Drug Carrier Systems

Horizons in Biochemistry and Biophysics Series

Editorial Board

Horizons in Biochemistry
and Biophysics

VOLUME 9

Drug Carrier Systems

Volume Editors
F.H.D. Roerdink and A.M. Kroon
Laboratory of Physiological Chemistry
University of Groningen, The Netherlands

Series Editors
E. Quagliariello, *Editor-in-Chief*
and
F. Palmieri, *Managing Editor*
Department of Biochemistry, University of Bari

A Wiley—Interscience Publication

JOHN WILEY & SONS

Chichester · *New York* · Brisbane · *Toronto* · *Singapore*

Copyright © 1989 by John Wiley & Sons Ltd.

British Library Cataloguing in Publication Data:
Drug carrier systems
 1. Medicine. Drug therapy. Drug therapy
 I. Roerdink, F.H.D. II. Kroon, A.M.
 (Albertus Maria), 1937–   III. Series
 615'.7

ISBN 0 471 92317 6

Printed in Great Britain

Contents

List of Contributors

ARNON, R.

Department of Chemical Immunology
Weizmann Institute of Science,
Rehovot, Israel

BLACKSHEAR, P. J.

Duke University Medical Centre
P.O. Box 3297
Durham, NC 27514, USA

DAVIS, S. S.

The University of Nottingham
University Park,
Nottingham, NG7 2RD, United Kingdom

DERKSEN, J. T. P.

Laboratory of Physiological Chemistry
University of Groningen,
Bloemsingel 10, 9712 KZ Groningen, The Netherlands

FEIJEN, J.

Department of Chemical Technology,
Twente University of Technology, P.O. Box 217,
7500 AE Enschede, The Netherlands

FIDLER, I. J.

The University of Texas System Cancer Centre,
M.D. Anderson Hospital and Tumor Institute, Texas
Medical Center, 6723 Bertner Avenue, Houston,
TX 77030, USA

FURR, B. J. A.

Imperial Chemical Industries PLC, Pharmaceuticals
Division, Mereside, Alderley Park, Macclesfield,
Cheshire, SK10 4TG, United Kingdom

GABIZON, A.

Cancer Research Institute, University of California,
San Francisco, and Liposome Technology, Inc.,
Menlo Park, California, USA

GREGORIADES, G.

Medical Research Council Group, Academic
Medicine, Royal Free Hospital School of Medicine,
London, NW3 2QG, United Kingdom

HOES, C. J. T.   Department of Chemical Technology, Twente
                      University of Technology, P.O. Box 217,
                      7500 AE Enschede, The Netherlands

HURWITZ, E.   Department of Chemical Immunology,
                      Weizmann Institute of Science,
                      Rehovot, Israel

HUTCHINSON, F. G.   Imperial Chemical Industries PLC, Pharmaceuticals
                      Division, Mereside, Alderley Park, Macclesfield,
                      Cheshire, SK10 4TG, United Kingdom

IHLER, G. M.   Department of Medical Biochemistry and Genetics,
                      Texas A&M College of Medicine,
                      College Station, TX 77840, USA

ILLUM, L.   The University of Nottingham,
                      University Park,
                      Nottingham, NG7 2RD, United Kingdom

JULIANO, R. L.   Department of Pharmacology, School of Medicine,
                      The University of North Carolina at Chapel Hill,
                      Chapel Hill, NC 27599-7365, USA

LAZAR, G.   Laboratory of Physiological Chemistry,
                      University of Groningen,
                      Bloemsingel 10, 9712 KZ Groningen, The Netherlands

ROERDINK, F. H.   Laboratory of Physiological Chemistry,
                      University of Groningen,
                      Bloemsingel 10, 9712 KZ Groningen, The Netherlands

ROHDE, T. D.   Duke University Medical Centre,
                      P.O. Box 3297,
                      Durham NC 27514, USA

SCHECHTER, B.   Department of Chemical Immunology,
                      Weizmann Institute of Science,
                      Rehovot, Israel

SCHERPHOF, G. L.   Laboratory of Physiological Chemistry,
                      University of Groningen,
                      Bloemsingel 10, 9712 KZ Groningen, The Netherlands

SPANJER, H. H.   *Laboratory of Physiological Chemistry,*
University of Groningen,
Bloemsingel 10, 9712 KZ Groningen, The Netherlands

Preface

SOME CONSIDERATIONS AND PROPOSALS

Detailed understanding of cell- and molecular biological processes has opened new ways for the interference with these processes. It is possible to test a wide variety of compounds as to their effects on enzymatic reactions, transport mechanisms, cell proliferation and cell differentiation-related activities in either subcellular systems or whole cells. Synthetic agents and also natural substances, although foreign to the organism under study, can be investigated. In addition, body-specific macromolecules such as hormones, plasma proteins and even intracellular enzymes or cell constituents can be isolated and purified or produced at large scale with the aid of modern biotechnological methods. Also these agents can be used for therapeutic purposes. Not to speak of the remote possibilities for causal treatments of inherited diseases by gene replacement or gene correction.

The availability of these therapeutic possibilities has once more pointed to the necessity that the relevant tissues, cells or subcellular targets should be reached. The beneficial effects fully depend on this prerequisite, especially since the therapeutic potential for certain cells and tissues coincides in most cases with toxicity of the same agents for other cells and tissues. The most striking example of this dilemma is shown by the anticancer drugs, which in general match a cytocidal effect with severe side effects on unaffected organs, not necessarily only those organs characterized by a high turnover rate of part of their cells. Another point of interest is that drug therapy is often more effective if the drug levels remain more or less constant over a large period of time. Extreme initial levels, which provoke severe side effects without improving the therapeutic potential, should be avoided.

These considerations call for drug carrier systems, which address the drugs specifically to the tissues and cells they should interfere with and/or release the active drug in a sustained way.

In the present volume the main developments of the last 20 years in the field of drug carrier systems and drug targeting are reviewed. Attention is paid to antibodies, polymers, liposomes, erythrocytes and erythrocyte ghosts and to implantable infusion pumps. From the data presented it is clear that considerable progress has been made over the past years. This progress not only led to a number of clearcut examples of the potency of the drug-targeting concept, but also pointed out the limitations of the available systems. The natural barriers which have to be overcome before the sensitive targets are really reached, are significant. Particle traffic and transport in the whole body is much more complicated than was expected when the various carrier systems were firstly introduced. In this volume problems of biodistribution and biodegradation are discussed. Moreover, the selectivity and the toxicity of the drugs carried by the various systems is given much attention. This especially holds for a variety of anticancer drugs.

From the data presented it follows that the cells of the reticuloendothelial system (RES) preferentially capture macromolecular drug-carrier conjugates. On the one hand this explains some disappointing results in reaching target organs other than the RES. On the other hand this circumstance has led to powerful approaches to the treatment of diseases specifically localized in the RES. The fight against a number of parasitic infections, which form a major healthcare problem worldwide, may benefit from these new developments. The odds are certainly warranting the onset of clinical trials in this field on a larger scale.

With respect to the various aspects of sustained release the developments are highly promising. Both implantable pumps, which can be refilled, and erythrocytes, in vitro loaded with certain drugs, resealed and then reinfused, appear strong tools in this respect.

In the near future progress may be expected in the field of immunotargeting. Reagents like immunotoxins and antibody-drug conjugates are macromolecules with molecular weights and dimensions, which are relatively low and small as compared to particle-type carriers such as liposomes and the various microspheres. For this reason they will escape easier from the phagocytotic activity of the macrophages in the various body compartments. Also the endothelial lining of the blood vessels can be passed easier by such molecules. The availability of monoclonal antibodies against a growing number of tissue-specific antigenic determinants is an important development in this respect. The binding of drugs such as cytostatic agents to the antibodies can be accomplished by various methods and the mechanisms for internationalization and liberation of the drugs in the target cells are currently under investigation. Encouraging results have already been obtained. As far as peptide drugs or enzymes are concerned future developments on hybrid proteins seem promising. Also the use of ternary complexes, which combine a drug with a targeting principle and a carrier system, may be expected to be developed further in the coming years.

Relatively little attention has yet been aid to the oral route of administration. In the near future this route too may benefit from the knowledge obtained in the various areas of drug targeting research described in this volume. It is obvious that carrier systems are able to protect drugs from hydrolytic breakdown in the gastrointestinal tract. By combining these carriers with appropriate bioadhesive reagents, it can be envisaged that proteins and other macromolecules prone to hydrolysis by the digestive enzymes, can be transported to the distal parts of the intestinal tract and can interfere with the cells of the immune system in that area of the bowel. This approach may open new possibilities for vaccination.

The volume editors: Frits H. Roerdink
Albert M. Kroon

Drug Carrier Systems
Edited by F.H.D. Roerdink and A.M. Kroon
© 1989 John Wiley & Sons Ltd.

TARGETING OF DRUGS: IMPLICATIONS IN MEDICINE

Gregory Gregoriadis,
Medical Research Council Group, Academic Medicine,
Royal Free Hospital School of Medicine,
London, NW3 2QG, UK

INTRODUCTION
It is widely accepted that the usefulness of many
drugs in therapeutic, diagnostic and preventive medicine
would be enhanced if drugs were to exert their desired
effect selectively on the target site. With few
exceptions, however (eg. interference of antibacterial
agents with metabolic pathways in bacteria, not shared
by the host), drug selectivity ranges from modest to
nil. The consequences of poor selectivity in
pharmacological action are clearly reflected in cancer
treatment where cytostatic agents will also kill or
damage normal rapidly proliferating cells. Lack of
selectivity is only one, albeit major, of the obstacles
to optimizing drug action. Others are inaccessibility
of target where, for instance, antimicrobial drugs fail
to enter intracellular sites harbouring microorganisms,
drug vulnerability in certain biological milieus leading
to or surrounding the target, premature drug secretion
or allergic reactions due to the molecular nature of
drugs.
In spite of the very considerable recent developments
in molecular and cell biology, progress in the design of
selective drugs has been meagre. On the other hand,
some of these developments including the advent of
monoclonal antibodies and better understanding of
ligand-receptor interactions and ensuing intracellular
events, have been the protagonists in the rapid progress
of an alternative approach to conferring selectivity on
drugs, namely targeting. The drug targeting concept is
based on the use of carrier systems to deliver drugs to
the intended area of action. Carriers can do so because
of an inherent or acquired ability to interact
selectively with respective cells (Gregoriadis, 1981).
Thus, glycoproteins bind through specific terminal
groups to receptors expressed on cell surfaces,
antibodies interact with cell surface antigens and
colloid particles (eg. liposomes) are endocytosed avidly
by macrophages. Loading of carriers with drugs can
occur through covalent and other types of bonding, or

passive entrapment as is usually the case with colloids.
Injected drug-loaded carriers are expected, once in the
vicinity of the target, to associate with it on contact.
Subsequently, through a variety of scenarios, the drug
is freed to act. However, accumulated experience
(Gregoriadis, 1988a; Davis et al, 1984; Tirrell et al,
1985; Gregoriadis and Poste, 1988) during the last
decade or so has shown that targeting in vivo can be
interfered with by the biological milieu within which
targeting occurs.

THE ROLE OF THE BIOLOGICAL MILIEU
    The biological milieu, for the purpose of this
discussion, includes the body fluids and tissues which
the carrier may encounter en route to its destination
and the target. Thus, blood which is normally where the
carrier finds itself after injection can, through some
of its cell or plasma components, alter the circulating
carrier to a state incompatible with drug retention or
divert it to irrelevant normal tissues which may suffer
as a result. Colloid systems for instance, are treated
as foreign by the body, opsonized and end up in the
reticuloendothelial system (RES) (Davis et al, 1984;
Gregoriadis and Poste, 1988). Further, natural
macromolecular carriers, such as antibodies,
lipoproteins, glycoproteins, etc. will be taken up by
tissues (eg. liver) and catabolized during the normal
process of the protein turnover (Allison, 1976). These
events may be so rapid as not to allow the carrier to
associate with the target significantly. In addition,
foreign or altered protein carriers (eg. mouse
monoclonal antibodies and partly denatured proteins
respectively) may elicit immune responses which could,
on repeated treatment, lead to allergic reactions or
neutralization of the carriers (Gregoriadis and Poste,
1988) (It is anticipated that use of monoclonal
antibodies of human origin will remove the threat of
immunological response; Crawford et al, 1983). On the
other hand, there may be slow, limited or no access of
the carrier to a target separated from the blood by
capillary and other (eg. basement) membranes. This is
especially true for some of the larger colloid or
polymeric type of carriers (Gregoriadis and Poste, 1988;
Poste and Kirsh, 1983), although it also applies to some
extent to proteins such as antibodies (Cobb et al,
1987).
    A potential source of problems is also the way by
which the carrier moiety mediates drug action on
association with the target. When cell death or some
kind of modification of its metabolic function are
required, it may be necessary for the carrier to enter
the cell. For ligands this is usually achieved by
receptor-mediated endocytosis (Gregoriadis et al, 1984;

Kolata, 1983) which introduces the ligand and drug moiety into cell compartments where the drug is freed to an active state through the action of the microenvironment. A good example is the use of asialoglycoproteins for the delivery of drugs (Fiume et al, 1984) to the hepatic parenchymal cells expressing the galactose receptor (Ashwell and Harford, 1982). Here, uptake of the glycoprotein by the cells is rapid, highly specific and leads to its localization in the lysosomal apparatus (Gregoriadis et al, 1970) within which lysosomal hydrolases will sever the bond between the drug and the carrier. Depending on the case, the drug could either act locally or, following diffusion, in other cell compartments or even extracellularly. It is possible however, that some cells will not interiorize bound ligands, bonds will not be hydrolysed and drugs will lose their activity or fail to reach their target. Most importantly, problems relating to the biological milieu could be aggravated in the presence of disease where blood composition, tissue microanatomy, membrane structure, receptors and other entities could be altered in ways that are detrimental to effective carrier function as established under normal conditions (McIntyre, 1986).

Appropriate choice of the carrier-drug unit and/or structural manipulations of it, may help to circumvent difficulties presented by the biological milieu. In terms of blood interference with carrier behaviour, the following examples come into mind: It has been shown (McIntosh and Thorpe, 1984) that terminal galactose residues of plasma or cell surface glycoproteins bind antibody-toxin conjugates via the galactose-recognizing sites of the B chain of the toxin (eg. ricin). Such non-specific binding can induce toxicity or severely limit the number of conjugate molecules available for distribution to tumour tissues. This can be now abbrogated through competitive antagonism with excess free galactose or lactose, chemical modification of the toxin to delete its galactose recognizing properties and also by using conjugates with A chains only (McIntosh and Thorpe, 1984). Plasma high density lipoproteins (HDL) on the other hand, destabilize liposomes (Krupp et al, 1976; Scherphof et al, 1978) by removing phospholipid molecules from their bilayers (Kirby et al, 1980). HDL action on liposomes however, can be reduced or abolished altogether by enriching the bilayers with cholesterol and/or phospholipids with high gel-liquid crystalline transition temperatures (Tc) (Kirby et al, 1980; Allen, 1981; Senior and Gregoriadis, 1982). These lipids promote bilayer packing and rigidity, in turn inhibiting HDL insertion (Gregoriadis, 1985).

A number of approaches have been adopted for the reduction of carrier interception by the RES. Recent

examples are the deglycosylation of the A chains of
ricin in immunotoxins, which enables the latter to avoid
the relevant sugar receptors in the RES (Blakey et al,
1987) and the co-administration of (ricin) immunotoxin
with yeast mannan which competes for uptake by the liver
(Bourrie et al, 1986). Generally, protein carriers must
obviously be prepared in a purified, non-denatured form.
Also, the load and type of linked drug must be such that
there is only minimal change of the protein's native
state (Matzku et al, 1985). To that end, conventional
cytotoxic drugs, of which relatively large quantities
must be linked to the protein so as to achieve the
required dosage, have been replaced with highly potent
toxins (eg. ricin, abrin; for reviews see Gregoriadis et
al, 1984). With colloid and polymeric carriers,
effective RES avoidance is more problematic. Since
recognition of most colloids by the RES is the result of
their opsoninization, attempts have been made to reduce
this by rendering colloids refractory to protein
adsorption (eg. a highly hydrophilic surface). As an
example, the rate of clearance of injected polystyrene
microspheres coated with the hydrophilic block copolymer poloxamine 908 was significantly reduced (Illum
et al, 1987). An even greater reduction of clearance
has been achieved in the case of liposomes by simply
diminishing vesicle size. Thus, small unilamellar
liposomes (SUV) with a 30-60um diameter exhibit half-
lives of several hours (Gregoriadis, 1988b).
Interestingly, half-lives can be extended greatly (up to
20 hours in mice) by incorporating into the SUV large
proportions of cholesterol and phospholipids (Senior and
Gregoriadis, 1982) with high Tc. It has been suggested
(Gregoriadis, 1985) that in addition to preventing HDL
insertion, a packed, rigid bilayer may also interfere
with opsonin adsorption (for further discussion see
Gregoriadis, 1988b).

As already stated, interception of (altered) protein
and colloid carriers by the RES reduces their value in
terms of quantitative contact with alternative targets.
This, however, is far from being a disadvantage when
drugs are to be delivered to the RES which is involved
in many microbial and other diseases and also
constitutes an important component of the immunological
response (for a discussion and relevant applications,
see later). It should also be pointed out here that, as
in most cases receptors on macrophages recognize foreign
carriers through plasma "ligand" components adsorbing
onto their surface, interaction of such carriers with
the RES is a form of receptor-mediated targeting
(commonly referred to as "passive"), in the sense that
selectivity is acquired on contact of carriers with the
biological milieu.

Although receptor-mediated targeting is potentially a most effective approach to optimizing drug action, some of the problems mentioned earlier, especially target inaccessibility, may prove insurmountable thus cancelling out sophisticated carrier-drug chemistry or ways of controlling other aspects of carrier behaviour in vivo. In addition, not all sites in need of treatment possess (or are known to possess) specific receptors and when they do, corresponding ligands may not be readily available. From the practical point of view, intravenous targeting which is considered ideal because blood will transport carriers to, or near, all areas in the body, is not a favoured route of drug administration in chronic treatment. For these and other reasons, an auxillary "targeting" concept has emerged, now generally known as the controlled release approach. Controlled release, as challenging as and in some ways perhaps more realistic than receptor-mediated targeting, employs a variety of microscopic or macroscopic implantable devices which will release their drug content, often at predetermined rates. Drug release is effected either by continuous degradation of the implant or by the efflux of the drug from the intact implant. Microscopic devices in the form of colloids can also be injected directly into the circulation to be subsequently "implanted" in tissues. It is thought (Storm et al, 1987) that the considerable success of certain liposomal cytostatic drugs in experimental cancer chemotherapy is probably the result of slow drug release from the liver and spleen where liposomes end up after injection. For further discussion on controlled release, which falls outside the scope of this review, see Leong et al, 1986.

IMPLICATIONS IN MEDICINE
Remarkable successes with drug targeting in the treatment or prevention of a wide spectrum of diseases in experimental animals and small clinical trials (eg. Gregoriadis, 1981, 1988a; Davis et al, 1984; Tirrell et al 1985; Gregoriadis and Poste, 1988) suggest that routine clinical applications may be forthcoming. To that end, the first and obvious consideration is that a carrier-drug unit designed to treat a particular disease has clear advantages over the conventional use of the therapeutic agent. Advantages may include a lower dosage at which the agent is effective (preferably in a single bolus thus diminishing both cost and toxicity), prevention of drug loss through premature excretion or inactivation as a result of carrier-mediated altered pharmacokinetics and improved access to the target because of the ability of the carrier to, for instance, enter intracellular sites. Also, novel drug toxicities as a result of altered pharmakinetics should, if

present, be minimal as not to outweigh advantages gained from targeted drug delivery. All these must be proven at the various experimental and clinical stages necessary for the final approval by regulatory bodies.

Historically, initial evidence for the effectiveness of a drug-carrier system in the treatment of a particular disease has been typically provided by academic researchers usually "disillusioned" with the performance of drugs available for this purpose. In some cases, however, researchers are unfamiliar with the clinical and/or industrial demands which must be satisfied before the system can be applied clinically. As a result, some of the ideas and concepts propagated are rather unrealistic and eventually fall by the wayside. This can be accelerated by premature claims which generate scepticism in those to whom the claims are targeted or are not substantiated when tested clinically. In others, carriers are solely designed to succeed in an in-vitro environment which has very little to do with the behaviour of the carrier in vivo or with the properties of the target as expressed in situ. Thus, fine work producing elaborate drug-carrier systems can become redundant when the systems are tried in vivo and their effectiveness shown to be disappointing.

Much of the success with drug targeting has originated from experiments performed with the biological milieu in mind and with the simultaneous involvement of key staff (eg. physicians, cell and molecular biologists, pharmacists, immunologists and biophysicists), occasionally including industrial workers who will infuse into the work their point of view. Recently, progress toward clinical applications of targeted drug delivery has gained new momentum thanks to the efforts of drug delivery research units set up by pharmaceutical industries and of new biotechnology companies the survival of which depends on the early development of drug delivery products. Such companies will employ a wealth of specialist scientists, have close links with hospital physicians and be advised by individuals knowledgeable of intricate legal, financial and marketing aspects. In this way, the combined efforts of academic and industrial workers are likely to lead in the near future to a number of products expected to optimize drug action in a variety of clinical uses. Some of these, together with applications of longer-term promise will be discussed here briefly and the role of individual carriers evaluated. More detailed treatment of ideas will no doubt be given in the various chapters of this book devoted to specific topics.

CANCER TREATMENT

Much of the research on targeted drug delivery has been focussed on cancer treatment not only because

cancer affects great numbers of people but also because
it is a major disease in the western world where science
is most developed and resources most concentrated. Yet,
conventional cancer treatment (chemotherapy, radiation
and surgery, alone or combined), especially of solid
tumours has been less than satisfactory for a large
number of patients: None of the drugs developed so far
is selective enough and as a result dosages cannot be
adjusted to levels required to kill all tumour cells
without being toxic at the same time. The discovery of
antigenic sites specific for some tumour cells raised
hopes (Bagshawe, 1983) that antibodies against such
antigens could serve to deliver cytostatic drugs
(including radionuclides) to the cells. Indeed, related
experimental work (see Bagshawe, 1985; Gregoriadis and
Poste, 1988; Tirrell et al, 1985; Gregoriadis et al,
1984, for numerous examples) with animals bearing
tumours established the feasibility of the approach.
Enhanced progress in this area, however, did not occur
until monoclonal antibodies became available and the
replacement of antibody-bound drugs with immunotoxins
(Marx, 1982; Vitetta and Uhr, 1985). The first,
provided an unlimited source of highly specific
molecules for tumour antigenic sites and the second,
extremely potent cell killer molecules (Olsnes and Pihl,
1982). Many laboratories have produced evidence of
therapeutic advantages (Baldwin and Byers, 1986) for
immunotoxins in animals, for instance nude mice bearing
human tumours (Griffin et al, 1987; Johnson and Laguzza,
1987; Gregoriadis, et al, 1984) and encouraging results
have been obtained in patients (Baldwin and Byers, 1986;
Beverley and Riethmuller, 1987). Clinical work
(Frankel, 1985; Douay et al, 1985) also suggests that
allergic reactions from mouse monoclonal antibodies are
either not apparent or can be tolerated reasonably well.
There is, nonetheless, considerable uneasiness (Garnett
and Baldwin, 1986) regarding the potential toxicity of
toxins to normal cells, such as those of the RES which
will take up much of the injected immunotoxins.
Moreover, toxins being foreign, will elicit immune
responses and are likely, on chronic treatment, to be
neutralized by the antibodies formed (Marx, 1982).
Researchers are, therefore, re-examining the use of
antibody-bound conventional cytostatic drugs (Garnett
and Baldwin, 1986; Diener et al, 1986; Manabe et al,
1984). Even more promising results in animals and
patients (Bagshawe, 1985; Chan et al, 1986; Carrasquillo
et al, 1984; Buraggi et al, 1985) have been obtained
with radiolabelled monoclonal antibodies in tumour
imaging where problems (Epenetos et al, 1986) are of
lesser magnitude compared to those encountered in
therapy.

The use of colloids, in cancer treatment has also
been considered (eg. Gregoriadis, 1988a; Tirrell et al,
1985; Davis et al, 1984) in spite of the obvious
difficulties (Poste and Kirsh, 1983) of particle
penetration to areas where tumours reside. However,
there is indirect evidence from animal work that tumour
cells might take up small liposomes (eg. Gregoriadis et
al, 1977; Ogihara et al, 1986; Patel et al, 1985) or
nanoparticles (Grislain et al, 1983) to a greater extent
than cells of neighbouring normal tissue. Whether these
results can be explained on the basis of a higher
endocytic activity of some tumour cells combined with
increased local permeability of capillaries, marker
diffusion from circulating vesicles or nanoparticles
followed by marker localization in the tumour area
because of increased blood flow in that area, or indeed,
as a result of migration of monocytes (with the engulfed
carrier) to tumours, remains an open question
(Gregoriadis, 1980). Unfortunately, preferential uptake
of (liposomal) markers by tumours in man has not as yet
been convincingly demonstrated (Gregoriadis et al, 1974;
Ryman and Tyrrell, 1980; Perez-Soler et al, 1985).
    The case of using colloids in cancer chemotherapy
appears stronger in work aimed at reducing toxicity
while at the same time maintaining the tumourcidal
effect of the drug. Experiments with liposomes
containing anthracycline cytostatics (eg. Gregoriadis,
1988a; Forssen and Tokes, 1983; Fichtner et al, 1984;
Gabizon, et al, 1986; Mayhew et al, 1987) have
unequivocally shown reduction of cardiotoxicity and
dermal toxicity and prolonged survival of tumour-bearing
animals compared to controls receiving the free drug.
Interestingly, there have been already promising results
with at least two of the ongoing related clinical
trials. In phase I work, for instance, the maximum
tolerated close for daunomycin-containing liposomes was
much greater (90_2mg drug per m^2) than that of the free
drug (30-45 mg/m^2 body surface). When the same
liposomes were used to treat, in a limited phase II
trial, five patients with recurrent breast cancer, three
of the patients became complete responders at 75 mg
daunomycin/m^2 (A. Rahman, personal communication). In a
second study with liposomal daunomycin (Sells et al,
1987) the drug was also well tolerated by six patients
with hepatic metastases from primary gastrointestinal
adenocarcinomas and there was evidence of significant
reduction of malignant hepatomegaly in one of them. It
has been postulated (for discussion see Gregoriadis,
1988a) that liposomal drug taken up by the RES tissues
is released to penetrate malignant cells and exert its
effect. If so, it is possible that other drug-loaded
colloid carriers including albumin microspheres,
nanoparticles, and erythrocyte ghosts (all of which

localise in the RES) may mediate a similar drug action.
Besides conventional liposomes, antibody-coated vesicles
have also been investigated (Gregoriadis et al, 1977;
Hashimoto et al, 1987; Connor et al, 1985) for drug
delivery to tumours. The advantage of using liposomes
over antibodies as such is that the former can
accomodate much larger quantities of drug per antibody
molecule linked to their surface (Wolff and Gregoriadis,
1984). However, in view of the limited, if any, access
of liposomes to extravascular sites, their carrier value
(with or without antibodies) for the direct delivery of
drugs to solid tumours is doubtful (Poste and Kirsh,
1983).
    Regardless of the delivery system used in cancer
treatment, one must consider a number of difficulties
(Marx, 1982) which targeting will probably not solve:
(a) many human tumour cells may turn out to be lacking
antigens exclusive to the tumour; (b) binding of certain
antibodies to tumour cells could promote the
disappearance of corresponding antigens from their
surface, thus minimizing the effectiveness of successive
treatments; (c) the heterogeneity of tumour cell
populations could leave much of the tumour unscathed,
although this may be circumvented to some extent by the
use of "cocktail" immunotoxins made of antibodies raised
against a spectrum of antigens covering the various cell
populations of the tumour. Many of these difficulties
could be circumvented however, by a much more promising
approach, namely activation of macrophages to a
tumourcidal state (Fidler, 1985). This is an area where
passively targeted liposomes hold much promise (Fidler,
1985). Indeed, extensive research with liposomes
containing macrophage activation agents (Fidler, 1985;
Phillips et al, 1985; Daemen et al, 1986) has
established that the system works in eradicating
metastases in experimental animals (eg. Fidler, 1985;
Philips et al, 1985).
    What is then the future of drug carriers available
for targeted cancer therapy? Great hopes have been
pinned onto monoclonal antibodies, especially
immunotoxins and much energy and resources are directed
towards their development. It should not be surprising
if in the next few years patients with certain tumours
do benefit from immunotoxin treatment (especially in
combination with conventional procedures) and live
longer. Optimism is much more justified (eg. Bregni et
al, 1986), however, in the case of some forms of
leukaemia treated by the ex-vivo purging of clonogenic
malignant cells using immunotoxins. Where, for reasons
already discussed, the use of monoclonal antibodies is
not possible, colloid systems such as liposomes acting
as depots of drug release from RES tissues (unaffected
by the presence of the drug), may prove of some value

provided that they contribute to significant
prolongation of life, hopefully in the absence of side
effects.

TREATMENT OF MICROBIAL INFECTIONS
    The need for site-specific drug delivery in the
treatment of microbial diseases has arisen mostly
because of the inability of many otherwise potent
agents, to effectively enter intracellular sites
harbouring microorganisms. Additional reasons (also
valid for other chemotherapeutic agents) include
premature agent excretion and/or inactivation, both of
which can be prevented or diminished by the carrier. As
a large variety of microorganisms reside in the liver
and spleen, especially the RES component of these
tissues, microparticles such as liposomes and niosomes
(Baillie et al, 1986) are the obvious choice of carrier.
Numerous publications since 1973 (Gregoriadis, 1973)
have shown that liposomes are superior to the free
agents both in terms of distribution to the appropriate
intracellular sites and therapeutic efficacy (eg.
Gregoriadis, 1988a; Bakker-Woudenberg and Roerdink,
1986; Alving, 1986).
    Many of the microbial diseases affect vast numbers of
people in third world countries. They include the
parasitic diseases visceral and cutaneous leishmaniasis,
malaria and trypanosomiasis. Although proper hygiene
and eventually vaccination constitute the best
approaches to mass eradication of such diseases,
chemotherapy is the only alternative at present,
especially for individuals already infected. The best
example of attempts to treat parasitic diseases via
targeted drug delivery is the use of liposome- (or
niosome-) entrapped Pentostam, glucantine, muramyl
dipeptide analogs and other agents in the treatment of
experimental leishmaniasis. Extensive work (eg. Alving,
1986; Berman et al, 1986; Baillie et al, 1986; Adinolfi
et al, 1985) has shown that much smaller amounts of
drugs (than used conventionally) can prolong the
survival of, or cure infected animals with little or no
apparent toxicity. These results, often achieved with a
single dosage of drug, strongly support the use of
appropriate preparations in clinical trials. Single
dosage treatment (and cure) of patients with
leishmaniasis would be a major advantage, both
economically and logistically. An array of other
antimicrobial drugs have also been incorporated in
liposomes for the treatment of a variety of
intracellular infections (for reviews see Gregoriadis,
1988a; Bakker-Woudenberg and Roerdink, 1986). It is
anticipated that related veterinary products will be
marketed in the near future.

A role for liposomes in antimicrobial therapy for which other colloids may not be suitable, is in the treatment of fungal diseases often seen in immunosuppressed patients. Amphotericin B used for the treatment of fungal diseases such as candidiasis, acts by binding to the ergosterol of fungi membranes thus creating channels through which vital molecules leak out from the cells and cause their death. Unfortunately, amphotericin B also binds to the cholesterol of mammalian cells, hence its toxicity. It has been demonstrated (eg. Lopez-Berestein et al, 1983) that disseminated candidiasis in animals can be treated successfully with amphotericin B incorporated into liposomes composed of saturated phospholipids and similarly encouraging observations were made with patients suffering of fungal disease (Lopez-Berestein et al, 1985). As a result, a multicentre clinical trial in collaboration with a pharmaceutical industry is to be carried out, apparently in the near future (Raymond, 1987). The beneficial effect of liposomal amphotericin B has been attributed (Juliano et al, 1987) to its considerably lower toxicity as it does not affect mammalian cells to the same extent it affects fungi. Recent experiments (Juliano et al, 1987) in vitro with mammalian cells exposed to liposomal amphotericin B has indicated the importance of saturated phospholipids as the carrier's components in reducing toxicity: preparations made of unsaturated phospholipids were found almost as toxic as the free drug.

Work with carrier-mediated use of agents for the direct killing of microbes is now being supplemented with attempts (eg. Adinolfi et al, 1985) to deliver (via liposomes) immunostimulating agents such as muramyl dipeptide and derivatives in order to activate macrophages to a microbiocidal state. Thus, experiences accumulated from the use of liposomes have not only injected considerable faith into the future of targeted antimicrobial therapy, they have also provided useful information as to what can be expected from the approach when testing other colloid systems, for instance niosomes, nanoparticles, albumin microspheres, etc.

An alternative class of carriers for the delivery of antimicrobial drugs to tissues is macromolecular ligands with specific affinity for cell receptors (Gregoriadis et al, 1984). The best examples here are galactose-terminating (eg. desialylated glycoproteins, polypeptides derived thereof, neoglycoproteins such as albumin coupled to galactose residues) or mannose-terminating macromolecules for the introduction of drugs, through receptor-mediated endocytosis, into the hepatic parenchymal and Kupffer cells respectively expressing the receptors for the sugars (Gregoriadis, et al, 1984). Targeting with this attractive systems was

proposed (for asialoglycoproteins) over seventeen years
ago (Rogers and Kornfeld, 1971). Although pursued to
some extent since then (eg. Fiume et al, 1984) it is
surprising that ligands have not been as yet adopted
widely, in view of their availability (eg. desialylated
plasma glycoproteins) and the considerable body of
accumulated knowledge of cell and molecular biology of
ligand-receptor interactions (Ashwell and Harford, 1982;
Gregoriadis, 1975). Such ligands have also been coupled
or otherwise incorporated (eg. Gregoriadis and
Neerunjun, 1975; Gregoriadis and Senior, 1984; Spanjer
and Scherphof, 1983; Szoka and Mayhew, 1983) onto the
surface of liposomes which, as already stated, can
accomodate much larger quantities of drug (per ligand
molecule). As the drug is associated with the liposome
moiety rather than with the ligand itself, it cannot
interfere with the receptor recognizing properties of
the latter, a situation which could occur with drugs
directly coupled to it. For the treatment of diseased
hepatic parenchymal cells with liposomes bearing
terminal galactose residues, vesicles must be small
enough to be able to pass through the fenestrations
(average size of 100 nm diameter). However, following
the discovery (Teradaira et al, 1983) of the galactose
receptor on fixed macrophages as well (for galactose
bearing particles), it would appear that Kupffer cells
of the liver are also candidates for galactose-bearing
larger liposomes. Another ligand (mannosylated
phospholipid), specific for the mannose receptor on
macrophages, has also been incorporated into liposomes
which could serve to transport immunomodulators into
macrophages and render the cells microbiocidal (Barratt
et al, 1986).

TREATMENT OF STORAGE DISEASES
     Two groups of diseases will be discussed under this
heading: (a) Inherited metabolic disorders, namely
lysosomal storage diseases in which deficiency of a
hydrolase in the lysosomes leads to the accumulation of
respective substrates in the organelles and, in turn,
tissue enlargement and malfunction;   (b) metal storage
diseases in which iron, copper, zink, plutonium, etc.
accumulate in tissues under a variety of circumstances.
Iron overload, for instance, occurs after frequent blood
transfusions or thalassaemia.   Other metals (eg. zink
and cadmium) can accumulate as a result of environmental
pollution.

     Early attempts to replace the missing enzyme in
patients with lysosomal storage diseases were largely
unsuccessful, even in cases where enzymes from human
sources were used (Brady et al, 1982).   Among
contributing factors, scarcity of patients (which

precluded the use of non-treated controls), limited amount of enzyme available and uncertainty as to whether the injected enzyme retains its activity in the circulating blood or reaches the site of intended action, were prominent. On the basis of experimental evidence from animal (Gregoriadis and Ryman, 1972a,b) and tissue culture (Gregoriadis and Buckland, 1973) work, it was thought that enzyme protection, enzyme access to intracellular sites and subsequent in-situ substrate hydrolysis, could be achieved in man by using liposomes as an enzyme delivery system. Unfortunately, a large number of intravenous injections (over 5 years) of liposome-entrapped glucocerebroside β -glucosidase purified from human placentas, into a splenectomised Gaucher's disease patient failed to drastically reduce the size of her liver (Gregoriadis et al, 1982). On the other hand, clinical observations suggested that treatment stabilized the patient, whose general condition did not deteriorate (it had done so during the years before treatment began; Gregoriadis et al, 1982). The patient died two years after cessation of treatment. In another attempt (Beutler et al, 1977) with human erythrocytes as the vehicle for the placental enzyme, some reduction of liver size was claimed. Reasons which could account for the failure of the two "trials" have been discussed elsewhere (Gregoriadis et al, 1982). They include insufficient amount of (entrapped) enzyme given, its inability to reach the majority of lysosomes because of reduced phagocytic activity in the diseased liver and poor enzyme action on a substrate which may have accumulated in a form not easily amenable to hydrolysis, in a microenvironment where conditions for such hydrolysis may be unfavourable.

The uncertainty as to the efficacy of targeted enzyme delivery could be resolved if and when a clinical trial is conducted in a reasonably large number of patients with similar severity of disease and supplemented with adequate controls, perhaps injecting larger amounts of enzyme. In this respect, targeted liposomes could also be considered in view of recent findings (Das et al, 1985) that glucocerebroside- β -glucosidase entrapped in glycosylated liposomes can be delivered selectively and at a more rapid rate to hepatic cell populations than by using non-glycosylated preparations. Since most, if not all, knowledge from the small scale trials with lysosomal storage disease patients has been amassed using liposomes or erythrocytes as enzyme carriers, it may be unwise to replace these two systems in larger trials with other colloids, or indeed, with macromolecular carriers coupled to the enzyme. The latter will probably add to the problems as the enzyme may be altered in the course of coupling and the enzyme to carrier mass ratio would be much lower than that

achieved with colloids or cells. It should also be
emphasized that the rarity of lysosomal storage disease
patients and, perhaps, the generally inherent
instability of enzymes will limit targeted delivery,
even if found to be successful, to experimental trials
only.

Correction of the malfunctioning gene, or its
replacement when absent, would be the ideal approach to
the management of enzyme deficiencies including
lysosomal storage diseases. Although genetic
engineering in man may be forthcoming sooner than
earlier predicted (Cline, 1984), it will probably begin
with ex-vivo manipulations of bone marrow stem cells.
It appears, for instance (discussed in Jones and
Summerfield, 1982), that suppression of Kupffer cell
function by blockade or X-irradiation stimulates
immigration of (precursor) Kupffer cells from the bone
marrow to the liver. Also, following transplantation of
a liver from a donor of the opposite sex, the
transplanted liver became populated with Kupffer cells
with the (male) sex caryotype of the host (Jones and
Summerfield, 1982). It is, therefore, possible that
insertion of appropriate genes in stem cells will lead
to the cure of lysosomal storage diseases affecting the
fixed macrophages of the RES. It is conceivable, on the
other hand, that genetically altered replacement cells
from the bone marrow may not be capable of establishing
stable intercellular contacts with endothelial cells and
hepatocytes in grossly malformed RES tissues as seen,
for example, in Gaucher disease patients.   Targeted
gene delivery is another attractive possibility and
success in the in-vivo (transient?) expression of genes
has been claimed by groups using liposomes (Nicolau et
al, 1983) or asialoglycoproteins (Wu and Wu, 1987) as
the carrier vehicle.

Turning onto metal storage diseases, mobilization of
excess metals by relevant chelating agents constitutes
the only treatment developed to date (Rahman, 1988).
However, the effectiveness of chelating agents,
demonstrated under in-vitro conditions, does not always
apply in vivo: because of complex interactions of stored
metals with biological ligands, the chelator may have to
compete against these ligands and it will, thus, be
required to possess strongly (anionically) charged
groups. These are likely to reduce the chelator's
ability to cross cell membranes. In addition, most
chelators being non-specific, are toxic as they will
interact with other non-target metals. Although some
chelators (eg. deferoxamine) lack toxicity, they are
rapidly degraded once introduced into the blood
circulation. There have been several attempts to
improve the efficacy of metal chelators by, for
instance, rendering these lipophilic so that they can

penetrate cells, adopting continuous infusion into
patients or the combined use of two chelators. With the
exception of infusion, success with such attempts has
been limited (Rahman, 1988).

Targeted delivery of chelators could provide
alternative means of improving their therapeutic
efficacy. Taking into consideration all major
prerequisites for an ideal chelator, ie. retention of
its structural integrity in blood, avoidance of
interaction with other, irrelevant metals and
accessibility to the intracellularly stored target
metal, usually within phagocytic cells, it would appear
that liposomes (or other colloids) and erythrocyte
ghosts rather than macromolecular ligands are the
vehicles of choice. Indeed, as early as 1973 (see
Rahman, 1988) liposomes were tested as a carrier of
chelating agents in plutonium-poisoned mice and
subsequently, by the same group and others (Rahman,
1988; Blank et al, 1984; Behari et al, 1986) in animal
models loaded experimentally with such metals as iron,
aluminum, ytterbium and cadmium. With most studies, the
effect of chelator in removing deposited metals from
tissues was greater than that of the free agent. In
addition to using chelator-containing conventional
liposomes for passive targeting to the tissues of the
RES, the system has also been applied in ligand-mediated
targeting. For instance, vesicles incorporating
galactocerebroside in the bilayers and containing
deferoxamine were more effective in removing iron from
hepatic parenchymal cells (expressing the galactose
receptor) than similar liposomes devoid of the
glycolipid (Rahman et al, 1982). To my knowledge, no
studies with any of the liposomal chelators has been so
far carried out in patients.

An alternative carrier for chelators for the removal
of stored metals from the liver and spleen macrophages,
is erythrocyte ghosts. The main attraction of this
system has always been that physicians are familiar with
erythrocytes (Deloach and Sprandel, 1985). However, the
system's credibility has been recently damaged by the
threat of viral (including HIV) infections from outdated
blood, normally an ideal source for ghosts. One is
therefore, limited to the use of the patient's own
blood. Although promising results were obtained in a
small clinical trial with iron-loaded patients treated
with ghost-entrapped deferoxamine (Green, 1985), no
further attempts have been reported.

VACCINES
Vaccination, chiefly responsible for the eradication
of smallpox and the control of polyomyelitis and German
measles in man and of foot-and-mouth, Marek's and
Newcastle disease in domestic animals, remains the best

answer to infectious diseases. Early vaccines were live
wild type organisms but these have been largely replaced
by attenuated or killed organisms or by purified
components (subunits) thereof. More recently,
developments in recombinant DNA techniques, monoclonal
antibodies and our understanding of the immunological
structure of proteins and the ways by which cells and
mediators are involved in the induction of immune
responses, have laid the foundations for a new
generation of vaccines (Liew, 1985). For instance,
subunit vaccines (eg. hepatitis B surface antigen) have
been produced through gene cloning and a number of
peptides mimicking very small regions of proteins on the
outer coat of viruses and capable of eliciting virus
neutralizing antibodies, have been synthesized (Liew,
1985). Such vaccines are defined at the molecular
level, can elicit immune responses controlling specific
infectious organisms and are, thus, potentially free of
the various problems inherent in conventional ones.
Unfortunately, subunit and peptide vaccines are only
weakly or non-immunogenic in the absence of
immunological adjuvants, a diverse array of agents that
increase specific immune responses to antigens (Allison
and Byars, 1986). They include aluminium salts, saponin
complexes with virus envelope antigens (immunostimulating complexes, ISCOM), bacterial products such
as lipopolysaccharide, lipid A, and mycobacteria cell
walls, minimal subunits of the latter (eg. muramyl
dipeptide and derivatives), physiological mediators (eg.
interleukins 1 and 2, interferon-γ), etc. Many of
these adjuvants produce side reactions ranging from
tissue damage, systemic effects and acute or chronic
inflammation at the site of injection to pyrogenicity,
arthritis and anterior uveitis. Only aluminium salts
have been licenced for use in man (Allison and Byars,
1986).
    Adjuvants induce humoural and/or cell-mediated
immunity (CMI). In humoural immunity antibodies with
sufficient affinity for the relevant antigens can
neutralize bacterial toxins and protect the host. CMI
on the other hand (necessary for protection against
intracellular bacteria, some viruses, protozoa and many
fungi), must be induced by adjuvants for protection from
related infections. There is already evidence (Allison,
1984) that adjuvants exert their action by activating
macrophages to release mediators such as interleukin 1,
which stimulate the proliferation of T-lymphocytes. An
additional (indirect) mechanism by which antigen
incorporating adjuvants may exert their action, is slow
degradation at the site of injection, possibly
facilitating uptake of the antigen by the regional lymph
nodes (Allison and Byars, 1986). Interestingly,
targeting antigens (via appropriate adjuvants) to highly

efficient presenting cells such as interdigitating and
follicular dendritic cells, may lead to the preferential
induction of CMI or humoural response respectively
(Allison and Byars, 1986).
It is perhaps not surprising, that colloid systems
such as liposomes known to interact with macrophages
avidly and to persist at the site of injection,
potentiate strong immune responses to entrapped
antigens. Ironically, it was originally hoped
(Gregoriadis and Ryman, 1972a,b) that liposomes may
actually prevent immune response to entrapped proteins
such as foreign enzymes in enzyme replacement therapy.
However, using diphtheria toxoid as a model protein, the
opposite was found (Allison and Gregoriadis, 1974) ie.
immune responses were enhanced. This was later
confirmed for a large number of antigens relevant to
human and veterinary immunization programmes. Liposomal
antigens studied to date (see Gregoriadis, 1988a;
Gregoriadis et al, 1988) include tetanus toxoid,
Streptococcus pneumoniae serotype 3, Salmonella
typhimurium lipopolysaccharide, cholera toxin,
adenovirus type 5 hexon, simplex virus type 1 antigens,
hepatitis B surface antigen, Epstein-Barr virus gp 340
protein, synthetic peptides of foot-and-mouth disease
virus and rat spermatozoal polypeptide fraction. In
several studies, protection of animal models was
achieved by immunization with the liposome-incorporated
antigens. Further, availability of liposomes with
variable structural characteristics and mode of antigen
accomodation (eg. entrapped versus surface-linked)
suggest versatility in immunoadjuvant action and vaccine
design. Recent work has, to some extent, established the
nature of immune responses (Davis et al, 1987) to
antigen elicited via liposomes and the role of liposomal
structural characteristics (Davis and Gregoriadis, 1987;
Gregoriadis et al, 1987) in optimizing such responses.
Targeted liposomal adjuvanticity has also been
demonstrated (Gregoriadis et al, 1988) in experiments in
which liposomes containing tetanus toxoid were coated
with mannosylated albumin.
Another adjuvant formulation with targeting potential
and shown to be effective in inducing immune responses
to a wide range of antigens, is based on the non-anionic
pluronic block polymer surfactants (Allison and Byars,
1986). The triblock polymers of these surfactants are
single unbranched chains made of a central hydrophobic
polymer of polyoxypropylene and two polymers of the
hydrophilic polyoxyethylene on either side. These high
molecular weight polymers (and several other variations)
increase immune responses, apparently because of their
ability to retain soluble macromolecules (eg. antigens)
on the surface of oil drops such as those of mineral
oil. A formulation has been recently (Allison and

Byars, 1986) developed consisting of materials
authorised for use in pharmaceutic and cosmetic
products. Components include squalene, pluronic L121
and tween 80, supplemented with threonyl muramyl
dipeptide characterized by low pyrogenicity. The
product, named Syntex adjuvant formulation (SAF-1)
appears in electron micrographs in the form of
microscopic spherules with antigens (labelled with
colloidal gold) visualized on their surface. Studies
(Allison and Byars, 1986) have shown that SAF-1
activates the alternative pathway of complement and C3b,
which is retained at the surface of the spherules,
probably facilitates interaction with interdigitating
and follicular dendritic cells expressing the C3b
receptors. Following this, antigens can be readily
transferred from the surface of the spherules to the
surface of antigen presenting cells. Vaccines based on
SAF-1 formulations have given promising results with
protection against feline leukaemia virus, simian AIDS
retrovirus and B-lymphoma in mice and humans (see
Allison and Byars, 1986).

Liposomes, pluronic block polymers and ISCOM
(structures derived from the saponin Quil A and virus
envelope antigens) formulations are all hopeful
immunoadjuvant candidates for the new generation of
vaccines. However, their widespread adoption will
depend not only on lack of toxicity, adjuvant efficacy
and its reproducibility in clinical trials, but also on
simplicity of technology and cost. Liposomes, when of
certain lipid compositions, have not shown overt
toxicity in experimental animals or after chronic
administration in man (see Gregoriadis, 1988a). The
system's efficacy in promoting primary and secondary
immune responses to incorporated antigens is well
documented in animal work (Gregoriadis, 1985) and it
would be of great interest to see whether this can be
confirmed in human studies. With regard to liposome
technology in general, previously identified potential
problems (Gregoriadis, 1984) of large scale production
and reproducibility have or are being resolved by
pharmaceutical industries devoted to these tasks
(Gregoriadis, 1984). Moreover, a recently developed
method (Kirby and Gregoriadis, 1984) for efficient
solute entrapment into liposomes may prove particularly
suitable for the production of liposomal vaccines
(Gregoriadis et al, 1987). The method involves
dehydration of a mixture of "empty" liposomes and solute
(eg. antigen) followed by controlled rehydration to give
the so-called dehydration-rehydration vesicles (DRV)
incorporating up to 80% of the solute. Sonication,
detergents and organic solvents, all of which may damage
or alter proteins, are not employed in this method
(Kirby and Gregoriadis, 1984). Further, antigen-

incorporating DRV can be freeze-dried for storage with
only minor loss of the entrapped material on
reconstitution with saline (Gregoriadis et al, 1987).

DIAGNOSTIC IMAGING
   Recent developments in diagnostic technology, namely
radionuclide imaging, magnetic resonance imaging (MRI)
and transmission computed tomography (CT) have created
the need for the development of relevant targeted agents
which would highlight differences between normal and
abnormal tissues. This could be achieved by enhancing
either the image of the target tissue or that of
neighbouring normal tissues. "Contrast"-generating
agents (eg. radionuclides, iodinated molecules for CT
and paramagnetic elements for MRI) can be toxic, non-
specific, unable to penetrate cells efficiently or
rapidly excreted. For instance, thorotrast and
iodinated starch particles used for the opacification of
the RES are not eliminated from the tissues and are,
therefore, unacceptable for clinical use. Other agents
such as those used in MRI have equal access to the
extracellular spaces of both normal and diseased tissues
(other than the brain) and, thus, lack specificity (for
extended discussion see Seltzer, 1988).
   Much of the work in imaging with drug delivery
systems has been carried out with antibodies (Epenetos
et al, 1986) and with liposomes (Seltzer, 1988).
Experimental and clinical evidence with radionuclide-
labelled antibodies has been sufficiently strong to
support their application in the imaging of tumours
((Epenetos et al, 1986). Of interest are also attempts
to target imaging agents to the hepatic parenchymal
cells by the use of galactose-terminating macromolecules
(Birken and Canfield, 1974).
   Liposomes, on the other hand, have been employed as a
carrier of contrast agents to portray the RES and a
number of non-RES tissues (Seltzer, 1988). Thus,
liposome-entrapped agents have given promising results
with CT (water and lipid soluble contrast materials,
radio-opaque lipids) and MRI (paramagnetic elements and
lipid derivatives thereof) in enhancing the image of the
normal liver and spleen and also with radionuclide
imaging (using various radioisotopes) in identifying not
only RES tissues but also lesions such as myocardial
infarction, abscesses and tumours in experimental
animals and clinically (Seltzer, 1988). It is
anticipated that recent advances in liposome technology
(Gregoriadis, 1984), including an improved imaging
agent/liposomal lipid mass ratio (Seltzer et al, 1988)
will strengthen the case of liposomes in diagnostic
imaging.

CONCLUSIONS

The need to endow specificity in most conventional drugs has led to the search of carrier systems which would transport drugs, directly or indirectly, from the site of administration to the site of action. If required, such systems could also serve to protect drugs from the non-target biological milieu (and vice-versa) or reduce premature excretion. In short, drug carriers are intended to, one way or another, optimize drug action. Two major categories of drug delivery systems have emerged: the first refers to systems such as antibodies and other ligands which exhibit specific affinity for receptors on cells. In this case, the drug is linked to the carrier (usually through covalent bonding), an approach which imposes the requirement of a drug activating hydrolytic reaction, normally within the target cells and, also, leaves the drug exposed to the biological milieu. Other possible adverse features are drug-induced masking of receptor recognizing sites on the ligand and modification of the latter's tertiary structure leading to altered pharmakinetics. In spite of these actual or potential difficulties, macromolecular carriers, notably antibodies, have been already applied successfully in the treatment of experimental malignancies. At present there is growing evidence from clinical work to support optimism in the future of monoclonal antibodies as a drug delivery system in the treatment of some forms of cancer in man. There is greater confidence, however, for the future of antibodies in the ex-vivo treatment of bone marrow and in tumour imaging.

The second category of drug delivery systems encompassing a variety of colloids, is more versatile both in terms of type of systems and applications. Although colloids as such are limited in tissue selectivity (predominantly the RES and hepatic parenchymal cells), they can nevertheless be targeted to accessible cells (eg. circulating cells) by the use of ligands attached on to their surface. In the latter case, colloids have certain advantages over ligands used as such in that (a) the attached ligand is not interfered with by the drug, safely entrapped within the carrier moiety of the complex; (b) large amounts of drug can be targeted by fewer ligand molecules. A major disadvantage of colloids as delivery systems is, of course, their size-imposed inability to cross most normal membrane barriers. Work with colloids, notably liposomes, suggests a versatility of applications in areas as diverse as antimicrobial and cancer therapy (especially through the activation of microbiocidal and tumourcidal macrophages respectively), storage diseases, blood surrogates, vaccines and diagnostics. Other applications, based on properties peculiar to colloids and related more to optimal drug release than to direct targeting, are topical (eg. skin and eyes) and oral therapy (for reviews see Gregoriadis, 1988a). With the exception of cancer chemotherapy (where antibodies have a greater by far potential) therapies and treatments with colloid-entrapped drugs appear, on the basis of experience with animals and patients, realistic enough to warrant further efforts from academic and industrial workers alike. Fortunately, invested interests of relevant biotechnology companies will ensure that everything possible is done to successfully transform laboratory and clinical experiences in drug targeting to products for routine use in the clinic.

ACKNOWLEDGEMENTS
I thank Mrs Angela Massaro for excellent secretarial assistance.

REFERENCES

Adinolfi, L.E., Bonventre, P.F., Vander-Pas, M. and Eppstein, D.A. (1985). 'Synergistic effect of glucantine and a liposome-encapsulated muramyl dipeptide analog in therapy of experimental visceral leishmaniasis', Infect. Immun,, 48, 409-416.

Allen, T. (1981). 'A study of phospholipid interaction between high-density lipoproteins and small unilamellar vesicles', Biochim. Biophys. Acta, 640, 385-397.

Allison, A.C. (Ed.) (1976). Structure and Function of Plasma Proteins, vol. 2, Plenum, New York.

Allison, A.C. (1984). 'Immunological adjuvants and their mode of action', in New Approaches to Vaccine Development (Eds. R. Bell and G. Torrigiani), pp. 133-166, Schwabe, Basel

Allison, A.C. and Byars, N.E. (1986). 'An adjuvant formulation that selectively elicits the formation of antibodies of protective isotypes and of cell-mediated immunity', J. Immunol. Meth., 95, 157-168.

Allison, A.C. and Gregoriadis, G. (1974). 'Liposomes as immunological adjuvants', Nature, 252, 252.

Alving, C.R. (1986). 'Liposomes as drug carriers in leishmaniasis and malaria', Parasitology Today, 2, 101-107.

Ashwell, G. and Harford, J. (1982). 'Carbohydrate-specific receptors of the liver', Ann. Rev. Biochem., 51, 531-554.

Bagshawe, K.D. (1983). 'Tumour markers - Where do we go from here?', Br. J. Cancer, 48, 167-175.

Bagshawe, K.D. (1985). 'Cancer drug targeting', Clin. Rad., 36, 545-551.

Baillie, A.J., Coombs, G.H., Dolan, T.F. and Laurie, J. (1986). 'Non-ionic surfactant vesicles, niosomes, as a delivery system for the antileishmanial drug sodium stibogluconate', J. Pharm. Pharmacol., 38, 502-505.

Bakker-Woudenberg, I.A.J.M. and Roerdink, F.H. (1986). 'Antimicrobial chemotherapy directed by liposomes', J. Antimicrob. Chemoth., 17, 547-552.

Baldwin, R.W. and Byers, V.S. (1986). 'Monoclonal antibodies in cancer treatment', The Lancet, i, 603-605.

Barratt, G., Tenu, J-P., Yapo, A. and Petit, J-F. (1986). 'Preparation and characterisation of liposomes containing mannosylated phospholipids capable of targeting drugs to macrophages', Biochim. Biophys. Acta, 862, 153-164.

Behari, J.R., Varghese, Z. and Gregoriadis, G. (1986). 'Use of liposome-entrapped diethylene-triamine pentaacetic acid in the treatment of mice loaded with cadmium', Biochem. Soc. Trans., 14, 1197-1198.

Berman, J.D., Hanson, W.L., Chapman, W.L., Alving, C.R. and Lopez-Berestein, G. (1986). 'Antileishmanial activity of liposome-encapsulated amphotericin B in hamsters and monkeys', Antimicrob. Agents Chemother., 30, 847-851.

Beutler, E., Dale, G.L., Guinto, E. and Kuhl, W. (1977). 'Enzyme replacement therapy in Gaucher's disease: Preliminary clinical trial of a new enzyme preparation', Proc. Natl. Acad. Sci. USA, 74, 4620-4623.

Beverley, P.C.L. and Riethmuller, G. (1987). 'Immunological intervention with monoclonal antibodies', Immunology Today, 8, 101-102.

Birken, S. and Canfield, R.E. (1974). 'Labelled asialo-human chorionic gonadotropin as a liver- scanning agent', J. Nucl. Med., 15, 1176-1178.

Blakey, D.C., Watson, G.J., Knowles, P.P. and Thorpe, P.E. (1987). 'Effect of chemical deglycosylation of ricin A chain on the in-vivo fall and cytotoxic activity of an immunotoxin composed of ricin A chain and anti-Thy 1.1 antibody', Cancer Res., 47, 947-952.

Blank, M.L.H, Byrd, B.L., Cress, E.A., Washburn, L.C. and Snyder, F. (1984). 'Liposomal preparations of calcium- or zink-DTPA have a high efficacy for removing colloidal ytterbium-169 from rat tissues', Toxicology, 30, 275-281.

Bourrie, B.J.P., Casellas, P., Blythman, H.E. and Jansen, F.K. (1986). 'Study of the plasma clearance of antibody-ricin-A-immunotoxins', Eur. J. Biochem., 155, 1-10.

Brady, R.O., Barranger, J.A., Furbish, F.S., Stowens, D.W. and Ginns, E.I. (1982). 'Prospects for enzyme replacement therapy in Gaucher disease', in Gaucher Disease: A Century of Delineation and Research (Eds. R.J. Desnick, S. Gatt and G.A. Grabowski), pp. 669-680, Alan R. Liss, New York.

Bregni, M., DeFabritiis, P., Raso, V., Greenberger, J., Lipton, J., Nadler, L., Rothstein, L., Ritz, J. and Bast Jr, R.C. (1986). 'Elimination of clonogenic tumour cells from human bone marrow using a combination of monoclonal antibody ricin A chain conjugates', Cancer Res., 46, 1208-1213.

Buraggi, G.L., Callegaro, L., Mariani, G., Turrin, A., Cascinelli, N., Attili, A., Bombardieri, E., Terno, G., Plassio, G., Dovis, M., Mazzuca, N., Natali, P.G., Scasselati, G.A., Rosa, U. and Ferrone, S. (1985). 'Imaging with ^{131}I-labelled monoclonal antibodies to a high-molecular weight melanoma-associated antigen in patients with melanoma: Efficacy of whole immunoglobulin and its $F(ab)_2$ fragments', Cancer Res., 45, 3378-3387.

Carrasquillo, J.A., Krohn, K.A., Baumier, P, McGuffin, R.W., Brown, J.P., Hellstrom, K.E., Hellstrom, I. and Larsen, S.M. (1984). 'Diagnosis of and therapy for solid tumours with radiolabelled antibodies and immune fragments', Cancer Treat. Rep., 68, 317-328.

Chan, S.Y.T., Evan, G.I., Ritson, G.I., Watson, J., Wraight, P. and Sikora, K. (1986). 'Localisation of lung cancer by a radiolabelled monoclonal antibody against the C-myc oncogene product', Br. J. Cancer, 54, 761-769.

Cline, M.J. (1984). 'Gene therapy', Mol. Cell. Biochem., 59, 3-10.

Cobb, L.M., Humphreys, J.A. and Harrison, A. (1987). 'The diffusion of a tumour-specific monoclonal antibody in lymphoma infiltrated spleen', Br. J. Cancer, 55, 53-55.

Connor, J., Sullivan, S. and Huang, L. (1985).
'Monoclonal antibody and liposomes', Pharmac. Ther., 28,
341-365.

Crawford, D.H., Huehns, E.R. and Epstein, M.A. (1983).
'Therapeutic use of human monoclonal antibodies', The
Lancet, i, 1040.

Daemen, T., Veninga, A., Roerdink, F.H. and Scherphof,
G.L. (1986). 'In vitro activation of rat liver
macrophages to tumoricidal activity by fate of liposome-
encapsulated muramyl dipeptide', Cancer Res., 46, 4330-
4335.

Das, P.K., Murray, G.J., Zirzow, G.C., Brady, R.O. and
Baranger, J.A. (1985). 'Lectin-specific targeting
of β-glucocerebrosidase to different liver cells via
glycosylated liposomes', Biochem. Med., 33, 124-131.

Davis, D. and Gregoriadis, G. (1987). 'Liposomes as
adjuvants with immunopurified tetanus toxoid: Influence
of liposomal characteristics', Immunology, 61, 229-234.

Davis, D., Davies, A. and Gregoriadis, G. (1987)
'Liposomes as adjuvants with immunopurified tetanus
toxoid: The immune response', Immunol. Lett., 14, 341-
348.

Davis, S.S., Illum, L., McVie, J.G. and Tomlinson, E.
(Eds.), (1984). Microspheres and Drug Targeting,
Elsevier, Amsterdam.

Deloach, J.R. and Sprandel, U. (Eds.), (1985). Red
Blood Cells as Carriers for Drugs, Karger, Basel.

Diener, E., Diner, U.E., Sinha, A., Xie, S. and
Vergidis, R. (1986). 'Specific immunosuppression by
immunotoxins containing daunomycin', Science, 231, 148-
150.

Douay, L., Gorin, N.C., Lopez, M., Casellas, P., Liance,
M.C., Jansen, F.K., Voisin, G.A., Baillou, C., Laporte,
J.P., Najman, A. and Duhamel, G. (1985). 'Evidence for
absence of toxicity of T101 immunotoxin on human
hematopoietic progenitor cells prior to bone marrow
transplantation', Cancer Res., 45, 438-441.

Epenetos, A.A., Snook, D., Durbin, H., Johnson, P.M. and
Taylor-Papadimitriou, J. (1986). 'Limitations of
radiolabelled antibodies for localization of human
neoplasms', Cancer Res., 46, 3183-3191.

Fichtner, I., Arndt, D., Elbe, B. and Reszka, R. (1984). 'Cardiotoxicity of free and liposomally encapsulated rubomycin in mice', Oncology, 41, 363-369.

Fidler, I.J. (1985). 'Macrophages and Metastasis: A biological approach to cancer therapy', Cancer Res., 45, 4714-4726.

Fiume, L., Mattioli, A. Busi, C. and Accorsi, C. (1984). 'Selective penetration and pharmacological activity of lactosaminated albumin conjugates of adenine arabinoside 5-monophosphate (ara-AMP) in mouse liver', Gut, 25, 1392-1398.

Forssen, E.A. and Tokes, Z.A. (1983). 'Attenuation of dermal toxicity of doxorubicin by liposome encapsulation', Cancer Treat. Rep., 67, 481-484.

Frankel, A.E. (1985). 'Antibody-toxin hybrids: A clinical review of their use', J. Biol. Resp. Mod., 4, 437-446.

Gabizon, A., Meshorer, A. and Barenholz, Y. (1986). 'Comparative long-term study of the toxicities of free and liposome-associated doxorubicin in mice after intravenous administration', JNCI, 77, 459-469.

Garnett, M.C. and Baldwin R.W. (1986). 'An improved synthesis of a methotrexate-albumin-791T36 monoclonal antibody conjugate cytotoxic to human osteogenic sarcoma cell lines', Cancer Res., 46, 2407-2412.

Green, R. (1985). 'Red cell ghost-entrapped deferoxamine as a model clinical targeted delivery system for iron chelators and other compounds', in Red Blood Cells as Carriers for Drugs (Eds. J.R. Deloach and U. Sprandel), pp. 25-35, Karger, Basel

Gregoriadis, G. (1973). 'Drug entrapment in liposomes', FEBS Lett., 36, 292-296.

Gregoriadis, G. (1975). 'Catabolism of glycoproteins. A possible role for sugars', in Lysosomes in Biology and Pathology (Eds. J.T. Dingle and R.T. Dean), pp. 265-294, Elsevier, Amsterdam.

Gregoriadis, G. (1980). 'The liposome drug-carrier concept. Its development and future', in Liposomes in Biological Systems (Eds. G. Gregoriadis and A.C. Allison), pp. 25-86, Wiley, Chichester.

Gregoriadis, G., (1981). 'Targeting of Drugs: Implications in Medicine', Lancet, ii, 241-247.

Gregoriadis, G. (Ed.) (1984). Liposome Technology, CRC Press, Boca Raton.

Gregoriadis, G., (1985). 'Liposomes as carriers of drugs and vaccines', Trends Biotechnol., 3, 235-241.

Gregoriadis, G. (Ed.) (1988a). Liposomes as Drug Carriers: Recent Trends and Progress, Wiley, Chichester.

Gregoriadis, G., (1988b). 'Fate of injected liposomes: Observations on entrapped solute retention, vesicle clearance and tissue distribution', in Liposomes as Drug Carriers: Recent Trends and Progress (Ed. G. Gregoriadis), pp. 3-18, Wiley, Chichester.

Gregoriadis, G. and Buckland, R. (1973). 'Enzyme-containing liposomes alleviate a model for storage disease', Nature, 244, 170-172.

Gregoriadis, G. and Neerunjun, D. (1975). 'Homing of liposomes to target cells', Biochem. Biophys. Res. Comm., 65, 537-544.

Gregoriadis, G. and Poste, G. (Eds.), (1988). Targeting of Drugs: Anatomical and Physiological Considerations, Plenum, New York.

Gregoriadis, G. and Ryman, B.E. (1972a). 'Fate of protein-containing liposomes injected into rats. An approach to the treatment of storage diseases', Eur. J. Biochem., 24, 485-491.

Gregoriadis, G. and Ryman, B.E. (1972b). 'Lysosomal localization of β - fructofuranosidase-containing liposomes injected into rats. Some implications in the treatment of genetic disorders', Biochem. J., 129, 123-133.

Gregoriadis, G. and Senior, J. (1984). 'Targeting of small unilamellar liposomes to the galactose receptor in vivo', Biochem. Soc. Trans., 12, 337-339.

Gregoriadis, G., Neerunjun, D. and Hunt, R. (1977). 'Fate of a liposome-associated agent injected into normal and tumour-bearing rodents. Attempts to improve localization in tumour tissues', Life Sci., 21, 357-370.

Gregoriadis, G., Davis, D. and Davies, A. (1987). 'Liposomes as immunological adjuvants: Antigen incorporation studies', Vaccine, 5, 143-149.

Gregoriadis, G., Morell, A.G., Scheinberg, H.I. and Ashwell, G. (1970). 'Catabolism of desialylated ceruloplasmin in the liver', J. Biol. Chem., 245, 5833-5837.

Gregoriadis, G., Swain, C.P., Wills, E.J. and Tavill, A.S. (1974). 'Drug-carrier potential of liposomes in cancer chemotherapy', The Lancet, i, 1313-1316.

Gregoriadis, G., Weereratne, H., Blair, H. and Bull, G.M. (1982). 'Liposomes in Gaucher type 1 disease: Use in enzyme therapy and the creation of an animal model', in Gaucher Disease: A Century of Delineation and Research (Eds. R.J. Desnick, S. Gatt and G.A. Grabowski), pp. 681-701, Alan R. Liss, New York.

Gregoriadis, G., Poste, G., Senior, J. and Trouet, A. (Eds.) (1984). Receptor-mediated Targeting of Drugs, Plenum, New York.

Gregoriadis, G., Garcon, N. Senior, J. and Davis, D. (1988). 'The immunoadjuvant action of liposomes: Nature of immune responses and influence of liposomal characteristics', in Liposomes as Drug Carriers: Recent Trends and Progress (Ed. G. Gregoriadis) pp. 279-307, Wiley, Chichester

Griffin, T.W., Richardson, C., Houston, L.L., LePage, D., Bogden, A. and Raso, V. (1987). 'Antitumour activity of intraperitoneal immunotoxins in a nude mouse model of human malignant mesothelioma', Cancer Res., 47, 4266-4270.

Grislain, L., Couvreur, P., Lenaerts, V., Roland, M., Deprez-Decampeneere, D. and Speiser, P. (1983). 'Pharmacokinetics and distribution of a biodegradable drug-carrier', Int. J. Pharm. (Amst.), 15, 335-345.

Hashimoto, Y., Sugawara, M., Kamija, T., Suzuki, S., Tanaka, T. and Hojo, H. (1987). 'Basic approach to application of chemoimmunoliposomes for cancer therapy', NCI Monogr., 3, 89-94.

Illum, L., Davis, S.S., Muller, R.H., Mak, E. and West, P. (1987). 'The organ distribution and circulation time of intravenously injected colloidal carriers sterically stabilized with a blockcopolymer poloxamine 908', Life Sci., 40, 367-374.

Johnson, D.A. and Laguzza, B.C. (1987). 'Antitumour xenograft activity with a conjugate of a Vinca derivative and squamous carcinoma-reactive monoclonal antibody PF1D', Cancer Res., 47, 3118-3122.

Jones, E.A. and Summerfield, J.A. (1982). 'Kupffer Cells', in The Liver: Biology and Pathobiology (Eds. I. Arias, H. Popper, D. Schachter and D.A. Shafritz), pp. 507-523, Raven Press, New York.

Juliano, R.L., Grant, C.W.M., Barber, K.R. and Kalp, M.A. (1987). 'Mechanism of the selective toxicity of amphotericin B incorporated into liposomes', Mol. Pharmacol., 31, 1-11.

Kirby, C. and Gregoriadis, G. (1984). 'Dehydration-rehydration vesicles (DRV): A new method for high yield entrapment in liposomes', Biotechnology, 2, 979-984.

Kirby, C., Clarke, J. and Gregoriadis, G. (1980). 'Cholesterol content of small unilamellar liposomes controls, phospholipid loss to high density lipoproteins in the presence of serum', FEBS Lett., 111, 324-328.

Kolata, G. (1983). 'The magic in magic bullets', Science, 222, 310-312.

Krupp, L., Chobanian, A.V. and Brecher, J.P. (1976). 'The in-vivo transformation of phospholipid vesicles to a particle resembling HDL in the rat', Biochem. Biophys. Res. Comm., 72, 1251-1258.

Leong, K.W., Kost, J., Mathiowitz, E. and Langer, R. (1986). 'Polyanhydrides for controlled release of bioactive materials', Biomaterials, 7, 364-371.

Liew, F.Y. (1985). 'New aspects of vaccine development'. Clin. Exp. Immunol., 62, 225-241.

Lopez-Berestein, G., Mehta, R., Hopfer, R.L., Mills, K., Kasi, L., Mehta, K., Fainstein, U., Luna, M., Hersh, E.M. and Juliano, R.L. (1983). 'Treatment and prophylaxis of disseminated Candida albicans infections in mice with liposome-encapsulated amphotericin B', J. Infect. Dis., 147, 939-945.

Lopez-Berestein, G., Fainstein, U., Hopfer, R., Sullivan, M., Rosenblum, M. Mehta, R., Luna, M., Hersh, E., Reuben, J., Mehta, K., Juliano, R.L. and Bodey, G. (1985). 'A preliminary communication: Treatment of systemic fungal infections in cancer patients with liposome-encapsulated amphotericin B', J. Infect. Dis., 151, 704-710.

Manabe, Y., Tsubota, T., Haruta, Y., Kataoka, K., Okazaki, M., Haisa, S., Nakamura, K. and Kimura, I. (1984). 'Production of a monoclonal antibody-methotrexate conjugate utilizing dextran T-40 and its biological activity', J. Lab. Clin. Med., 104, 445-454.

Marx, J.L. (1982). 'Monoclonal antibodies in cancer', Science, 216, 283-285.

Matzku, S., Kirchgessner, H., Dippold, W.G. and Bruggen, J. (1985). 'Immunoreactivity of monoclonal anti-melanoma antibodies in relation to the amount of radioactive iodine substituted to the antibody molecule', Eur. J. Nucl. Med., 11, 260-264.

Mayhew, E.G., Goldrosen, M.H., Vaage, J. and Rustum, Y.M. (1987). 'Effects of liposome-entrapped doxorubicin on liver metastases of mouse colon carcinomas 26 and 38', JNCI, 78, 707-713.

McIntosh, D. and Thorpe, P. (1984). 'Role of the B-chain in the cytotoxic action of antibody-ricin and antibody-abrin conjugates', in Receptor-Mediated Targeting of Drugs (Eds. G. Gregoriadis, G. Poste, J. Senior and A. Trouet), pp.105-118, Plenum, New York.

McIntyre, N. (1986). 'Hepatic functions in health and disease: Implications in drug carrier use', in Targeting of Drugs with Synthetic Systems (Eds. G. Gregoriadis, J. Senior and G. Poste), pp. 87-96.

Nicolau, C., LePape, A., Soriano, P., Fargette, F. and Juhel, M-F. (1983). 'In-vivo expression of rat insulin after intravenous administration of the liposome-entrapped gene for rat insulin I', Proc. Natl. Acad. Sci. USA, 80, 1065-1072.

Ogihara, I, Kojima, S. and Jay, M. (1986) 'Tumour uptake of ^{67}Ga-carrying liposomes', Eur. J. Nucl. Med., 11, 405-411.

Olsnes, S. and Pihl, A. (1982). 'Chimeric toxins', Pharmac. Ther., 15, 355-381.

Patel, K.R., Tin, G.W., Wiliams, L.E. and Baldeschwieler, J.D. (1985). 'Biodistribution of phospholipid vesicles in mice bearing Lewis lung carcinoma and granuloma', J. Nucl. Med., 26, 1048-1055.

Perez-Soler, R., Lopez-Berestein, G., Kasi, L.P., Cabanillas, F., Jahns, M., Glenn, H., Hersh, E.M. and Haynie, T. (1985). 'Distribution of Technetium-99m-labelled multilamellar liposomes in patients with Hodgkin's disease', J. Nucl. Med.,26, 743-749.

Phillips, N.C., Moras, M.L., Chedid, L., Lefrancier, P. and Bernard, J.M. (1985). 'Activation of alveolar macrophage tumorcidal activity and eradication of experimental metastases by freeze-dried liposomes

containing a new lipophilic muramyl dipeptide
derivative', Cancer Res., 45, 128-134.

Poste, G. and Kirsh, R. (1983). 'Site-specific
(targeted) drug delivery in cancer therapy',
Biotechnology, 1, 869-878.

Rahman, Y.E. (1988). 'Use of liposomes in metal
poisonings and metal storage diseases', in Liposomes as
Drug Carriers: Recent Trends and Progress (Ed. G.
Gregoriadis), pp. 461-472, Wiley, Chichester.

Rahman, Y.E., Cerny, E.A., Patel, K.R., Lau,. E.H. and
Wright, B.J. (1982). Differential uptake of liposomes
ranging in size and lipid composition by parenchymal and
Kupffer cells of mouse liver, Life Sci., 31, 2061-2071.

Raymond, C.A. (1987). 'Liposomes embark on rescue
mission to make highly toxic drugs more useful', JAMA,
257, 1143-1144.

Rogers, J.C. and Kornfeld, S. (1971). 'Hepatic uptake
of proteins coupled to fetuin glycopeptides', Biochem.
Biophys. Res. Comm., 45, 622-629.

Ryman, B.E. and Tyrrell, D.A. (1980). 'Liposomes -bags
of potential', Essays Biochem., 16, 49-98.

Scherphof, G., Roerdink, G., Waite, M. and Parks, J.
(1978). 'Disintegration of phosphatidylcholine
liposomes in plasma as a result of interaction with
high-density lipoproteins', Biochim. Biophys. Acta, 542,
296-307.

Sells, R.A., Owen, R.R., New, R.R.C. and Gilmore, I.T.
(1987). 'Reduction in toxicity of doxorubicin by
liposomal entrapment', The Lancet, ii, 624-625.

Seltzer, S.E. (1988). 'Liposomes in diagnostic
imaging', in Liposomes as Drug Carriers: Recent Trends
and Progress (Ed. G. Gregoriadis), pp. 509-525, Wiley,
Chichester.

Seltzer, S.E., Gregoriadis, G. and Dick, R. (1988).
'Evaluation of the dehydration-rehydration method for
production of contrast-carrying liposomes', Invest.
Radiol., In press.

Senior, J. and Gregoriadis, G. (1982). 'Is half-life of
circulating small unilamellar liposomes determined by
changes in their permeability?', FEBS Lett., 145, 109-
114.

Spanjer, H.H. and Scherphof, G. (1983). 'Targeting of lactosylceramide-containing liposomes to hepatocytes in vivo', Biochim. Biophys. Acta, 734, 40-47.

Storm, G., Roerdink, F.H., Steerenberg, P.A., deJong, W.H. and Crommelin, D.J.A. (1987). 'Influence of lipid composition on the antitumour activity exerted by doxorubicin-containing liposomes in a rat solid tumour model', Cancer Res., 47, 3366-3372.

Szoka, F.C. and Mayhew, E. (1983). 'Alteration of liposome disposition in vivo', Biochem. Biophys. Res. Comm., 110, 140-146.

Teradaira, R., Kolb-Bachofen, V., Schlepper-Schafer, J. and Kolb, H. (1983). 'Galactose-particle receptor on liver macrophages: Quantitation of particle uptake', Biochim. Biophys. Acta, 759, 306-310.

Tirrell, D.A., Donaruma, L.G. and Turek, A.B. (Eds.) (1985). Macromolecules as Drugs and as Carriers for Biologically Active Materials, Ann. N.Y. Acad. Sci., vol.446.

Vitetta, E. and Uhr, J.W. (1985). 'Immunotoxins: Redirecting nature's poisons', Cell, 41, 653-654.

Wolff, B and Gregoriadis, G. (1984). 'The use of monoclonal anti-Thy$_1$ IgG$_1$ for the targeting of liposomes to AKR-A cells in vitro and in vivo', Biochim. Biophys. Acta, 802, 259-273.

Wu, G.Y. and Wu, C.H. (1987). In-vivo targeted gene delivery and expression in rat liver, AASLD, 38th Annual Meeting, Abstract 338.

Drug Carrier Systems
Edited by F.H.D. Roerdink and A.M. Kroon
© 1989 John Wiley & Sons Ltd.

THE USE OF ANTIBODIES AND POLYMERS AS CARRIERS OF CYTOTOXIC DRUGS
IN THE TREATMENT OF CANCER

R. Arnon, E. Hurwitz and B. Schechter

Department of Chemical Immunology,
Weizmann Institute of Science, Rehovot, Israel

I. INTRODUCTION

The idea of immuno-targeting of antineoplastic drugs to tumor cells, by their attachment to specific anti-tumor antibodies attracted many scientists. The prospect is that such reagents would recognize the cancerous cells and would thus selectively increase the local concentration of the drug, leading to site-directed killing of the tumor cells, while a lower systemic drug concentration is maintained. In this way the selectivity of the drug toxicity for the tumor cells might be enhanced. While at the beginning of this research the possibility of obtaining antibodies capable of specifically recognizing tumor cells was rather limited, the development of the hybridoma technique has greatly overcome this difficulty. Today, numerous monoclonal antibodies directed to a large variety of tumor cells have been elicited, both against experimental tumors in animals and against primary human tumors. Hybridomas can now be made also human-human (Abrams et al, 1986; Andreason et al, 1986) or even custom designed, by genetic engineering, to contain a human Ig-Fc portion with a mouse antibody binding site (Tan et al, 1985; Sahagan et al, 1986; Nishimura et al,1987). Such antibodies maintain their antigenic binding portion without some of the undesirable side effects of the Fc, such as its antigenicity which may lead to the production of anti-antibodies (Klein et al, 1986).

In spite of this enormous advancement, it is still not clear whether antibodies could become effective and practical as specific drug carriers. Out of many recent studies (Biddle et al, 1986; Bumol et al, 1986; Dillman et al, 1986; Endo et al, 1987; Garnett et al, 1986; Lee et al, 1986; Nepo et al, 1985; Rowland et al, 1986; Smyth et al, 1986; 1987a, 1987b), only a few indicate unequivocally an advantage to the use of drug-antibody conjugates. Some results suggest that the simultaneous delivery of drug and antibody, possibly as non-covalent complexes (Dillman et al, 1986; Schecter et al, 1987a; Smyth et al, 1986), or even their injection at the different times and routes (Hurwitz et al, 1986; 1987), may be at least as beneficiary against the tumors, as the covalent antibody-drug conjugates. The antibody, in some of these studies, had by

33

itself a therapeutic effect which may have been indirect, involving the recruitment of cells or factors cytoxic to the tumor, while the drug exerted its effects independently. In other cases, the drug and the antibody had a synergistic effect and were, in combination, even more effective than the sum of their separate activities. In these cases, one could assume either an in-vivo interaction between the drug and the antibody, or an indirect mutual effect, whereby one of them renders the tumor cells more susceptible to the other.

The nature of tumor-associated antigens to which the antibodies were directed varied in different systems. In many cases these were cell membrane antigens while in others they were secreted antigens, only transiently expressed on the surface of the tumor target cells. Studies in the past have suggested that it was possible to use antibodies directed against such secreted antigens and that antigen-antibody complexes in circulation did not necessarily interfere with the activity of drug-antibody conjugates (Tsukada et al., 1982b).

It should be mentioned that the mere macromolecularization of the drug might also have an effect. On one hand, the chemical binding of the drug could result in a loss of its activity; on the other hand, if the resulting drug complex or conjugate still retains its activity, this might differ in some respects from that of the low molecular weight substance. Macromolecular conjugates of anti-tumor drugs could affect their distribution in the body and facilitate slow and continuous release, while ensuring better stability of the active substances. This attachment could thus reduce the overall toxic effects of these compounds without impairing their beneficial activity. This approach has been demonstrated with some anti-cancer drugs (e.g. Szekerke et al, 1972; Chu and Whiteley, 1977; Bernstein et al, 1978) and is further explored by us as will be evidenced in the following.

II. CONSIDERATIONS IN DRUG TARGETING

A) The Antibody: In order for an antibody to be a suitable drug carrier it has to be highly specific and to possess high binding affinity and avidity towards the tumor target cell. When conservative polyclonal antibodies, prepared against tumor-associated antigens, were employed, exhaustive absorption by normal tissue was required to ensure adequate specificity. Monoclonal antibodies have both advantages and disadvantages over the polyclonal ones. They are advantageous since they can be readily obtained by immunization with whole tumor cells, including neoplasms which do not bear a known tumor marker. They can also be prepared from naturally occurring antibody-producing peripheral blood lymphocytes by in-vitro culturing procedures, thus leading to human-human hybridomas. They are uniformly specific and, if directed against an abundant antigen, may show very high efficiency. Their uniformity in some cases can be of a disadvantage, for example if they are directed toward a scarce antigen on the tumor cell, or if the drug-binding modification leads to loss of their reactivity.

Polyclonal antibodies may be advantageous, since they often show higher avidity to the tumor cells due to their simultaneous reactivity with many antigenic determinants. Their polyclonality also makes them less sensitive to loss of antibody activity during the drug binding procedures. It is thus apparent that both polyclonal and monoclonal antibodies should be considered in future efforts toward immunotargeted chemotherapy.

The antibody isotype may also be of significance. In various studies in-vitro (Herlyn et al, 1985) and in-vivo (Steplewski, et al, 1985; Denkers et al, 1985; Seto et al, 1986; Lowder et al, 1987) different Ig isotypes were evaluated for their cytotoxic efficacy against the respective tumor cells. In-vitro, all IgG sub-classes were able to cause antibody-mediated cell-cytoxicity. Complement-mediated cytoxicity was mostly restricted to IgG2a and IgM, the latter being the most effective. On the other hand, IgM could be limited in its ability to cross blood barriers or to penetrate a tumor growth and was shown unable to localize one tumor tested (Pimm et al, 1985). IgG was found to have a faster clearance rate (Denkers et al, 1985) which for targeting could be advantageous. Earlier results in-vivo claimed effectively of IgG2a only (Miller et al, 1983). However, recent studies were not as conclusive and no direct correlation was found between the isotypes (IgG2a vs IgG1) and their in-vivo anti-tumor reactivity (Lowder et al, 1987). It is thus apparent that for their use as targeting devices the criteria for choice should probably be the binding affinity of the antibody rather than its isotypic characteristics.

The most common experimental tumor systems for studying targeting to human cancers are those xenografting in nude mice. This is a very artificial model and possibly some of the "failures" in drug-targeting by antibodies may have arisen from the use of nude mice. While these mice lack the thymus, they still have a vigorous host versus graft reaction, (by non T-cell mediated immune mechanisms), which in some cases had to be suppressed prior to the tumor implantation. This was done by irradiating the mice at 7-10 days prior to the experiment (Shouval et al, 1981). It is thus possible that in the nude mice model an antibody alone can elicit an anti-tumor effect by recruiting some of the cells, which already have a strong anti-tumor effect. Some results showing an anti-tumor effect by mixtures of drugs and antibodies may be explained on this basis.

It has been shown that repeated injections of antibodies to tumor cells may modulate the antigenic expression of the cells (Ritz et al, 1982; Galun et al, 1987). The effect of antigenic modulation was enhanced if the treatment was with an immunotoxin rather than by the respective antibody alone (Manske et al, 1986). It is therefore apparent that for repeated treatments by drug-antibody conjugates, more than one type of antibody should be employed (Ceriani et al, 1987).

B.   The Choice of Drugs: Many anti-cancer drugs are available and could be potential candidates for the drug targeting approach.   The prerequisite is that the drug will have a functional group by which it could be either linked or complexed to the carrier, without a drastic loss of its pharmacological activity.   We have concentrated our efforts mostly on two types of drugs.   The anthracycline analogue and platinum derivatives.   Of the anthracycline drugs, adriamycin (doxorubicin) has the widest spectrum of antitumor activity among chemotherapeutic agents and is used with a high degree of efficacy in the treatment of many human cancers worldwide (Weiss et al, 1986).   The major obstacle to its use is its cumulative cardiotoxicity.   New derivatives, potent and less toxic have been developed but are still not in clinical use yet.   At least one of them, epirubicin, an isomer of adriamycin could be attached to antibodies and polymers by the same procedures as adriamycin.

The platinum derivatives we used are cis-diclorodiammine-platinum(II) (cis-DDP) and cis-diaquodiammine platinum(II) (cis-aq). Cis-DDP is a highly effective antitumor agent currently used in the treatment of various human tumors mainly testicular and ovarian carcinomas (Einhorn and William, 1979).   This drug, however, is very toxic causing pronounced nausea, vomiting and in particular damage to the kidney which may result in long or irreversible renal disfunction (Groth et al., 1986).   Its structural analogue cis-aq is similar in its properties to cis-DDP, but its toxic effects are even more pronounced, and is therefore not used clinically. One possible approach for decreasing the toxic side effects of the cis-platinum derivatives and increasing their antitumor efficacy is to use them in association with high molecular weight carriers that will alter their differential delivery in-vivo.

III. THE CHEMISTRY OF DRUG CONJUGATES WITH ANTIBODIES

Most of the coupling procedures used by us involved the binding of the drug to the antibody via a spacer, which enabled the attachment of a high number of drug molecules to each antibody molecule without affecting its activity.   We have used a variety of dextrans and their derivatives or poly-glutamic acid for this purpose.

Dextran is considered as a most appropriate drug carrier to be used either as such or as an intermediate carrier in the drug targeting (Bernstein et al, 1978).   It has the advantage of being inexpensive, non-degradable, highly soluble, non-toxic, and has been used already in man as a plasma volume expander (Rowland, 1983).   In order to use dextran for targeting of drugs it is necessary to introduce in it functional groups, which are essential for both drug attachment and linkage to antibodies.

The first derivative we have used is oxidized dextran. Dextrans of M.W. 10,000 and 40,000 (pharmacia, T-10 and T-40, respectively) were oxidized by sodium periodate at a 0.03 M concentration.   Periodate was added at half equal amounts to glucose residues (which was considered a 50% oxidation).   The drugs adriamycin, daunomycin, (Hurwitz et al, 1975) and cytosine arabinoside (Hurwitz et al, 1985) were bound through their amino group to the aldehydic functions of oxidized dextran, whereas the antibodies were attached to the same 24 hours later (Hurwitz et al., 1978a). The Schiff bases thus formed were stabilized by reduction with sodium borohydride or cyanoborohydride, depending on the drug sensitivity to the reducing agent. The cytosine arabinoside-dextran conjugate was completely reduced (namely in equimolar amounts of the reducing agent and the periodate used for the dextran oxidation). The daunomycin-dextran conjugate was partially reduced, whereas the adriamycin-dextran conjugate was not reduced at all. These latter two derivatives were stable in phosphate buffered saline over long periods of time and were only partially unstable at acidic conditions. It is, therefore, assumed that at least part of the bonds formed were oxazolidine rings between the amino group and its vicinal hydroxy group (Hurwitz et al, 1980). The antibody bound to periodate-oxidized -dextran, through the formation of Schiff bases between its free amino groups and the dextran's aldehyde functions, is stable without reduction, possibly because of the multiplicity of the binding sites.

Another derivative of dextran was carboxymethyl dextran (CM-dex) which was obtained by reaction of dextran with monochloroacetic acid (Hurwitz et al., 1980). CM-dex was used as such for complexing cis-platinum drugs. It was also further modified by the binding of hydrazine or adipic dihydrazine, which resulted in the formation of dextran-hydrazide. The dextran-hydrazide conjugate reacted with the carbonyl group of daunomycin or adriamycin, each residue capable of binding one moiety of the drug (Hurwitz et al, 1980; 1983a). The resulting adriamycin-dextran-hydrazone could be bound to antibodies, for example anti-neuroblastoma, by cross linking with glutaraldehyde. The same type of conjugation was utilized for the binding of a 5-fluorouridine derivative to antibodies. Fluorouridine was oxidized by periodate in order to open the sugar ring and form two aldehyde groups, which in turn bound to dextran-hydrazide. The antibody was then cross-linked to the spacer, as above, by glutaraldehyde (Hurwitz et al, 1985). In a similar procedure, dextran-adipic dihydrazone was utilized with even better yields (unpublished data).

The other spacer we used for binding drugs to antibodies is poly-L-glutamic acid (PG). This polymer is a non-toxic poly-anion that had also been suggested as a blood volume expander and was used in various studies as a drug carrier (Rowland, 1983). In our studies PG was used for attachment of platinum derivatives and also for linking of adriamycin to antibodies. In the latter case, PG was derivatized with adipic dihydtrazide (similarly to CM-dex). The same chemical group was used for the attachment of the antibody. The antibody's polysaccharide chain was oxidized by periodate prior to the reaction, to form the aldehyde groups necessary for its binding. The oxidation of the sugar moiety of the antibody and its binding to the polymer had no effect on its antigen binding capacity (Galun et al, 1987).

IV.   DAUNOMYCIN AND ADRIAMYCIN CONJUGATES

In our early studies, we attached daunomycin and/or adriamycin either directly (Hurwitz et al, 1975; Levy et al, 1975) or via a dextran bridge (Hurwitz et al, 1978b) to antibodies against a B leukemia, a plasmacytoma (RPC5), a T lymphoma (YAC, induced by Moloney virus) and a carcinoma (3LL) (Hurwitz et al, 1979). The resulting conjugates possessed both antibody and drug activity, and were cytoxic in-vitro. Some of them exhibited differential in-vitro and in-vivo activities (Hurwitz et al, 1978), and compared favourably with controls which consisted of free drug, drug bound to non-specific immunoglobulin G, antibody alone or mixtures of drug and antibody. These results were described in several previous review articles (Arnon and Sela, 1982a and 1982b) and will not be further discussed. More recent studies have indicated that similar conjugates can be prepared with antibodies more relevant to human tumors, as is described in the following.

A.   Conjugates with Anti-Alpha-Fetoprotein: Alpha-fetoprotein (AFP)-secreting hepatoma cell lines (human or rat) are capable of binding antibodies specific to this protein, in-vitro (Koji et al, 1980; Tsukada et al, 1982b; 1985). Furthermore, anti-rat-AFP was shown to localize in-vivo on the tumor cells (Koji et al, 1980), and to localize in liver cancer without interference of circulating AFP (Goldenberg, 1985). The administration of the antibodies to tumor-bearing rats reduced AFP-secretion and affected tumor development (Tsukada et al., 1982b). This experimental tumor system seemed, therefore, suitable for drug immunotargeting studies.

Daunomycin was linked via oxidized dextran T-10 to both monoclonal and polyclonal antibodies against rat AFP (Tsukada et al 1982a, 1982b). In different preparations, the drug activity of the conjugates, as determined by the inhibition of [^{3}H]thymidine incorporation into the cells, amounted to 60-100% of the free drug activity. The antigen binding activity of the antibody was also fully retained in the conjugates, as determined by their reaction with ^{125}I-labeled AFP. The conjugates of daunomycin with the polyclonal horse anti-AFP and the mouse monoclonal anti-AFP antibodies showed similar and specific cytotoxicity towards the hepatoma cells in-vitro, both demonstrating higher efficacy than free daunomycin. High AFP levels, which accompany tumor growth, were reduced by the specific drug-antibody conjugate even more efficiently and over a longer period of time than by anti-AFP itself.

Rats challenged intraperitoneally (i.p.) with 10^4 rat hepatoma (AH66) cells were treated with a multiple dose administration of the specific drug-antibody conjugates of both monoclonal and polyclonal antibodies, and with the controls which included antibody and drug alone, mixture of the two, as well as a conjugate of drug-dextran with a non-relevant immunoglobulin (Fig.1). The treatments were given 5 times, on alternative days, starting from the third day after the tumor injection, and

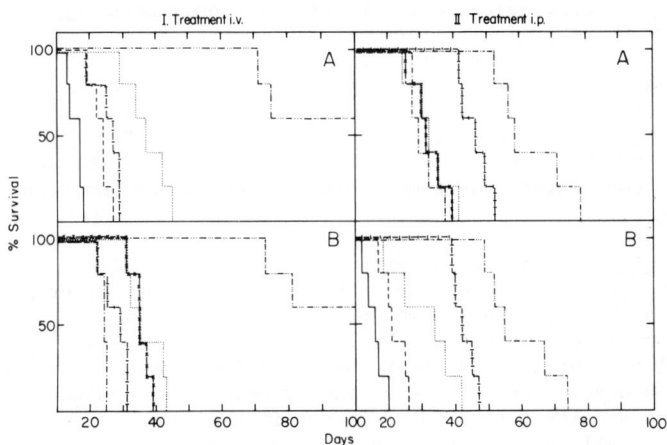

Fig 1. The therapeutic effectivity of daunomycin conjugates with polyclonal (A) and monoclonal (B) antibodies against rat AFP. (-), PBS control; (---), normal horse IgG; (-.-), free daunomycin; (x-x-x), daunonomycin dextran; (...), antibody alone; (-1-1-), a mixture of daunomycin and antibody; (-....-), specific daunomycin antibody conjugate. From: Ysukada et al (1982b).

were administered intravenously (i.v). or i.p. (Tsukada et al, 1982b). It is of interest that the efficacy of the monoclonal anti-AFP and its drug conjugate was similar to that of conventional horse anti-rat AFP and its drug derivative. Both conjugates were much more effective than the free drug and antibody, or even than mixtures of daunomycin and antibody, which caused only slight delay in the median survival time by both routes of treatment. The specific conjugate injected i.p. prolonged the survival time but did not prevent death from the tumor. The intravenously administered specific conjugate prevented tumor development in most of the mice. The reason for the lower effectivity of the i.p. treatment is not clear, particularly in view of the fact that material injected via the i.p. route could have reached the tumor directly in this case. A possible explanation is that intraperitoneally the drug was more toxic, and consequently less effective at the lower doses administered, that could be tolerated by the mice. The high efficacy demonstrated by the antibody-drug conjugates when injected intravenously may indicate specific homing to the target.

B. Studies with Anti-Human AFP and anti-hepatitis B surface antigen: Several drug conjugates of mouse monoclonal anti-human AFP were prepared. Cytosine arabinoside (Shouval et al, 1986), daunomycin and adriamycin (Galun et al, 1987) were each attached to these antibodies, either via polyglutamic acid or via dextran T-10 or T-40 spacers. The general toxicity and the specific activity of these conjugates were tested with a human hepatoma cell line PLC/PRF/5, which is AFP-secreting. The most effective conjugates were those with adriamycin.

The pharmacological activity of the conjugates <u>in-vitro</u> was evaluated by the inhibition of the incorporation of either [^{3}H] leucine or [^{3}H] thymidine into the cells. [^{3}H] thymidine incorporation was shown to be in good correlation with clonogenic ability of human tumor cells (Jones et al, 1985). Incidentally, anti-AFP by itself had an enhancing effect on the DNA and protein synthesis by the tumor cells. The drug activity of the adriamycin-antibody conjugates was comparable or even higher than that of the free drug.

<u>In-vivo</u>, in the xenograft model in nude mice, 5-10x 10^6 PLC/PRF/5 were injected subcutaneously 10 days after the mice received a 400 rad irradiation (Shouval et al, 1981). The effect of the adriamycin conjugates was tested by multiple i.v. injections, started 1 to several days after the tumor inoculation and repeated every 3-4 days. The effect of the treatment was evaluated by the tumor size and its rate of development in the mice. As shown in Figure 2, all drug conjugates tested, whether specific, namely antibody bound, or just bound to the respective polymer spacer (poly glutamic or dextran) were more effective than the free drug. In the first part of the experiment, the specific drug-anti-AFP conjugate via the PG spacer was the most effective treatment. Yet, later on, the results obtained with adriamycin-dextran or adriamycin-PG were similar or even slightly better than those obtained with the specific antibody conjugates (Fig 2). This could have been due either to the enhancing effect the antibody itself had on the tumor growth, or to the changes in the tumor cell antigenicity. This assumption of tumor-associated antigen

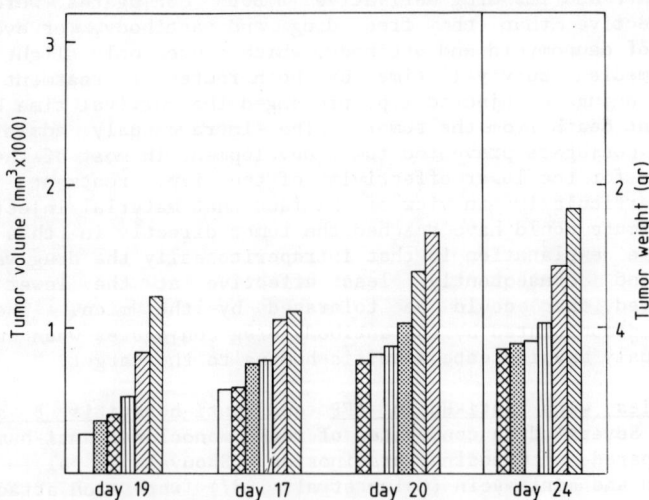

Fig 2. The effect of adriamycin-anti-AFP conjugates on the growth of the hepatoma PLC/PRF/5 in nude mice.
▢ Adr-PG-anti-AFP; ▨ adr-dex-anti-AFP; ▨ adriamycin-extran; adr-PG; ▨ control; ▨ anti-AFP.

modulation is sustained by the finding that the percent of cells which were able to bind anti-AFP at the end of the experiment was lower than that of the hepatoma cells grown in culture (about 40%). Results obtained by Ohkawa et al, 1986 and Tsukada et al., 1985, using anti-human AFP for daunomycin targeting against two types of human tumors, hepatoma and yolk sac tumors, were much more encouraging. In those studies the conjugates, which were prepared similarly to ours, were very effective in reducing tumor growth. These results were not entirely comparable to ours for several possible reasons - different tumor lines were used, the antibody was polyclonal, the treatment was given intraperitoneally and the drug used was daunomycin. The latter drug as such, injected intraperitoneally, was in our hands inferior to adriamycin.

In order to overcome the problem of antibody induced tumor-associated antigens modulation, it was considered useful to employ more than one type of antibody in each tumor system. The hepatoma PLC/PRF/5 contains hepatitis B virus (HBS) DNA and consequently is secreting hepatitis B surface antigen. It was shown that anti-HBS could bind to this hepatoma cells and to affect the development of the tumor xenograft nude mice (Shouval et al, 1982). Conjugates of adriamycin-dextran-anti-HBS (monoclonal, IgG_2a and IgG_1) were prepared and tested for their cytotoxicity and antigen binding activities in this tumor system (Shouval et al, 1987). Anti-hepotoma-drug-conjugates, with the adriamycin attached to either anti-AFP or anti-HBS, showed a different behaviour in-vitro than the effect observed previously with other antibody-drug conjugates and their respective tumor cells. Whereas drug-antibody conjugates with YAC cells, with a thymoma or with a human neuroblastoma were shown capable of penetrating into their respective target cells, the anti-hepatoma conjugates, showed almost no internalization capability. In preliminary experiments the drug itself, adriamycin, was taken up by the cells and their nuclei, but the drug conjugates, after an incubation of 4 hr at room temperature, showed only a very low level of penetration into the cells. The extent of penetration was measured by either optical density for monitoring the drug, or ^{125}I-radioactivity for labeled antibody. A control hepatoma cell line non-secreting either AFP or HBS, did not differ significantly from the PLC/PRF/5 cells in this respect. Possibly, the toxic activity of adriamycin in the anti-hepatoma conjugates is delivered by interactions on the cell membrane without requiring cell penetration. Such an action of adriamycin was previously observed in other tumor systems (Bredehorst et al, 1986).

In-vivo, administered intraveneously, both anti-HBS alone, and the adriamycin-anti-HBS conjugates were effective against the hepatoma in experiments (still in progress) similar to those described with anti-AFP-drug conjugates. Both types of adriamycin conjugates were inactive when injected intraperitoneally.

C. **Daunomycin Conjugates with Antibodies to a B-Lymphoma (38c 13)**:
Idiotypic determinants on the immunoglobulin that is expressed on B leukemia or lymphoma cells have been used by us (Perek et al, 1983) and by others (Maloney et al, 1985; Vitteta et al, 1979) as tumor-

specific antigens.   In our studies we used anti-idiotypic
antibodies, both   polyclonal prepared in goats and monoclonal,
specific against   the IgM molecule on the B lymphoma 38c 13 were
prepared (Maloney al, 1985).   This specific polyclonal idiotypic
antibody was used as daunomycin carrier via oxidized dextran
(Hurwitz et al, 1983b).   The conjugated antibodies maintained their
original antigen-binding activity and their specificity as shown by
binding to the purified IgM 38c 13, as described by Perek et al
(1983).   The pharmacological activity of the drug in-vitro was
likewise maintained in the conjugate at its original level.   The
chemotherapeutic efficacy of the daunomycin-antibody conjugates was
studied by their administration i.v. or i.p. into mice inoculated
with the 38c 13 tumor cells.   In this case the best results were
achieved when the treatment was administered i.p.   Under these
conditions, the free daunomycin was of no therapeutic value, while
daunomycin conjugates with the specific antibodies, effected
complete cures in most of the treated mice.   As shown in Figure 3,
the specific conjugate was effective but not more than drug and
antibody injected by this route together or at different times
(Hurwitz et al, 1986, 1987).   The idiotypic antibody by itself was
also partially effective but did not cure more than 40% of the mice.
The addition of various anti-neoplastic drugs (daunomycin, cytosine
asabinoside, 5-fluorouracil, cis-platinum and cyclophosphamide) at
doses   at which they were inactive by themselves, increased
enormously the survival rates of the tumor-injected mice.   Both
polyclonal and monoclonal anti- idiotypic antibodies were effective
in these experiments, leading to almost total cure in combination
with the various drugs.   The monoclonal antibodies in these mixtures
were more effective than the polyclonal ones and only 1/100 of the
amount was needed in order to exert the same effects.

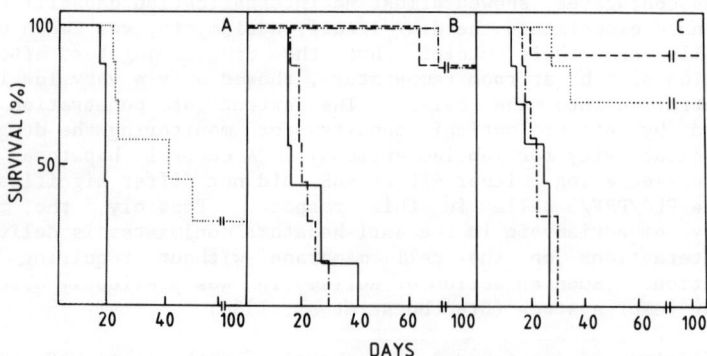

Fig 3.   The therapeutic synergistic effect of the 2B6 monoclonal
anti-38c Id and several anti-neoplastic drugs.
(A) Controls: No treatment (—); anti-38c ascites, 5 ul (·⋯).   (B)
Antibody in combination with Cis-platinum and 5-fluotouracil.   Cis-
platinum, 100 ug (—); Cis-platinum and antibody, 100 ug and 5 ul,
respectively (--), 5-fluorouracil, 750 ug (-·-); 5-fluorouracil and
antibody, 750 ug and 5 ul, respectively (-⋯-).   (C) Antibody in
combination with daunomycin and adriamycin.   Daunomycin, 50 ug (-·-);
daunomycin and antibody, 50 ug and 5 ul, respectively (--).
Adriamycin, 25 ug (-); adriamycin and antibody, 25 ug and 5 ul,
respectively (·⋯).

D. Daunomycin - Dextran Conjugates: An interesting finding was that a conjugate containing daunomycin covalently linked to dextran, which served as a control for the drug-dextran-antibody conjugates had by itself a higher antitumor activity than free daunomycin (Bernstein et al, 1978). This was evident mainly at high drug concentrations in which free daunomycin had a toxic effect. Considering the crucial importance of the issue of toxicity and its ramifications to the applicability of anti-cancer drugs, we compared the toxicity of the free and macromolecular bound daunomycin in a more detailed manner.

Our studies have demonstrated, (Levi-Schaffer et al, 1982) that both the acute and subacute toxicities of daunomycin-dextran conjugate are much lower than that of the free drug. Thus, as shown in Figure 4 the LD_{50} of the dextran conjugate is almost three times higher than that of free daunomycin and so are the values of the therapeutic indices. Consequently, in the case of daunomycin-dextran conjugate there is a range of drug concentration (between 22 and 24 mg/kg) in which there is no mortality of mice either from drug toxicity or from the tumor, in mice inoculated with YAC cells. This could be considered as a "safe region".

The lower toxicity was manifested also after multiple (four) administrations, of the optimal therapeutic dose, which resulted in 100% survival of the mice treated with the daunomycin-dextran conjugate as compared with very high mortality (90-100% of the mice similarly treated with the free drug. Furthermore, in histological examinations, which demonstrated that free daunomycin was extremely detrimental and caused massive atrophy of spleen and bone marrow, and also damages liver and heart tissue, the daunomycin-dextran conjugate had hardly any damaging effect, either immediately after or even 2 months following four injections of the therapeutic dose.

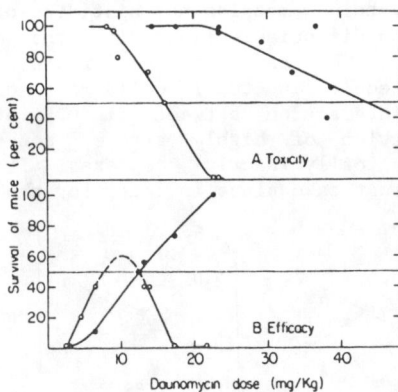

Fig 4.   Toxicity and efficacy of free daunomycin (open circles) and its dextran conjugate (solid circles).   In the toxicity experiments the drug was injected to untreated mice, and in the efficacy experiment to mice prechallenged with 10^5 cells of YAC lymphoma.

Thus, although lacking the feature of specificity this drug-carrier conjugate may still manifest an important advantage for treatment of cancer.

V. CIS-PLATINUM COMPLEXES AND CONJUGATES

Due to its chemical properties, cis-DDP is a highly suitable candidate to be used in drug-carrier systems. This drug which consists of a coordination complex (Fig 5), serves as a bifunctional reagent with both chlorine atoms open to substitution by other ligands that are either in large excess, or that form thermodinamically more stable links with platinum (II). The cis-DDP structural analog cis-aq (Fig 5) is similar in its properties to cis-DDP, except that the two water molecules replacing the chlorine atoms in cis-DDP are more reactive in exchange reaction. As such, cis-aq is too toxic for clinical use, but, as will be demonstrated in the following, in a complexed form it definitely demonstrated therapeutic potential.

A. Preparation of Cis-Platinum Complexes: The interaction of cis-DDP or cis-aq with reactive groups on macromolecules is effected through the exchange of one or both chlorine atoms or water molecules, and the thermodynamic stability of the newly formed complexes is dependent on the chemical nature of the carrier and whether the interaction is mono- or bi-functional. A prerequisite for a macromolecule to serve as a carrier for cis-DDP or cis-aq in biological systems is the presence of low affinity binding sites that can form partially stable, or "reversible" links with these compounds. Proteins, for example, are inappropriate to serve as carriers since most of the interactions occur with amino acid side chains such as cystein, methionine, histidine or arginine, which form very stable, non-dissociable links with the drug (Howe-Grant and Lippard 1980). The same is true for DNA, which is the most relevant target molecule in tumor cells for cis-DDP activity. Consequently the interaction between cis-DDP or cis-aq and DNA results in the formation of highly stable, non-dissociable, and therefore pharmacologically inactive complexes. It thus seems that drug dissociation must take place in order for the complex to exert anti-tumor activity.

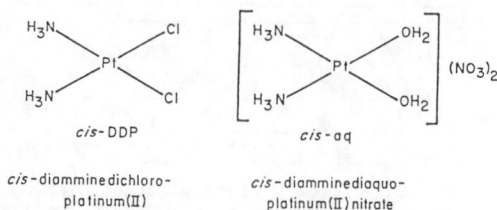

cis-DDP

cis-diamminedichloro-
platinum(II)

cis-aq

cis-diamminediaquo-
platinum(II)nitrate

Fig 5. Chemical structure of cis-DDP and cis-aq.

Polycarboxylates, such as carboxymethyl dextran or polyglutamic acid, on the other hand, form low affinity bonds with the two platinum (II) compounds and are potentially suitable to serve as drug carriers. In the course of our study we have tested the suitability of these polyanionic carriers to serve in the preparation of pharmacologically active cis-DDP or cis-aq multi-complexes.

B. **Complexes of cis-DDP or cis-aq with dextran derivatives**: The derivative we have used for the platinum (II) derivatives is CM-dex with which cis-DDP and cis-aq readily form reversible complexes. Complex preparation is simple and does not require the intervention of any coupling reagents. The platinum(II) compounds are allowed to interact with the carrier in an aqueous solution at 37°c for 17 to 24 hr, followed by dialysis to remove any uncomplexed drug. Quantitative determination of complexed drug was performed by a colorimetic reaction using the ortho-phenylenediamine (OPDA) reagent (Golla and Ayres, 1973). Complexed drug that was bound to the carrier with a relative low binding affinity would undergo a substitution reaction with the OPDA due to its higher affinity to the latter, and the light blue cis-DDP-OPDA complex is determined at 703nm. The OPDA reaction also provided a semi-quantitative measure for the dissociability of carrier-complexed drug, and hence, means for the prediction of its pharmacological activity. A distinction between high and low affinity bound drug in various complexes or within a given complex could be done by determining the OPDA-reactive platinum(II) before and after dialysis of the drug-carrier reaction mixtures. Thus, whereas in undialysed preparations both bound and unbound drugs are monitored, which in the case of complete dissociation should amount to 100% of the drug input, in dialyzed complexes only the reversibly bound cis-DDP is expressed.

Soluble complexes prepared with CM-dex T-10 and CM-dex T-40 could carry up to 10 or 40 mole drug per mole of dextran, respectively. Higher drug loads resulted in loss of solubility. At low cis-DDP or cis-aq loads, most of the drug was bound to the carriers and the majority of the bound drug was dissociable. At higher loads, the portion of reversibly bound drug decreased, indicating a saturation point at a molar ratio of around one platinum(II) per 5 glucose residues of the dextran. A small fraction of the drug became high affinity and hence irreversibly bound.

Due to the reactivity of cis-DDP or cis-aq with various ligands, complex formation was feasible only in an aqueous solution, in the absence of salts, buffers etc. Chlorine ions are especially effective in inhibiting the interaction between these platinum (II) compounds and other ligands. Since the complexes described here are designed to be used in a physiological milieu, it was important to establish the effect of NaCl on the complexes. Indeed, NaCl at physiological concentration inhibited the formation of drug CM-dex complexes, but it did not seem to cause dissociation of already formed complexes. The complexes were also shown to be stable in

water upon storage at 4°c for several weeks. Complexed cis-DDP
or cis-aq can, however, undergo substitution or exchange reactions
with ligands exhibiting higher affinity to the platinum (II)
compounds, such as the OPDA reagent, or more relevant, DNA
(Schechter et al., 1986a). These findings suggest that when
introduced into a biological system, the complexed drug would not be
dissociated by NaCl but rather by other substances that form
thermodinamically more stable links with the platinum(II)
compounds, among which DNA is apparently of the highest stability.

Since the interaction between cis-DDP or cis-aq and DNA is
considered to be the primary biochemical event leading to inhibition
of cellular DNA synthesis, this parameter was taken as a measure for
complex activity of the complexes when tested in-vitro against tumor
cells. Indeed, drug complexes of CM-dex T-10 and CM-dex T-40 were
shown to be active in inhibiting DNA synthesis in a variety of tumor
cells, some of which is illustrated in Figure 6. The complexes were
as active as the free drug, indicating that all of the complexed
drug was pharmacologically active. The inhibitory activity was
not restricted to DNA only, but also to another parameter of
cellular function such as protein synthesis, indicating a
substantial damage to the tumor cells, (Schechter et al., 1986b).
The low and high molecular weight complexes were active in-vitro to
a similar extent, and no major differences were noted between
complexes of cis-DDP and cis-aq.

The CM-dex complexes of cis-DDP and cis-aq were also tested
in-vivo against the cis-DDP sensitive F9 embryonal carcinoma in
Balb/c mice (Table 1). It should be indicated, that as with other
chemotherapeutic agents, that are inherently of cytotoxic nature,

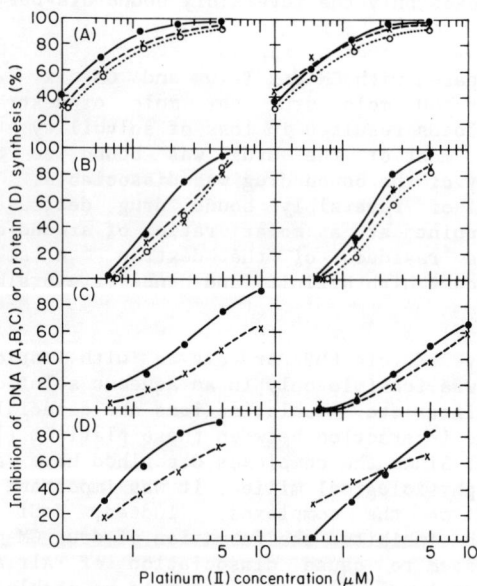

Fig. 6. Antitumor
activity in-vitro of
cis-DDP (left) and
cis-aq (right) (●)
and their complexes
of cm-dex T-10 (o)
or cm-dex T-40 (x)
against murine F9
embryonal carcinoma
(A and D), 38c-13 B
lymphoma (B) and
human osteogenic
sarcoma (C).

Table 1. The therapeutic effect of CM-DDP and CM-aq complexes of CM-dex[a].

Treatment	Drug (mg/kg)	cis-DDP ILS[b] (%)	cis-DDP S[c] (%)	cis-aq ILS (%)	cis-aq S (%)
Free drug	1.5	65	20		
(day 1)	2.0				80
	2.5	270	72	0	20 (20)
	3.75	–	80	0	0 (60)
	5.0	–	20 (40)	–	0 (100)
	7.5	–	0 (80)		
Drug-CM-dex T-10	2.5	45	23	30	20
(day 1)	5.0	100	57	78	0
	7.5	80	57	132	0
	10.0	–	80	70	40
Drug-CM-dex T-40	2.5	65	57	110	20
(day 1)	5.0	–	100	–	100
	7.5	–	43 (43)	–	100
	10.0	–	0 (80)	–	60 (40)
Drug-CM-dex T-10	5.0x3	–	100	–	100
(days 1,5,10)	7.5x3	–	80	–	100
Drug-CM-dex T-40[d]	2.5x2	–	100	–	100
	5.0x2	–	100	–	100

[a] BALB/c female mice (5-7 mice/group) were inoculated i.p. with 4x10[5] F9 cells. The mean day of survival (MDS) of control untreated mice was 24.
[b] ILS - increased life span
[c] S - Long term survivors. Numbers in parenthesis refer to the percentage of toxic death.
[d] Treatment by 2.5 or 5 mg/Kg were given at days 1 and 5 or 1 and 24, respectively.

treatment by free cis-DDP which is of relative high toxicity, was restricted to a narrow dose range. This often led to mis-treatment of the tumor-bearing mice leading to either sub-effectivity of the drug or death due to toxicity. The CM-dex T-10 complexes were less toxic than the free drug, but also less active at drug levels equivalent to the therapeutic doses of free cis-DDP. This complex was, however, highly effective in suppressing tumor growth when administered at much higher doses or upon repeated treatments that were well tolerated by the tumor-bearing mice and consequently led to 100% survival of the mice in most experiments. The CM-dex T-40 complexes were similar in their activity to the free drug, but their efficacy was somewhat higher. Hence it was easier to achieve 100% survival with these complexes than with the free drug. CM-dex T-40 complexes seem to be most appropriate for use in drug targeting mainly due to its high content of releasable drug. The attachment of 2-3 moles of cis-DDP-carrying CM-dex to one mole of immunoglobulin was shown to provide a high content of drug in the conjugate without impairing the antibody activity (see below).

C. **Complexes of Cis-DDP or Cis-aq with Poly-L-Glutamic Acid (PG)**: Polyglutamic acid (PG) was found very suitable for the attachment of cis-platinum compounds. Cis-DDP or cis-aq complexes with PG were prepared using two PG preparations, one of Mr-13000 and the other Mr-58000. Soluble complexes of the polymers could carry up to one mole drug per 5 mole of glutamyl residues, namely up to 20 or 90 mole drug per mole of the low and high molecular weight PG, respectively. However, in contrast to the CM-dex complexes, the interaction of the platinum(II) compounds with PG was of higher thermodynamic stability. First, under conditions leading to soluble complexes more than 90% of the drug in the reaction mixture was bound to the PG, and second, only 50-60% of the complexed drug was dissociable, i.e., reacted with OPDA. The increased stability of the PG-complexes was also apparent from their low anti-tumor activity in-vitro, that was equivalent to only 15-25% of the drug in the complex (Schechter et al, 1987a). Nonetheless, these complexes were better chemotherapeutic agents than free cis-DDP when tested against the F9 tumor in-vivo as reflected by their much lower toxicity, wide therapeutic range and increased efficacy. The cis-DDP or cis-aq complexes of PG were only partially active in-vivo at the equivalent therapeutic dose of the free drug (5mg/kg). However, due to their lower toxicity, they could be administered at higher doses, and at 8-12mg/kg (equivalent to OPDA reactive drug) they were as effective in preventing tumor growth as 5mg/kg of free cis-DDP. The marked difference between free and complexed drug was that, whereas cure of tumor-bearing mice by the free cis-DDP was restricted and not always successful, due to its narrow therapeutic dose range, the complexes were significantly less toxic, and their highest tolerated dose was 3-3.5 fold higher than their therapeutic level. Thus, at a concentration range of 10-35mg/kg for PG Mr-13000 and 10-30mg/kg for PG Mr-58000 no mortality occurred in the mice either from the tumor or due to drug toxicity (Fig 7). The PG complexes could be administered at relatively high doses and consequently their advantage was even more pronounced in mice carrying higher tumor loads when given either in a single or repeated doses (Table 2).

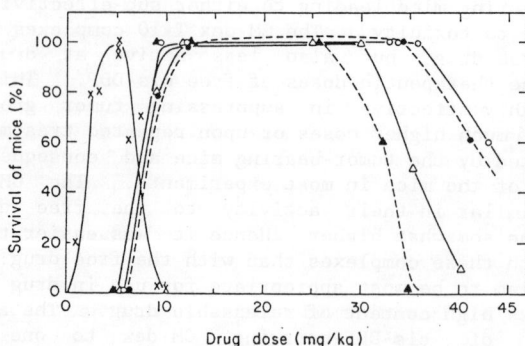

Fig 7. Therapeutic activity and toxicity of cis-DDP (X), cis-DDP-PG Mr-13000 (o), cis-DDP-PG Mr-58000 (▲), cis-aq PG Mr-13000 (●) and cis-aq-PG Mr-58000 (▲).

Table 2. The therapeutic effect of cis-DDP and cis-aq complexes of PG[a]

Treatment	Day of treatment	Drug (mg/kg)	ILS (%)	S
I cis-DDP	11	5	20	0/7
cis-DDP-PG Mr-13,000	11	32	70	1/7
cis-DDP-PG Mr-58,000	11	24	60	0/7
cis-aq-PG Mr-13,000	11	32	5	0/7
cis-aq-PG Mr-58,000	11	24	45	0/7
II cis-DDP	9,12,15	5.5x3	-	- (100)[b]
cis-DDP-PG Mr-13,000	9,12,15	32x3	20	3/7
cis-DDP-PG Mr-58,000	9,12,15	23x3	-	7/7
cis-aq-PG Mr-13,000	9,12,15	23x3	65	3/7
III cis-DDP	12	7.5	0	0/7 (28)[b]
cis-DDP-PG Mr-13,000	12,15,18	32x3	0	3/7
cis-DDP-PG Mr-58,000	12,15,18	23x3	0	3/7
cis-aq-PG Mr-58,000	12,15,18	23x3	40	2/7

[a] BALB/c female mice (7 mice/group) were inoculated i.p. with 4×10^5 tumor cells (MDS of control 22.5 days) in experiment I, and with 2×10^5 tumor cells (MDS of control 34 days) in experiments II and III.
[b] Percent toxid death.

D. Cis-DDP and Cis-aq Complexed to Antitumor Antibodies Via CM-dex:
In principle, unmodified IgG which contains carboxylic groups could serve as such for direct binding of cis-DDP or cis-aq. In practice, however, proteins in general, and immunoglobulin (IgG) is no exception, do not bind any significant amounts of releasable platinum (II). Direct binding of these compounds to IgG is also hampered by the low solubility of these proteins in aqueous solutions and the low solubility of the cis-DDP-IgG complexes that are formed. It is thus apparent that the introduction of an intermediate carrier is essential for drug complexing to antibodies. CM-dex T-40 was found to be suitable as an intermediate carrier for cis-DDP or cis-aq, thanks to its capacity to bind high contents of releasable drug. The antibodies used in this study were the idiotypic polyclonal antibodies (described in section IV C) that recognized a specific membrane IgM on B lymphoma cells (38c 13). Whereas in the previously described daunomycin conjugates the drug was first bound to the spacer, either oxidized dextran or PG, and then attached to the antibody, in the preparation of the cis-platinum complexes, CM-dex was first conjugated to the antibody and the drug was subsequently complexed to the conjugates.

The conjugation of CM-dex to IgG was carried out by a modified water soluble carbodiimide mediated coupling reaction, in which N-hydroxysuccinimide was used to increase the yields of this reaction (Staros et al). This procedure resulted in immunologically active CM-dex-IgG preparations that were highly soluble both in saline and in water (Schechter et al, 1987b). The solubility of the conjugates in distilled water was most important for drug-binding to the conjugated CM-dex, since binding in the presence of salts was

greatly diminished.   The CM-dex-IgG conjugates were then separated
from unconjugated CM-dex by ammonium sulfate at high saturation, and
from un-conjugated IgG   by   chromatography   on   DEAE-cellulose
(unconjugated IgG was eluted from the DEAE-cellulose by saline
whereas CM-dex-IgG was eluted only by a 0.5M NaCl   solution).   This
chromatography procedure enabled the removal of unmodified antibody
which might block the activity of the drug CM-dex-Ab complexes.   An
optimal   molar ratio of the reactants,   yielded a conjugate in which
around 3 mole of CM-dex were coupled per one mole of   IgG,   with   no
major   loss   of   antibody   activity.   Binding of cis-DDP to such
conjugates was performed by reacting 120 moles of drug per mole IgG
for 6 hr at 37°c.   These   conditions   were   found   optimum   for
ensuring   both   complex   solubility   and   preservation   of   antibody
activity.   The resultant drug CM-dex-Ab complexes carried up to   50
mole of releasable drug per mole IgG.

The complexes retained their original antigen binding capacity
as   demonstrated   by   an   indirect   immunofluorescence assay using a
fluorescence analyzer cell sorter.   The pharmacological activity of
these complexes as evaluated   by   an   in-vitro   assay   demonstrated
selective binding to the tumor cells.   The 38c 13 target cells were
treated for 1-2 hrs with the specific conjugates before culture.   As
shown,   the   specific   conjugates   exerted   preferential   cytoxicity
towards the B-lymphoma cells, in comparison to controls of the   free
drug, drug-CM-dex complex or non-specific conjugates with irrelevant
antibodies (Table 3).   It thus seems that the selective toxicity of
the   complexes   towards   the   tumor   cells   was   due to their specific
binding to the target cells.

Table 3.   Pharmacological activity of cis-DDP and cis-aq complexed to CM-
dex-g-anti 38C-13 antibodies[a].

Exp.	Drug complex to		50% inhibition at drugs concentration (x10^{-5}M)	
			cis-DDP	cis-aq
I.	(Free drug)		2.0	1.6
	CM-dex-g-anti-38C-13	(P.A.)	0.5	0.58
	CM-dex-g-anti-TNP	(P.A.)	2.2	1.6
	CM-dex		7.8	9.5
II.	(Free drug)		1.5	1.0
	CM-dex-g-anti-38C-13	(P.A.)	0.27	0.25
	CM-dex-g-anti-TNP	(P.A.)	2.5	3.3
	CM-dex-g-anti-RIgG	(IgG)	2.7	3.5
	g-anti-38C-13	(P.A.)	no inhibition	

[a] cis-DDP and cis-aq were complexes to CM-dex derivatives of specific and
non-specific purified antibodies (P.A.) or IgG as well as to unconjugated
CM-dex and to unmodified g-anti-38C-13 antibodies.   The complexes were
incubated with 38C-13 cells for 1 hr (Exp. I) or 2 hr (Exp. II) at 37°C.
Incubation was then continued in drug free medium for 22-23 hr.   A pulse
of [^{3}H] methylthymidine was given at the last 3 hr of incubation.

VI. CONCLUDING REMARKS

The problem we addressed ourselves to in this chapter is the potential use of drug-antibody conjugate, as well as macromolecular drug complexes, for targeted chemotherapy for the treatment of cancer.   In this approach the prerequisite is that the drug will retain its pharmacological activity in either conjugated or complexed form, and that the antibody binding activity will not be impaired. This prerequisite was indeed fulfilled and in most of our studies high _in-vitro_ activity levels were manifested by the conjugates and complexes towards the specific tumor cells, as well as _in-vivo_ activity that led to life prolongation and high percentage of long-term survivors in the treated animals.

Most of the results discussed in this article relate to two types of drugs - the antracyclines daunomycin and adriamycin, and cis-platinum compounds.   Both types of drugs are known for their high beneficial value in cancer chemotherapy.   However, their detrimental side effects cannot be ignored.   In the case of daunomycin and adriamicin the main problematic issue is their cardotoxicity, whereas in the case of cis-platinum it is the nauseating effect and nephrotoxicity which restrict their use.   In general, cancer chemotherapy is ultimately limited by the toxicity of the drugs to normal tissue, especially when applied in high dosages.   The finding that the mere macromolecularization of these drugs, both daunomycin and cis-platinum compounds, by their attachment to high molecular weight carriers, leads to a marked decrease in their general toxicity without interfering with their pharmacological activity is indeed encouraging.

The main conclusion from our studies, however, relates to the suitability of antibodies to serve as carriers for anti-cancer drugs for the purpose of immuno-targeting. Both polyclonal specific antibodies and monoclonal antibodies are adequate.   The rapid advance in the development of monoclonal antibodies and satisfactory procedures for drug conjugation gives credence to the hope that appropriate reagents for specific immunochemotherapy might indeed be within reach.

REFERENCES

Abrams, P.G., Rossio, J.L., Stevenson, H.C., and Foon, K.A. (1986).
'Optimal strategies for developing human-human monoclonal antibodies',
Method Enzymol. 121, 107-19.

Andreason, R.B., and Olsson, L. (1986). 'Antibody producing human-human
hybridoma III. Derviation and characterization of two antibodies with
specificty for human myeloid cells', J. Immunol., 137, 1083-1090.

Arnon, R and Sela, M., (1982a) 'In-vitro and in-vivo efficacy of
conjugates of daunomycin with anti-tumor antibodies'. Immunol. Rev,
62, 5-27.

Arnon, R. and Sela, M. (1982b) 'Targeted chemotherapy: Drugs conjugated
to anti-tumor antibodies', Cancer Surveys., 1, 429-449.

Bernstein, A., Hurwitz, E., Maron, R., Arnon, R., Sela, M., and Wilchek,
M. (1978). 'Higher anti-tumor efficacy of daunomycin when linked to
dextran: In-vivo and in-vitro studies', J. Natl. Cancer Inst, 60,
379-384.

Biddle, W.C., Papsidero, L. D., and Sarcione, E.J. (1986) 'An adriamycin-
monoclonal antibody immmune-conjugate effective against human mammary
carcinoma', Proc. Ann. Meet Am. Assoc. Cancer. Res., 27, 403.

Bredehorst R., Panneerselvam, M., Old L. J., and Vogel C.W. (1986). 'A
novel cell surface effect of adriamycin: Enhancement of tumor cell
killing by monoclonal antibody and complement', Proc. Ann. Meet. Am.
Assoc. Cancer Res., 27, 317.

Bumol, T.F., Andrews, E.L., Todd, G., Zimmerman, J.L., De Herot, S.V.,
Parish, J.E., Baker, A.L., Briggs, S., and Apelpren, L.D. (1986).
'Preclinical development of a murine monocolonal antibody-vinca
alkeloid conjugate targeted for human lung adenocarcinoma', Proc. Annu.
Meet, Am. Assoc., Cancer Res., 27, 278.

Ceriani, R.L., Blank E.W., and Peterson, J. A. (1987). 'Experimental
immunotherapy of human breast carcinoma implanted in nude mice with a
mixture of monoclonal antibodies against human milk fat globule
components'. Cancer Research., 47, 532-540.

Chu, B.C.F., and Whiteley, J. M. (1977). 'High molecular weight
derivatives of methotraxate as chemotherapeutic agents'. Molecular
Pharmacology., 13, 80-88.

Denkers, E.Y., Badger, C.C., Ledbetter J.A., and Bernstein, I.D. (1985).
'Influence of antibody isotype on passive serotherapy of lymphoma', J.
Immunol., 135, 2183-6.

Dillman, R.O., Shawler, D.E., Meyer, D.L., Koziol, J.A., and Frincke, J.M.
(1986). 'Preclinical trials with combinations and conjugates of T101
monoclonal antibody and doxorubicin', Cancer Res., 46, 4886-4891.

Einhorn, L.H., and Williams, S.D. (1979). 'The role of cisplatin in solid
tumor therapy', New. Engl. J. Med, 300, 289-291.

Endo, Y., Kato, Y., Taked, M., Saito, N., Umemoto, K., Kishida, and Hara,
T. (1987). 'In-vitro cytotoxicity of a human serum albumin-mediated
conjugate of Methotrexate with anti-MM46 monoclonal antibody', Cancer
Res., 47, 1076-1080.

Galun, E., Shouval, D., Adler, R., Shahar, M., Sela, M., Wilchek, M., and
Hurwitz, E. (submitted for publication). The effect of
anti-α-fetorprotein-adriamycin conjugates on a human hepatoma.

Garnett, M.C., and Baldwin, R.W. (1986). 'An improved synthesis of a
methotrexate albumin-791T/36 monoclonal antibody conjugate cytotoxic to
human osteogenic sarcoma cells', Cancer Res., 46, 2407-2412.

Goldenberg, D.M., Deland, F.H., DeJager, R., Primus, F.J., Brennan, K, and
Ruoslahti, E. (1985). 'Imaging of liver cancer with ^{131}I monoclonal
and polyclonal antibodies against α-fetoprotein', Proc. Am. Soc. Clin.
Oncol. 4, 224.

Golla, E.D., and Ayres, G.H. (1973). 'Spectrophotometric determination of
platinum with O-phenylenodiammine', Talanta, 20, 199-210.

Groth, S., Nielson, H., Sorenson, J.B., Christensen, A.B., Pedersoen, A.G.,
and Rorth, M. (1986). 'Acute and long term nephrotoxicity of cis-
platinum in man', Cancer Chemother. Pharmacol. 17, 191-196.

Herlyn, D., Herlyn, M., Steplewski, Z., and Koprowski, H. (1985).
'Monoclonal anti-human tumor antibodies of six isotypes in cytotoxic
reactions with human and murine effector cells', Cell Immunol., 92,
105-14.

Howe-Grant, M.E., and Lippard, S.J., (1980). 'Aqueous platinum(II)
Chemistry; Binding to biological molecules', in Siegel. E. (Ed) Metal
Ions in Biological Systems.Vol 11, Marcel Dekker, N.Y., 63-125.

Hurwitz, E., Levy, R., Maron, R., Wilchek, M., Arnon, R., and Sela, M.
(1975). 'The covalent binding of daunomycin and adriamycin to
antibdoies with retention of both drug and antibody activity', Cancer
Res., 35, 1175-1181.

Hurwitz, E., Maron, R., Bernstein, A., Wilchek, M., Sela, M. and R. Arnon.
(1978a). 'The effect in-vivo of chemotherapeutic drug-antibody
conjugates in two murine experimental tumor systems', Int. J. Cancer,
21, 747-755.

Hurwitz, E., Maron, R., Arnon, R., Wilchek, M., and Sela, M. (1978b).
'Daunomycin-immunoglobulin conjugates, uptake and activity in-vitro',
Eur. J. Cancer, 14, 1213-1220.

Hurwitz, E., Schechter, B., Arnon, R., and Sela, M. (1979). 'Binding of
anti-tumor immunoglobulins and their daunomycin conjugates to the tumor
and its metastases. In-vitro and in-vivo studies with Lewis Lung
Carcinoma', Int. J. Cancer, 24, 461-470.

Hurwitz, E., Wilchek, M., and Pitha, J. (1980). 'Soluble macromolecules
as carriers for daunomycin', J. Applied Biochem, 2, 25-35.

Hurwitz, E., Arnon R., Sahar, E., and Danon, Y. (1983a). 'A conjugate of
adriamycin and monoclonal antibodies to Thy-1 antigen inhibits human
neuroblastoma cells in vitro', Ann. N.Y. Acad. Sci, 417, 125-136.

Hurwitz, E., Kashi, R., Burowski, D., Arnon R., and Haimovich, J. (1983b).
'Site-directed chemotherapy with a drug bound to anti-idiotypic
antibodies to a lymphoma cell-surface IgM', Int. J. Cancer, 31,
745-748.

Hurwitz, E., Kashi, R., Arnon, R., Wilchek, M., and Sela, M. (1985). 'The
covalent linking of two nucleotide analogues to antibodies', J. Med.
Chem, 28, 137-140.

Hurwitz, E., Burowoski, D., Kashi, R., Hollander, N. and Haimovich, J.
(1986). 'A synergistic effect between anti-neoplastic drugs in the
therapy of a murine B-cell tumor', Int. J. of Cancer., 37, 739-747.

Hurwitz, E., Burowski, D., and Haimovich, J., (1987). 'Treatment of an
experimental B cell tumor by a combination of tumor specific antibody
and the drugs: Daunomycin and cyclophosphomide', Cancer Ltrs.,
submitted for publication.

Jones, C.A., Tsukamoto, T., O'Brien, P.C., Uhl, C.B., Alley, M.C., and
Lieber, M.M. (1985). 'Soft agarose culture human tumor colony forming
assay for drug sensitivity testing: [^{3}H] Thymidine incorporation vs
colony counting', Br. J. Cancer., 52, 303-310.

Koji, T., Ishii, N., Munihisa, T., Kusumoto, Y., Nakamura, S., Tamenishi,
A., Hara, A., Kobayashi, K., Tsukada, Y., Nishi, S., and Hirai, H.
(1980). 'Localization of radioiodinated antibody to alpha-fetoprotein
in hepatoma transplanted in rats and a case report of alpha-fetoprotein
antibody treatment of a hepatoma patient', Cancer Res., 40, 3013-3015.

Klein, J.L., Sandoz, J.W., Kopher, K.A., Leichner, P.K., and Oroler, S.E.
'Detection of specific anti-antibodies in patients treated with
radiolabeled antibody'. Int. J. Radio. Oncol. Biol. Phys. 12,
939-43.

Lee, C.L., Deguchi, T., Horoszewicz, J.S., and Chu, T.M. (1986). 'Anti-
tumor reactivity of adriamycin and methotrexate conjugates with
monoclonal antibody on human prostate tumor xenograft', Proc. Ann.
Meet. Am. Assoc. Cancer Res., 27. 320.

Levi-Schaffer, F., Bernstein, M., Meshorer , A., and Arnon R. (1982).
'Reduced toxicity of daunorubicin by conjugation to dextran', Cancer
Treatment Reports, 66, 107-114.

Levy, R., Hurwitz,. E., Maron, R., Arnon, R., and Sela, M. (1975). 'The
specific cytotoxic effects of daunomycin conjugated to antitumor
antibodies', Cancer Res., 35, 1182-1186.

Lowder, J.N., Meeker, T.C., Campell, M., Garcia, C.F., Gralow, J., Miller,
R.A., Warnke, R., and Levy, R. (1987). 'Studies on B lymphoid tumors
treated with monoclonal anti-idiotype antibodies. Correlation with
clinical responses', Blood, 69, 199-210.

Maloney, D.G., Kaminskui, M.S., Burowski, D., Haimovich, J. and Levy, R.
(1985). 'Monoclonal anti-idiotype antibodies against the murine B cell
lymphoma 38c13: Characterization and use for the biology of the tumor
in-vivo and in-vitro', Hybridoma, 4, 191-209.

Manske, J.M., Buchsbaum, D.J., Azemore S.M., Hanna D.E., and Vallera D.A.
(1986) 'Antigenic modulation by anti-CD5 immunotoxins', J. Immunol.,
136, 4721-8.

Miller, R.A., Oseroff A.R., Stratte, P.T. and Levy R. (1983).
'Monoclonal antibody therapeutic trials in 7 patients with T-cell
lymphoma', Blood, 62, 988-995,

Nepo, A.G., Greaton, C.J., Taub, R.N., and Mesa-Tejada, R. (1985).
'Daunorubicin-conjugated monoclonal antibody CU18: In-vivo interaction
with T47D breast carcinoma cells', Biannual Int. Breast Cancer
Research Conference., p. 197.

Nishimura, Y., Yokoyama, M., Araki, K., Veda, R., Kudo, A., and Watanahe,
T. (1987). 'Recombinant human-mouse chimeric monoclonal antibody
specific for common acute lymphocytic leukemia antigen', Cancer Res.,
47, 999-1005.

Ohkawa, K., Hibi, N., and Tsukada, Y. (1986). 'Evaluation of a conjugate of purified antibodies against human AFP-dextran-daunorubicin to AFP-producing yolk sac cell lines', Cancer Immunol. Immunother., 22, 81-86.

Perek, Y., Hurwitz, E., Burowski, D., and Haimovich J. (1983). 'Immunotherapy of a murine B cell tumor with antibodies and F(ab')$_2$ fragments against idiotypic determinants of its surface IgM'., J. Immunol., 131, 1600-1603.

Pimm, M.V., and Baldwin, R.W. (1985). 'Distribution of IgM monoclonal antibody in mice with human tumor xenografts: Lack of tumor localization', Eur. J. Cancer Clin. Oncol., 21, 765-768.

Ritz, J., and Schlossman, F. (1982). 'Utilization of monoclonal antibodies in the treatment of leukemia and lymphoma', Blood, 59, 1-11.

Rowland, G.F. (1983). 'The use of antibodies in drug targeting and synergy' in Goldberg E.P. (Ed) Targeted drugs., Vol 2, 57-72.

Rowland, G.F., Simmonds, R.G., Gore, V.A., Marsden, C.H., and Smith, W. (1986). 'Drug localization and growth inhibition studies of vindesine-monoclonal anti-CEA conjugates in a human tumor xenograft', Cancer Immunol. Immunother, 21, 183-187.

Sahagan, B.G., Doroi, H., Saltzgaber-Muller, J., Toneguzzo,F., Guindon, C.A., Lilly, S.P., McDonald, K.W., Morrissey, D.V., Stone, B.A., and Davis, G.L. (1986). 'A genetically engineered murine/human chimeric antibody retains specificity for human tumor-associated antigens', J. Immunol., 137, 1066-74.

Schechter, B., Pauzner, P., Arnon R., and Wilchek, M. (1986a). 'Cis-platinum(II) complexes of carboxymethyl-dextran as potential anti-tumor agents. I. Preparation and characterization', Cancer Biochem. Biophys, 8, 277-287.

Schechter, B., Pauzner, R., Wilchek, M., and Arnon, R. (1986b). 'Cis-platinum(II) complexes of carboxymethyl-dextran as potential anti-tumor agents. II. In-vitro and in-vivo activity', Cancer Biochem. Biophys., 8, 289-298.

Schechter, B., Wilchek, M., and Arnon R. (1987a). 'Increased therapeutic efficacy of cis-platinum complexes of poly-L-glutamic acid against a murine carcinoma', Int. J. Cancer., 39, 409-413.

Schechter B., Pauzner, R., Arnon, R., Haimovich, J., and Wilchek, M. (1987b). 'Selective cytotoxicity against tumor cells by cis-platinum complexed to anti-tumor antibodies via carboxymethyl dextran', Cancer Immunology Immunotherapy, (in press).

Seto, M., Takaha, Shi, Nakamura, S., Saito, M., Hara, T., and Nishizuka, Y. (1986). 'Effector mechanism in anti-tumor activity of monoclonal antibodies produced against an ascitic mouse mammary tumor', Cancer Res., 46, 2056-61.

Shouval, D., Reid, L.M., Chakroborty, P.R., Ruiz-Opazo, N., Morecki, R., Gerber, M.A., Thung, S.N., and Shafritz, D.A. (1981). 'Tumorigenicity in nude mice of a human hepatoma cell line containing hepatitis B virus DNA', Cancer Res., 41, 1342-1350.

Shouval, D., Shafritz, D.A., Zurawski, D.A., Jr., Isselbacher, K.J., and Wands, J.R. (1982). 'Immunotherapy in nude mice of human hepatoma using monoclonal antibodies against hepatitis B virus', Nature (London), 298, 567-569.

Shouval, D., Adler, R., Wands, J.R., and Hurwitz, E. (1986). 'Conjugates between monoclonal antibodies to HBsAg and cytosine arabinoside', J. Hepatology., 3, (Suppl.2), 87-95.

Shouval, D., Adler, R., Wands, J.R., Eliakim, M., Sela, M., and Hurwitz, E. (1987). 'Chemo-immunotherapy of human hepatoma by a conjugate between adriamycin and monoclonal anti-HBS', in Proc. Intern Symposium on viral hepatitis', London, May, Eds. A. Zuckerman, et al.

Smyth, M.J., Pietersz, G.A., Classon, B.J., and McKenzie, I.F.C. (1986). 'Specific targeting of chlormabucil to tumors with the use of monoclonal antibodies', J. Natl. Cancer Inst, 76, 503-510.

Smyth, M.J., Pietersz, G.A., and McKenzie, I.F.C. (1987a). 'Selective enhancement of antitumor activity of N-acetyl melphalan upon conjugation to monoclonal antibodies', Cancer Res., 47, 62-69.

Smyth, M.J., Pietersz, G.A., and McKenzie, I.F.C (1987b). 'The in-vitro and in-vivo anti-tumor activity of N-AcMEL F(ab')$_2$ conjugates', Brit. J. Cancer., 55, 7-11.

Staros, J.V., Wright R.W., and Swingle, D.M. (1986). 'Enhancement by N-hydroxysulfsuccinimide of water-soluble carbodiimide-mediated coupling reactions', Analytical Biochemistry., 156, 220-222.

Steplewski, Z., Spira G., Blasczyle, M., Lubeck, M.D., Radbruch, A., Illgs, H., Heryln, D., Rajewsky, K., and Scharff, M. (1985). 'Isolation and characterization of anti-monosialoganglioside monoclonal antibody 19-9 class switch variants', PNAS, 82, 8653-7.

Szekerke, M., Wade, R., and Whisson, M.E. (1972). 'The use of macromolecules as carriers of cytotoxic groups. I. Conjugates of nitrogen mustards with proteins, polypeptidyl proteins and polypeptides'. Neoplasma, 19, 199-210.

Tan, L.K., Oi, V.T., and Morrison, S.L. (1985). 'A human-mouse chimeric immunoglobulin is expressed on mouse myeloma cells', J. Immunol., 135, 3564-7.

Tsukada, Y., Bischof, W.K.-D., Hibi, N., Hirai, H., Hurwitz, E., and Sela, M. (1982a). 'Effect of a conjugate of daunomycin and antibodies to rat α-fetoprotein on the growth of α-fetoprotein-producing tumor cells', Proc. Natl. Acad. Sci. USA., 79, 621-625.

Tsukada, Y., Hurwitz, E., Kashi, R., Sela M.,Hibi, N., Hara, A., and Hirai, H. (1982b). 'Chemotherapy by intravenous administration of conjugates of daunomycin with monoclonal and conventional anti-rat α-fetoprotein antibodies', Proc. Natl. Acad. Sci., USA., 79, 7896-7899.

Tsukada, Y., Ohkawa, K., and Hibi, N. (1985). 'Suppression of human α-fetoprotein-producing hepatocellular carcinoma growth in nude mice by an anti-α-fetoprotein antibody-daunorubicin conjugate with a poly-L-glutamic acid derivative as intermediate drug carrier', Br. J. Cancer., 52, 111-116.

Vitetta, E.S., Yuan, D., Krolick, K., Isakson, P., Knapp, M., Slavin, S., and Strober, S. (1979). 'Characterization of the spontaneous murine B-cell leukemia (BCL$_1$) III. Evidence for monoclonality by using an anti-idiotype antibody', J. Immunol, 122, 1649-1654.

Weiss, R.B., Sarosy, G., Chagett-Car, K., Russo, M., and Leyland-Jones, B. (1986). 'Anthracyline analogs: The past, present and future', Cancer Chemother. Pharmacol., 18, 185-197.

Drug Carrier Systems
Edited by F.H.D. Roerdink and A.M. Kroon
© 1989 John Wiley & Sons Ltd.

THE APPLICATION OF DRUG-POLYMER CONJUGATES IN CHEMOTHERAPY

C.J.T. Hoes and J. Feijen
Department of Chemical Technology, Twente University
of Technology, P.O. Box 217, 7500 AE Enschede, The Netherlands

INTRODUCTION

Controlled release of drugs with polymeric systems

After a single dose of a drug by either parenteral or enteral routes of administration the drug concentration at the target organ increases up to a maximum value which then declines due to excretion and/or metabolic conversion. To achieve a therapeutically effective level of drug for an extended time period high doses are required. However the drug concentration has to remain below the threshold toxicity level. A better approach to a constant drug concentration can be obtained by repeated or continuous dosage of drug at set intervals. The use of these procedures is impractical for the application of many drugs. The application of polymeric systems offers an attractive alternative to achieve constant drug levels at the target organ. During the last decade many investigators have explored the development of polymeric drug delivery systems for application in both humans and animals and in agriculture (Anderson and Kim, 1986; Baker, 1987; Robinson and Lee, 1987).
The rate of drug release from a polymeric drug delivery device can be controlled either physically or chemically (Kim et al., 1980). In a physically regulated drug delivery system the drug is surrounded by an insoluble polymeric membrane (reservoir device) or is enclosed within an insoluble polymer matrix (monolithic device). The rate of drug release is primarily governed by diffusion through the polymeric barrier and can be kept constant for a reservoir device provided that the concentration of dissolved drug does not change during release. In monolithic devices the drug release rate decreases with time, usually with a square root dependence for dispersed systems. A constant rate can be achieved by proper choice of device geometry or by controlled erosion or biodegradation of the polymer. Variation in the characteristics of the device such as type and molecular weight of the polymer, geometry, swelling behaviour and porosity allows considerable flexibility in adjusting the drug release rate. The design of cost-effective physically regulated drug delivery systems is favoured by the very high cost associated with the development of new drugs.
In a different type of drug delivery device the chemical or enzymatic hydrolysis of the covalent linkage between a drug and a polymeric carrier is utilized. The rate of drug release from this type of conjugate may be regulated and adapted to the biological environment by proper choice of the drug-polymer bond. A general distinc-

tion in polymer-drug conjugates can be made on the basis of their
solubility in body fluids (blood, lymph and interstitial fluid). In
a soluble polymer-drug conjugate for systemic administration the
polymer functions as the carrier to transport the drug to target
cells. These target cells can be present either in the circulation
or in other body compartments which are only accessible after transendothelial passage of the conjugate from the general circulation.
The insoluble type of polymer-drug conjugate is suitable for the
local release of drug after implantation. When prepared as microspheres or nanocapsules insoluble conjugates can also be administered systemically. The covalently bound type of drug delivery
device is still in the development phase for a number of reasons.
The knowledge of the in vivo behaviour of polymers and drug conjugates including inflammatory and immunological reactions, biodistribution, binding with proteins and cells as well as drug release is
incomplete. Further, chemically linked drug-polymer conjugates are
considered as new drugs and approval for clinical application by
authorities requires extensive biological testing which is very
expensive. On the other hand, effective and selective chemically
linked polymer-drug conjugates should be readily accepted in the
treatment of several diseases, most notably, cancer.

Polymer-drug conjugates in cancer chemotherapy

Cancer is characterized by deregulated cell growth. In cancer chemotherapy the most obvious limitation is the lack of selectivity of
drugs for cancer cells compared with normal cells. Conjugates of
drugs bound to polymers have been advocated as a possible method to
increase the therapeutic effectiveness either by the sustained
release of drug preventing potentially harmful peak concentrations
of free drug or by effecting some specificity for tumor cells. Currently, the selective targeting of drugs to tumor cells attracts
considerable interest. Targeting may be accomplished by linking a
receptor-specific moiety such as an antibody, lectin, hormone or
carbohydrate to a soluble polymer-drug conjugate. The targeting
moiety is recognized by the appropriate receptor on the target cell
allowing a highly selective drug action. The polymer functions as an
intermediate carrier between drug and homing device to increase the
amount of transported drug. In other chapters in this volume the
selective targeting of drugs is discussed in detail.
The insoluble type of polymer-drug conjugate may be suitable for the
local release of drug notably in the treatment of solid tumors.
In this chapter the current status of soluble conjugates of anti-
tumor drugs bound to synthetic polymeric carriers is presented.
Preference is given to conjugates which are either at an advanced
stage of development or are promising. An exhaustive review is not
attempted and can be found elsewhere (Duncan and Kopecek, 1984;
Ferruti and Tanzi, 1986).

DESIGN CRITERIA FOR COVALENTLY BOUND POLYMER-DRUG CONJUGATES

General aspects

The pharmacological effect of a drug at the cellular level is generally dose-related. The local concentration profile depends on the dose of drug, the route of administration, the distribution among body compartments and the rate of elimination by excretion and/or metabolic conversion. Linkage of a drug to a macromolecular carrier will profoundly alter the pharmacokinetics while the bioavailability of the drug at the target site should be preserved, preferably with enhanced specificity and duration of action. Relatively few soluble polymers have been studied at the present time with respect to their fate in the body except those used as plasma expanders. Nevertheless, the available data on macromolecular cytostatic agents offer prospect in certain areas, especially the continuous release within body compartments and in targeted drug delivery.

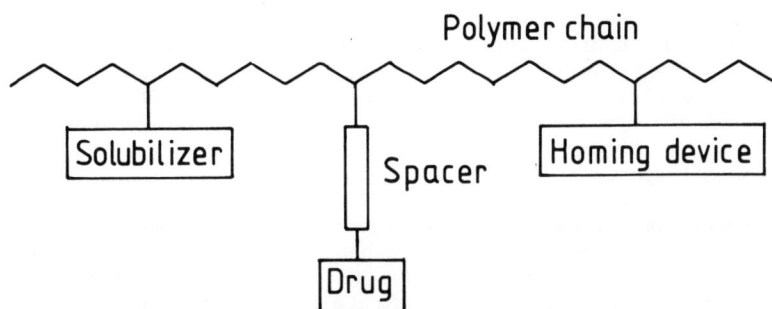

Fig. 1   General structure of polymer-drug conjugate (from
           Ringsdorf, 1975)

A generalized structure of a covalently bound macromolecular drug for systemic administration was first discussed by Ringsdorf (1975) and is given in Fig. 1. Many experimentally prepared conjugates conform to this scheme.

1. The polymeric carrier which renders the conjugate soluble contains functional groups for covalent linkage and may be biodegradable. The carrier can be either biologically inert or active.
2. Drug moieties are connected via a covalent linkage which may be cleaved hydrolytically or enzymatically.
3. A spacer may be present to increase the physical separation of drug and carrier which is advantageous for drug-receptor interaction or selectivity of cleavage of the linkage with the drug, for example by enzymes.
4. Additional pendant groups (solubilizers) may be present which affect the charge and hydrophilicity/hydrophobicity thereby regulating the solubility or exerting additional biological effects.
5. A homing device also called drug targeting moiety may be present to facilitate specific recognition by cells.

Pinocytosis, biodistribution and targeting of synthetic polymers

Macromolecular compounds including proteins and synthetic polymers cannot diffuse through cell membranes. However, these compounds can still enter the cell by a process called pinocytosis. Prior to pino-cytosis, the substrate may bind to the cell membrane either non-specifically due to hydrophobic or electrostatic interactions (ad-sorptive pinocytosis) or through receptors (receptor- mediated pino-cytosis). In the absence of binding between the macromolecule and the cell surface the uptake is mediated by fluid-phase pinocytosis (Fig. 2) (Lloyd, 1987; Ryser and Shen, 1986). Pinocytosis is the general process in which parts of the cell membrane containing extracellular soluble substrates, usually proteins, continuously invaginate yielding an intracellular vacuole called a pinosome. The pinosome migrates towards a vacuolar compartment known as the endo-some or CURL (compartment of uncoupling of receptors and ligands) where a segregation process occurs especially in the case of ligand-receptor complexes. The low pH (5–5.5) in the CURL effects the dis-sociation of ligand and receptor and the latter is recycled back to the cell membrane. Subsequent fusion of the ligand (substrate) with a lysosome exposes it to a large variety of hydrolytic enzymes at an acid pH, resulting in degradation. In all three forms of pinocytosis the protein or polymer substrate reaches the lysosome. The rate of uptake of substrates is strongly enhanced when adsorptive processes occur (Fig. 2). The rate of uptake is highest in receptor-mediated pinocytosis especially at low substrate concentration but levels off at higher substrate concentrations due to the saturation of recep-

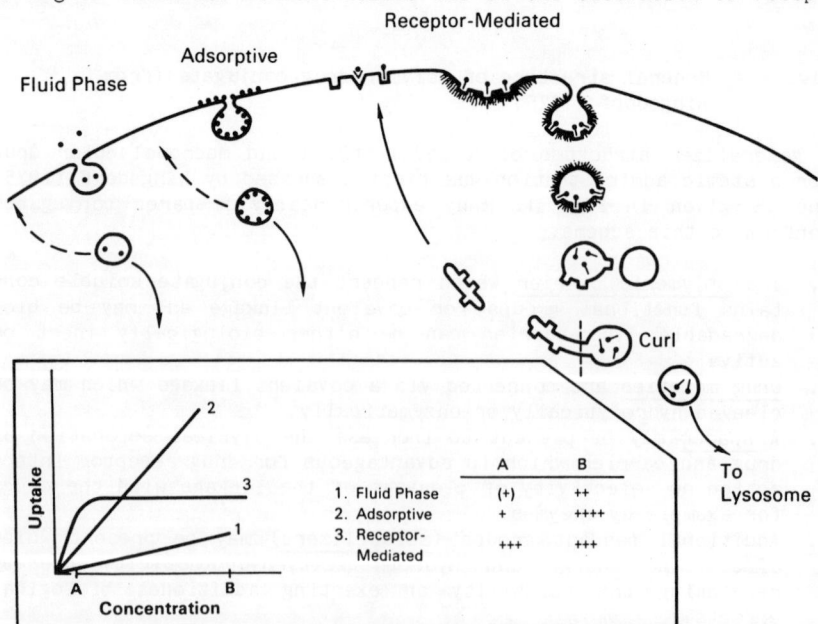

Fig. 2   The three main forms of pinocytosis (from Ryser and Shen, 1986)

tors on the cell surface (10^4–10^6/cell).
The mechanism of these cellular processes is currently being un-
raveled in detail (Pastan and Willingham, 1985). Soluble synthetic
polymers are taken up by cells either by adsorptive or fluid-phase
pinocytosis, dependent on their chemical structure and molecular
weight and the type of cell (Duncan and Kopecek, 1984).
Pinocytosis is clearly distinct from phagocytosis. Phagocytosis is
the mechanism by which particulate matter such as bacteria and cell
debris is captured and processed by specialized cells, i.e. the
macrophages of the reticulo-endothelial system (RES). The RES is the
collective name for the mobile phagocytic cells in the blood (monocytes) and the immobile phagocytes in certain organs, especially the
Kupffer cells of the liver and the macrophages of the kidneys,
spleen, lungs and bone marrow.
Two major advantages with respect to the biodistribution of soluble
polymers compared with that of particulate carriers such as liposomes, micro- and nanocapsules have been postulated. First, the
uptake of particulate carriers with a size between 100 and 400 nm
by the cells of the reticuloendothelial system via phagocytosis
readily occurs which renders it inherently difficult to target drugs
to other cells. Uptake by RES cells is determined both by the size
and surface properties of the carrier and the passive targeting of
colloidal carriers can be modified by altering therefore their surface properties (Davis and Illum, 1986).
Soluble polymers such as poly[N-(2-hydroxypropyl)methacrylamide]
(polyHPMA) discussed further below assume an unordered conformation
in solution with an average molecular diameter between 3 and 30 nm
dependent on molecular weight (10.000-800.000) and are taken up by
cells only by the process of pinocytosis which is slower than phagocytosis. Hence, polyHPMA and similar carriers do not show dominant
passive targeting to RES cells and offer a better prospect for targeting of drugs to other cells (Seymour et al., 1987a). Secondly,
soluble carriers are able to cross the endothelial barriers present
between various body compartments. This allows access to a greater
number of cell types and makes the intraperitoneal (i.p.) or subcutaneous (s.c.) routes of administration feasible, in contrast with
particulate carriers (Seymour et al., 1987a). On the other hand, the
effect of the drug on the biodistribution of polymeric conjugates is
not well known. Studies on this aspect have only been recently initiated (Ulbrich et al., 1987).
Currently the targeting of drugs with the aid of cell-specific molecules in the treatment of cancer and other diseases is attracting
more interest as demonstrated in other chapters in this volume. A
marked example is the presence of recognition systems on certain
cells for exposed carbohydrate residues of desialylated serum glycoproteins, neoglycoproteins, liposomes and synthetic polymers. This
results in increased clearance of these entities from the blood and
highly selective uptake by the cell. The galactose receptor on liver
hepatocytes, the mannose/N-acetylglucosamine receptor on liver
Kupffer cells and other macrophages and the mannose-6-phosphate
receptor on fibroblasts and other tissue cells are examples of such
recognition systems (Schneider et al., 1983; Seymour et al., 1987a).
Monoclonal antibodies directed against tumor-associated antigens
offer great potential in the diagnosis of tumor cells and their use

in the targeting of drugs is a logical extension towards selective chemotherapeutic treatment. When drugs are linked directly with antibodies the degree of drug substitution should be kept low to prevent loss of antigen-binding capacity. A much higher drug load can be achieved by the use of an intermediate polymeric carrier. Using soluble polymers which avoid RES uptake seems beneficial in preserving the targeting potential and selectivity in such conjugates (Rihova et al., 1986).

Choice of polymeric carrier

A multitude of carriers has been proposed, including natural proteins (albumin, immunoglobulins, lectins), polysaccharides (dextran and charged derivatives) and many synthetic polymers such as vinylic polymers, poly(α-amino acids) and polyethers. Essential properties are low toxicity and immunogenicity, biodegradability, the presence of functional groups for chemical coupling of the drug and in the case of synthetic polymers a narrow distribution of molecular weight.
Two fundamentally different approaches for the choice of a particular carrier can be distinguished.

1. The polymer is inert in the biological environment and thereby acts merely as a specific transport vehicle. The drug is then released systemically or locally and can exert its cell growth inhibiting action. Examples of carriers belonging to this class are poly[(2-hydroxypropyl)methacrylamide] (polyHPMA) and dextran, which are both discussed below.
2. The polymer is biologically active. A number of polyanions for example the polymer obtained from copolymerization of maleic anhydride and divinylether (DIVEMA), have strong biological effects including antitumor activity. Polycations such as poly-(L-lysine) are known to be toxic to cells due to strong membrane binding. These charged polymers have been proposed as drug carriers offering a synergistic effect caused by the enhanced uptake of the polymer by tumor cells and the cytotoxic action of the drug released locally. With these drug-polymer conjugates the in vivo mechanism of action is probably very complicated involving the activation of the immune and complement systems.

Once introduced into the body, polymers above a certain threshold molecular weight generally on the order of 40.000 are not excreted via the kidneys and are thus retained for a prolonged time. It is recognized that biodegradation yielding non-toxic low-molecular weight fragments or monomeric units is highly desirable. Biodegradability of the polymeric carrier requires the presence of hydrolyzable bonds as with proteins and poly(α-L-amino acids), which are degradable by proteolytic enzymes present inside cells or in plasma. Similarly some polysaccharides e.g. dextran are degradable by glycosidases. Vinyl polymers which contain carbon-carbon bonds in the main chain are not readily biodegradable but can be rendered excretable through the kidneys by crosslinking linear chains of relatively low degree of polymerization with biodegradable peptide-containing diamines. After in vivo degradation of the crosslinks the short linear chains are removed by renal excretion. It is also possible to incorporate hydrolytically labile ester bonds in vinyl polymers by ring-opening radical copolymerization, but this strategy has not yet been applied in polymer-drug conjugates (Bailey and Gapud, 1985).

Nature of drug-carrier linkage

Many different types of covalent linkages between the drug and the polymeric carrier are possible. These linkages affect both the rate and specificity of drug release. In general drug release from polymeric conjugates can be either extracellular or endocellular. Extracellular drug release appears to be relatively simple to accomplish but selectivity is limited. The drug is generally bound to the carrier via a hydrolytically labile bond and is released slowly in the circulation by chemical solvolysis or plasma enzymes. The pharmacological effects of the release of drug from polymeric conjugates by this mechanism are similar to those observed by slow parenteral infusion of the free drug.

In an endocellular drug release system the drug-polymer bond must be stable with respect to chemical or enzymatic hydrolysis during transport. The conjugate is taken up by cells via pinocytosis and the free drug is released within the lysosomes by enzymes or as a result of the low pH. The free drug may then exert its action provided that it remains stable inside the lysosomes and can diffuse out of the lysosomes. Both the pinocytic uptake of the conjugate and the enzymatic release should be efficient to accomplish a therapeutically effective drug concentration. For amino- or hydroxy-containing drugs bound across an amide or ester bond, respectively, efficient endocellular drug release by lysosomal enzymes from the polymeric carrier can be accomplished by the use of a peptide spacer having an appropriate length and sequence (Hoes et al., 1986; Kopecek, 1984a,b; Trouet et al., 1982).

The rate of drug release from macromolecular conjugates by enzymes also depends on the conformation of the polymeric carrier and/or on the electrostatic effects exerted by the carrier on enzyme-substrate binding (Pato et al., 1984). The release rate of the drug is also affected by the character of the drug itself (Duncan et al., 1987). The polymer-bound spacer-drug moiety is cleaved by the major lysosomal endopeptidases (cathepsin B, H and L) usually at different sites releasing the drug as well as aminoterminated peptide-drug fragments. The latter are degraded further by relatively aspecific lysosomal aminopeptidases. Less progress has been made in the search for peptide spacers between carboxyl-containing drugs and polymeric carriers. Although such spacers are also hydrolyzed by the action of cathepsins, the additional breakdown of carboxyl-terminated peptide-drug fragments requires the action of lysosomal carboxypeptidases which are more specific than aminopeptidases (Renard et al., 1986). Whether complete breakdown of spacer-drug fragments by lysosomal enzymes is necessary to achieve cytotoxic action is dependent on the specific drug. For instance, amino acid or peptide derivatives of the anthracyclines are much less active than the free drugs, while peptide conjugates of methotrexate have activities comparable with that of the parent drug (Ryser and Shen, 1986; Trouet et al., 1982).

Endocellular drug release has been achieved nonenzymatically by using an acid-labile linkage between drug and carrier which takes advantage of the low pH in lysosomes as well as in prelysosomal endosomes (Ryser and Shen, 1986). Thus, with polymer-drug conjugates which have been optimized for endocellular drug delivery, a highly specific tumor therapy may ultimately be achieved by incorporation of a tumorcell-specific targeting moiety.

To accomplish selectivity towards tumor cells with an extracellular polymer-drug conjugate, the prodrug concept has been proposed (Stella and Himmelstein, 1985). In a number of tumor cells elevated levels of certain hydrolytic enzymes have been observed which are either bound to the cell membrane or are released by the cells in the medium. For instance, plasmin, γ-glutamyl-transferase (γ-GT), collagenase and others have been found to be present in cancer cells (Quigley, 1979). Various low-molecular weight prodrugs have been demonstrated to act as substrates for such enzymes, but as yet few examples in the field of polymer-drug conjugates have been descri-bed. In a polymeric conjugate the drug-polymer bond may be selected to be susceptible to enzymatic hydrolysis by the use of an appro-priate spacer or by the choice of carrier. For instance, peptide-derivatives of drugs might be suitable as spacers in polymeric con-jugates for the plasmin-mediated release of drug (Chakravarty et al., 1983 a,b). Amide-bound conjugates of adriamycin (ADR) and poly-(α-L-glutamic acid) were designed in this laboratory as macromole-cular prodrugs for the selective release of drug at tumor cells having elevated levels of γ-GT (Van Heeswijk et al., 1984). However, the low in vitro activity of these conjugates does not yet corro-borate the validity of the prodrug approach in polymeric systems.

EXAMPLES OF POLYMER-DRUG CONJUGATES

Poly[N-(2-hydroxypropyl)methacrylamide], polyHPMA

Poly[N-(2-hydroxypropyl)methacrylamide] (polyHPMA) is a water-soluble polymer which is biocompatible, non-toxic and non-immuno-genic (Kopecek et al., 1973). Extensive studies by Duncan, Kopecek and co-workers on soluble conjugates of drugs and polyHPMA illu-strate the current status in endocellular drug release from poly-meric conjugates. Relevant parameters for the performance of a macromolecular endocellular drug release device such as uptake by cells, endocellular release of drugs and model compounds, body dis-tribution, stability in body fluids and cell-specific targeting of conjugates have been documented for polyHPMA (for reviews, see Duncan and Kopecek, 1984; Kopecek, 1984 a,b). The ultimate aim of Kopecek et al. with polyHPMA conjugates is to accomplish cell-speci-fic targeting of drugs, notably cytostatics. To this end appropriate ligands for instance carbohydrates, hormones or monoclonal anti-bodies are incorporated in the polymeric conjugates (Duncan et al., 1987; Kopecek et al., 1985; Rihova and Kopecek, 1985).

Synthesis and structure of polyHPMA conjugates

Soluble conjugates based on polyHPMA are prepared as follows (Fig.3

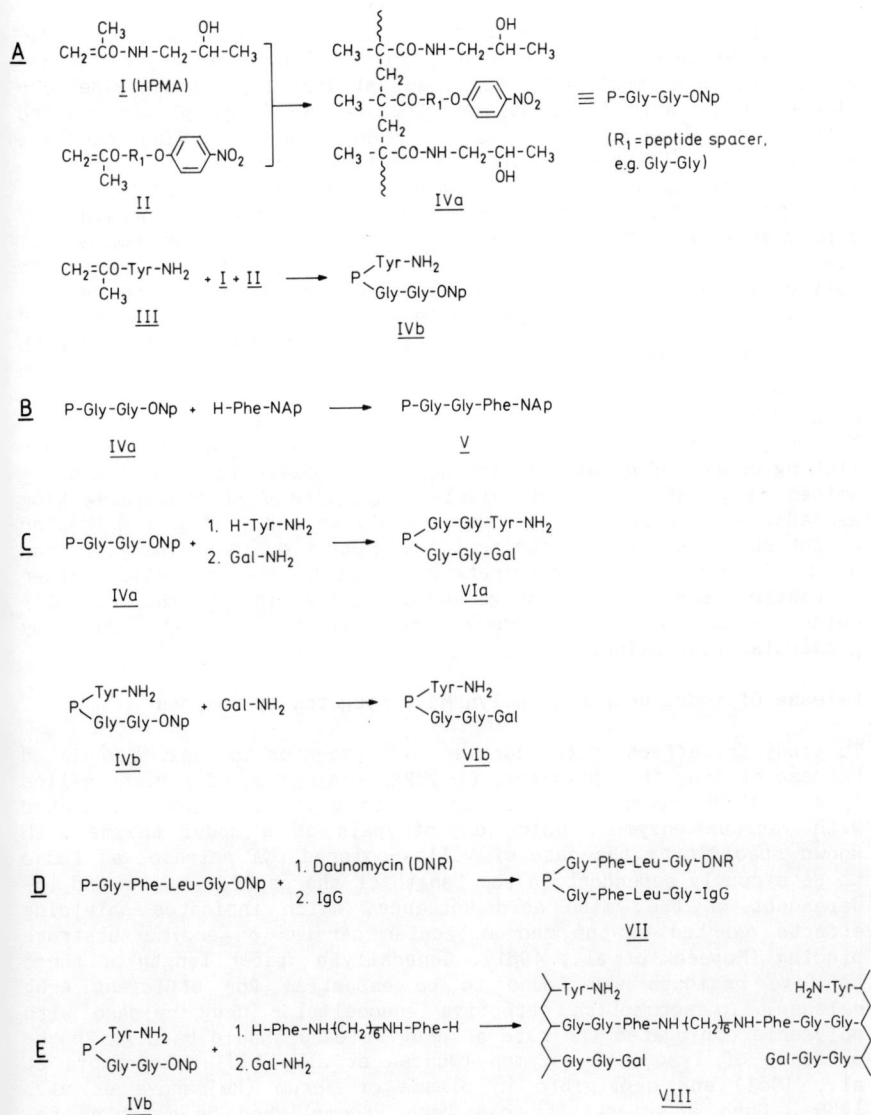

A

CH₃ = CH₂=CO-NH-CH₂-CH-CH₃ with OH
I (HPMA)

CH₂=CO-R₁-O-⟨⟩-NO₂ with CH₃
II

→

CH₃-C-CO-NH-CH₂-CH-CH₃ (OH)
CH₂
CH₃-C-CO-R₁-O-⟨⟩-NO₂
CH₂
CH₃-C-CO-NH-CH₂-CH-CH₃ (OH)
IVa

≡ P-Gly-Gly-ONp

(R₁ = peptide spacer, e.g. Gly-Gly)

CH₂=CO-Tyr-NH₂ with CH₃
III
+ I + II →

P ⟨ Tyr-NH₂ / Gly-Gly-ONp
IVb

B   P-Gly-Gly-ONp + H-Phe-NAp ⟶ P-Gly-Gly-Phe-NAp
     IVa   V

C   P-Gly-Gly-ONp + 1. H-Tyr-NH₂   ⟶ P ⟨ Gly-Gly-Tyr-NH₂ / Gly-Gly-Gal
     IVa   2. Gal-NH₂   VIa

P ⟨ Tyr-NH₂ / Gly-Gly-ONp + Gal-NH₂ ⟶ P ⟨ Tyr-NH₂ / Gly-Gly-Gal
IVb   VIb

D   P-Gly-Phe-Leu-Gly-ONp + 1. Daunomycin (DNR)   ⟶ P ⟨ Gly-Phe-Leu-Gly-DNR / Gly-Phe-Leu-Gly-IgG
                              2. IgG   VII

E   P ⟨ Tyr-NH₂ / Gly-Gly-ONp + 1. H-Phe-NH(CH₂)₆NH-Phe-H   ⟶   Tyr-NH₂ ... H₂N-Tyr / Gly-Gly-Phe-NH(CH₂)₆NH-Phe-Gly-Gly- / Gly-Gly-Gal ... Gal-Gly-Gly
     IVb   2. Gal-NH₂   VIII

Fig. 3. Examples of synthetic routes towards soluble conjugates of polyHPMA

The hydrophilic monomer (R,S) N-(2-hydroxypropyl)methacrylamide (I, HPMA) is copolymerized with C-terminal activated p-nitrophenylesters of methacryloylated L-amino acids or oligopeptides II using radical initiation (Fig. 3A). This yields linear random copolymers IVa containing generally 2-10 mole % of reactive ester groups. The activated intermediate is then reacted further with a variety of amino-group-containing compounds. For model studies a p-nitroanilide-terminated amino acid (Fig. 3B) or tyrosinamide (Fig. 3C) is reacted with IVa to yield conjugates of the model drug (V or VIa) bound via a potentially cleavable peptide spacer to the carrier. Amino-containing drugs (Fig. 3D) can be coupled similarly (VII). A slightly different activated polymer precursor IVb is obtained by copolymerization of I, II and methacryloylated tyrosinamide III yielding a non-cleavable tyrosinamide moiety bound directly to the polymer backbone as well as an activated spacer moiety for subsequent coupling. This strategy has been adopted to follow the fate of the polymer <u>in vivo</u> after radiolabeling of the tyrosyl moiety with ^{125}I. A targeting moiety such as a carbohydrate (Fig. 3C) or an antibody (Fig. 3D) may be linked in addition to the drug or model drug to yield conjugates (VIa, VIb, VII) with targeting capability. The drug-polymer conjugates can be rendered biodegradable by cross-linking chains of relatively low degree of polymerization with diamines (Fig. 3E) to yield soluble conjugates VIII containing biodegradable peptide bonds in the crosslinks and having a molecular weight sufficiently high for prolonged retention in the body. A drug or model drug and a carbohydrate may also be present (VIII). After enzymatic cleavage of the drugs and crosslinks <u>in vivo</u> the low-molecular weight vinylic polymers are removed from the body by glomerular filtration.

Release of model drug from polyHPMA conjugates by enzymes

To study the effect of the carrier and spacer on the enzyme-mediated release of drug from polymers, polyHPMA conjugates of p-nitroaniline (pNA) with different spacers similar to V (Fig. 3) were incubated with various enzymes. Using chymotrypsin as a model enzyme with known specificity the rate of yellow-colored pNA release was found to be strongly dependent on the length of the peptide spacer and independent of the amino acid sequence, which indicates shielding effects exerted by the macromolecular carrier on enzyme/substrate binding (Kopecek et al., 1981). Generally a spacer length of three to five residues was found to be essential for efficient drug release. To accomplish effective endocellular drug release with polymeric conjugates the rate of drug release should be high in the presence of lysosomal enzymes (Duncan et al., 1983a; Rejmanova et al., 1983) and negligible in plasma or serum (Rejmanova et al., 1985). Such a selectivity has been accomplished by matching the peptide spacer length and sequence in polyHPMA conjugates of pNA with the specificity of the thiol proteinases cathepsin B, H and L, the major endopeptidases present in lysosomal enzymes. Fig. 4 shows the effect of spacer length and sequence on pNA release by cathepsin

A

% NAp cleaved

8 –
6 –
4 –
2 –

P – Gly-Phe-Tyr-Ala-NAp
P – Gly-Phe-Leu-Gly –NAp

P – Gly-Phe-Ala-NAp

P – Gly-Leu-Ala-NAp

P – Ala-Val-Ala – NAp

P – Gly-Val-Leu-NAp

P – Ala-Gly-Val-Phe-NAp
P – Ala-Val-Phe-NAp

P – Gly –Phe-Leu-Gly-Phe-NAp

10   20   30   time (min)

B   P–Gly–Phe–Leu–Gly–NAp

% NAp released

40 –

30 –

20 –

10 –

0 –

TRITOSOMES

CATHEPSIN B

CATHEPSIN L

CATHEPSIN H

0   1   2   3   4   5   time, [h]

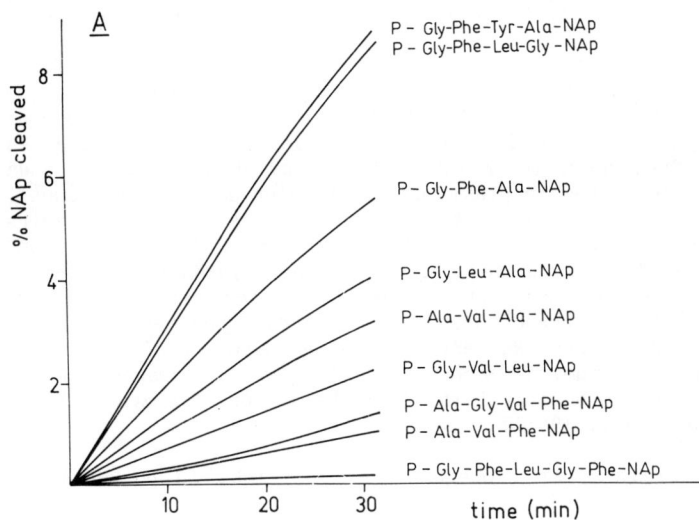

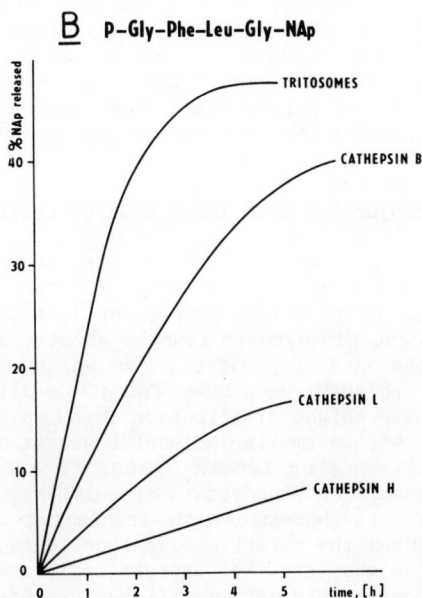

Fig. 4   The effect of spacer length and sequence on cathepsin B-
mediated release of pNA (A) and the release of pNA by
various enzymes from a polyHPMA conjugate (B) (from
Kopecek, 1984a)

B (Fig. 4A) and the dominant contribution from the individual cathepsins present in lysosomal enzymes isolated from the rat liver (Tritosomes) (Fig. 4B).

Some of the spacers, notably those composed of five amino acids, are cleaved at different bonds by these endopeptidases releasing drug as well as spacer drug fragments. These fragments are degraded further by the relatively aspecific aminopeptidases present in lysosomes to yield the free drug. The relative stability of drug-spacer bonds in poly HPMA conjugates in plasma or serum deserves further discussion. The major proteolytic activity in blood results from the proteinases which have a trypsin-like specificity different from the cathepsin-specific peptide spacer sequences used in polyHPMA conjugates. However lysosomal enzymes may be present in the blood and might in principle effect extracellular drug release. Experimentally insignificant levels of free pNA were found after exposure of conjugates to rat plasma or serum. This was explained by the presence of enzyme levels too low to be effective or by the action of enzyme inhibitors (Rejmanova et al., 1985).

Enzymatic cleavage of biodegradable crosslinks in polyHPMA conjugates

As mentioned above, the use of soluble polyHPMA conjugates containing biodegradable crosslinks e.g. VIII (Fig. 3E) offers a strategy to eliminate inherently nondegradable polyHPMA from the body after release of drug. In studies similar to those described above for pNA release the enzymatic degradability of crosslinks in poly-HPMA conjugates was found to depend on the length and amino acid sequence of the spacer. The rate of polymer excretion under in vivo conditions can thus be adjusted by varying these parameters (Cartlidge et al., 1987a; Subr et al., 1986).

Pinocytosis of polyHPMA and conjugates with model drug by cells

PolyHPMA is a hydrophilic polymer which does not bind with cell membranes and which is pinocytosed by cells cultured in vitro with a rate of uptake similar to that of poly(vinylpyrrolidone), a polymer used as a marker for fluid-phase pinocytosis (Duncan et al., 1981). A significant dependence of the rate of pinocytosis by various cells on the molecular weight of polyHPMA has been found. Crosslinked polyHPMA with a wide molecular weight distribution ($\bar{M}_w/\bar{M}_n$ 6.8) was fractionated to yield samples with a smaller molecular weight distribution ($\bar{M}_w/\bar{M}_n$ 1.8-2.5) ranging from $\bar{M}_w$ 34000 to > 400000 (Cartlidge et al., 1986). The rate of pinocytosis of polyHPMA by rat visceral yolk sacs was found to decrease with increasing weight average molecular weight, whereas the relative extent of degradation was independent of molecular weight (Fig. 5). Rat visceral yolk sacs are used as a model for a pinocytically active cell because they are not phagocytically active. The pinocytic process can be inhibited by adding 2,4-dinitrophenol while cathepsin-mediated degradation of substrates maintaining pinocytosis is inhibited in the presence of leupeptin (Fig. 5).

	$\bar{M}_w$	$\bar{M}_w / \bar{M}_n$
1	190 000	6·8
2	>400 000	—
3	400 000	2·5
4	155 000	1·9
5	110 000	1·7
6	34 000	1·8

Total uptake (µl/mg protein)

20

10

+ 2,4-DNP

+ leupeptin

1 2 3 4 5 6 6 6
fraction

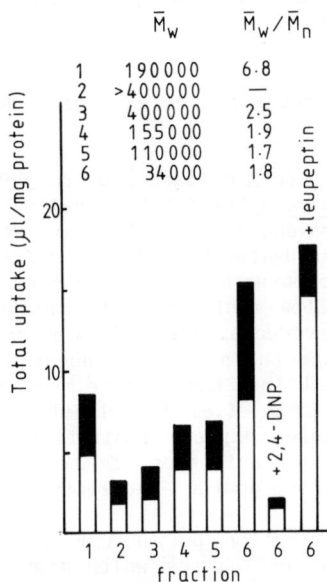

Fig. 5   Effect of molecular weight of crosslinked polyHPMA on tissue accumulation (□), degradation (■) and total uptake by rat yolk sacs cultured in vitro (from Cartlidge et al., 1986)

In adult rat intestine cells, however, the rate of uptake increased with increasing molecular weight. This was explained by an increase in non-specific adsorption of the polymer to the cell membrane resulting in an enhanced rate of pinocytosis. These results indicate the necessity of the use of polymer-drug conjugates with a narrow molecular weight distribution in vivo.
The rate of uptake of polyHPMA by yolk sacs is enhanced by incorporation of aromatic groups like tyrosinamide or cationic trimethylammonium groups. This enhancement is due to the increased nonspecific adsorption of conjugates to the cell membrane prior to pinocytosis. Increased levels of substitution with these groups effect higher rates of pinocytosis (Duncan et al., 1984; McCormick et al., 1986). From the adsorptive properties of polymers containing tyrosinamide as a model drug it has been inferred that hydrophobic drugs attached to soluble carriers can give rise to nonspecific cell binding even in the presence of cell-specific targeting moieties (Duncan et al., 1984).
Thus the maximum degree of substitution of soluble polymers with hydrophobic drugs is limited both to suppress the nonspecific cell binding as well as to preserve the solubility of the conjugates. Receptor-mediated pinocytosis may occur when ligands such as carbohydrates, hormones or antibodies bound to the drug-polymer conjugate are recognized by specific receptors on the cell membrane. Such specific pinocytosis which occurs with a rate increasing with in-

creased ligand substitution level has great potential for targeted drug delivery as will be discussed below (Chytry et al., 1987; Duncan et al., 1986; Seymour et al., 1987a).

In vitro and in vivo endocellular release of model drug from poly HPMA conjugates

The endocellular nature of drug release from polyHPMA conjugates has been confirmed both in vitro using rat visceral yolk sacs (Duncan et al., 1981) and after intravenous application in rats (Duncan et al., 1983b, 1986). PolyHPMA conjugates having four different spacers with tyrosyl-p-nitroanilide moieties were radiolabeled with ^{125}I at the tyrosyl side chain to allow monitoring of conjugate and low-molecular weight hydrolysis products. The accumulation of conjugates by rat visceral yolk sacs and the spacer-dependent amount of degradation products released back into the culture medium is shown in Fig. 6. The release of model drug (i.e. ^{125}I-labeled tyrosine) is evidently endocellular because complete inhibition of pinocytosis by adding 2,4-dinitrophenol or inhibition of the lysosomal cathepsins maintaining pinocytosis by adding leupeptin both abolish drug release into the medium.

After intravenous administration polyHPMA-drug conjugates will be exposed to many different cell types which might possibly abolish any selectivity in drug delivery. To facilitate the study of in vivo endocellular drug release using polyHPMA conjugates polymer-bound galactose (Gal) moieties were introduced. After i.v. administration to rats an almost selective (70-90% within 30 min) accumulation of the conjugate in hepatocytes was observed (Duncan et al., 1986). A polyPHMA copolymer containing both Gly-Gly-Gal (3.7 mole %) and a biodegradable Gly-Gly-Tyr-NH$_2$ (0.4 mole %) moiety was radiolabeled with ^{125}I, injected i.v. into rats, and the liver was removed after 20-30 min. The lysosomes were quickly isolated from hepatocytes and incubated in 0.25 M sucrose to minimize lysosomal breakage. The release of low-molecular weight degradation products (probably ^{125}I] iodo-L-tyrosine) into the medium was monitored for 4 h at 37°C. It was found that 50-60% of free model drug relative to the amount originally bound to the polymer and associated with the lysosomes was released over this period. It was also demonstrated that release of labeled compounds was not due to lysosomal breakage. Thus these data support the endocellular nature of drug release from polyHPMA conjugates under in vivo conditions.

Biodistribution and in vivo targeting of polyHPMA carriers

A polymer or polymer-drug conjugate administered by parenteral routes which include intravenous (i.v.), intraperitoneal (i.p.) or subcutaneous (s.c.) injection will show a different distribution between various body compartments and organs compared with low-molecular weight compounds. This is due to the diffusional limitations of the macromolecule or to alterations in specific interactions with cells in different compartments and organs (Drobnik and Rypacek, 1984). Similar to low-molecular weight compounds the distribution of

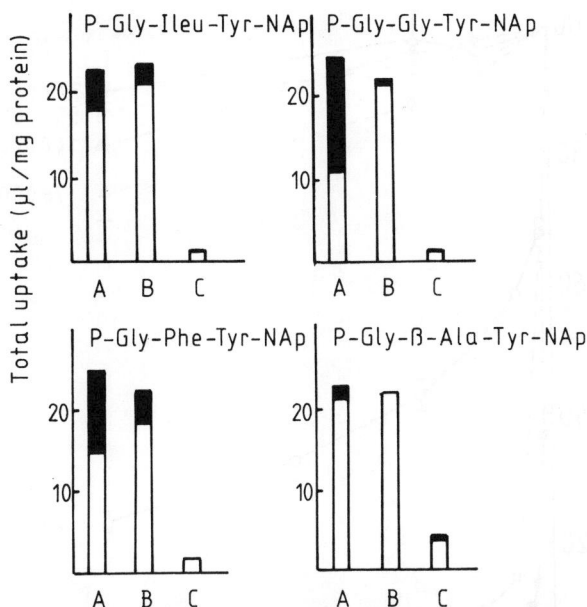

Fig. 6   Uptake ($\square$) and degradation ($\blacksquare$) of ^{125}I-labeled polyHPMA
conjugates (15 µg/ml) by rat yolk sacs in culture medium.
Effect of inhibition of cathepsins by leupeptin (B) and of
pinocytic inhibition by 2,4-dinitrophenol (C) (from Duncan
et al., 1981)

polymers can be described by a pharmacokinetic model involving the
rates of dosage, intercompartmental transfer, uptake by various
cells, degradation and excretion. PolyHPMA is one of the few polymers in addition to dextran and poly(vinylpyrrolidone) which are
used as plasma expanders of which the pharmacokinetics has been
studied. Actually, polyHPMA has also been proposed as a plasma expander (Kopecek et al., 1973). Knowledge of the biodistribution of
polyHPMA is considered essential to allow its use as the carrier in
targeted drug delivery by parenteral routs of administration
(Seymour et al., 1987a).
The biodistribution of ^{125}I-labeled polyHPMA carriers including
blood levels, the levels in various organs and other tissues and
excretion into the urine and faeces was evaluated in rats. The molecular weight and polydispersity of the polymer in addition to the
presence of biodegradable crosslinks, attachment of carbohydrate
residues for cell-specific targeting and the route of administration
affect the biodistribution. The molecular weight ($\bar{M}_w$) of polyHPMA
has a pronounced effect on the clearance from blood after i.v. administration in rats which is primarily because of the rapid urinary
excretion of polymer with molecular weights below the threshold
value for glomerular filtration (Fig. 7) (Seymour et al., 1987b).
The use of linear polyHPMA fractions each with a narrow molecular
weight distribution ($\bar{M}_w/\bar{M}_n$ 1.2) allows an accurate estimate of this

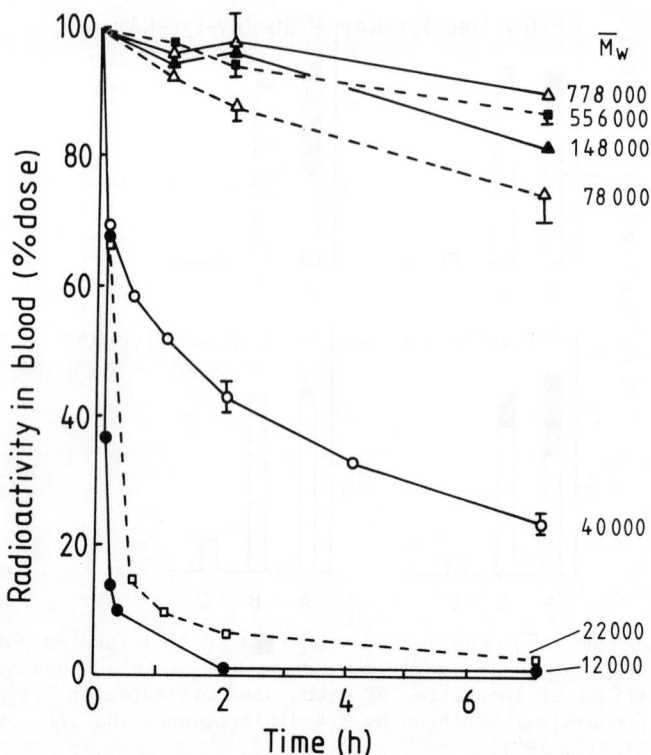

Fig. 7   Blood clearance of ^{125}I-labeled polyHPMA of different molecular weights after i.v. administration in rats (from Seymour et al., 1987b)

threshold value. For this particular type of polymer a value of 45000 ± 2500 was found. PolyHPMA fractions with molecular weights above this value are retained in the circulation for a much longer time period with the rate of clearance increasing with decreasing molecular weight of the conjugates. The polymer fractions not cleared by the kidneys can be removed from the circulation either in sinusoidal capillaries or via transendothelial passage. In the first case free exchange with the interstitial fluid is possible without size limitation (e.g. liver, spleen and bone marrow). In the latter case a size discrimination has been found explaining qualitatively the observed molecular weight dependence (Seymour et al., 1987b). The biodistribution of polyHPMA carriers among various organs has been compared with that of carriers having an additional galactos- amide moiety (Cartlidge et al., 1987b). Crosslinked conjugates 1 and 2 (Fig. 8A) were radiolabeled with ^{125}I, injected i.v. into rats and the blood clearance (Fig. 8B) as well as the level of radioactivity in major organs after 1 h (Fig. 9) was determined. Conjugate 2 with- out the carbohydrate ligand is cleared relatively slowly, with small amounts appearing in the liver, kidney, spleen and lungs while the major fraction (77%) is still in the blood.

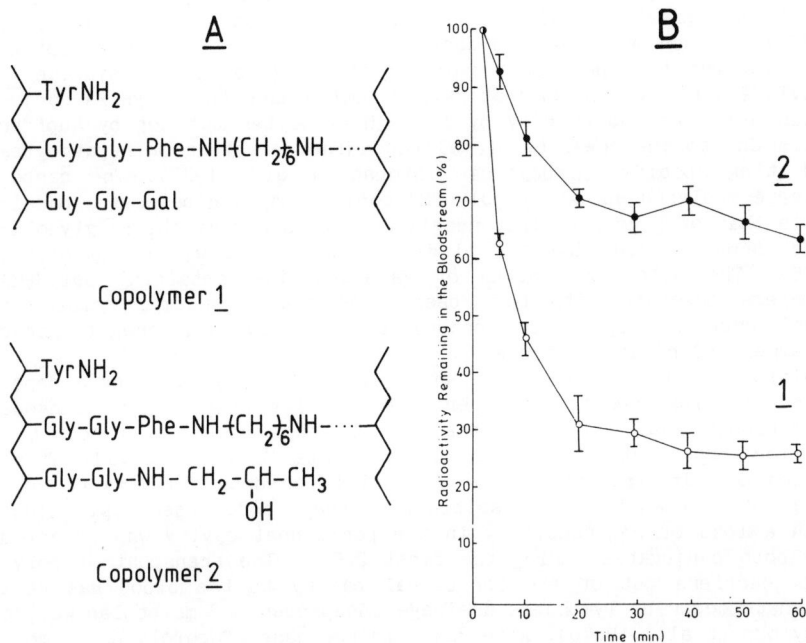

Fig. 8   Schematic structure (A) and blood clearance after i.v. administration in rats (B) of crosslinked polyHPMA conjugates (from Cartlidge et al., 1987b)

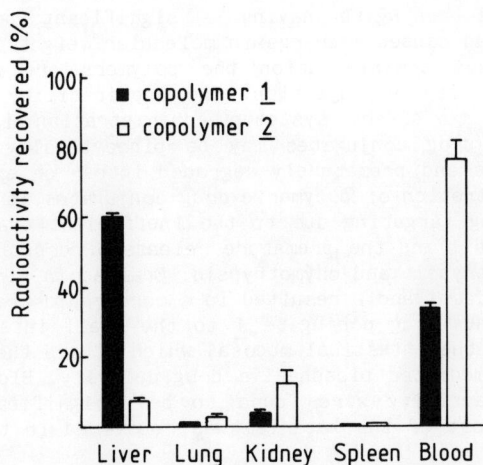

Fig. 9   Body distribution at 1 h after i.v. administration in rats of crosslinked polyHPMA conjugates 1 and 2 (from Cartlidge et al., 1987b)

The distribution of crosslinked polyHPMA is similar to that observed for linear polyHPMA conjugates when the molecular weight is higher than the threshold value for renal excretion (Seymour et al., 1987b). A much more rapid clearance and an almost selective uptake (60%) occurs for the galactosamide-containing polyHPMA conjugate 1. Previous data have shown that galactosamide-carrying polyHPMA conjugates are taken up quickly by liver hepatocytes but not by Kupffer cells due to the presence of galactosamine receptors on hepatocytes mediating specific pinocytosis (Duncan et al., 1986). The carbohydrate-mediated uptake of polyHPMA conjugates by hepatocytes parallells that of glycoproteins, neoglycoproteins and synthetic glycolipids described in numerous studies (for a review, see Barondes, 1986). The selective uptake of galactosamide-containing polyHPMA carriers together with the observation that lysosomal relase of model drug in hepatocytes occurs is promising for drug-targeted treatment of hepatoma tumors.

Administration of conjugates 1 and 2 by routes establishing a depot effect for instance from the peritoneal cavity or after subcutaneous injection, results in slow transfer to the blood. As a result a similar hepatocyte targeting of conjugate 1 is observed. These routes of administration offer potential for drug delivery to cell types other than those accessible from the blood. After i.p. injection a loss of radioactivity in the peritoneal cavity was observed for both conjugates during the first 2.5 h. The transport of poly-HPMA carriers out of the peritoneal cavity to the blood has been shown to occur by lymphatic drainage independent of molecular weight (Seymour et al., 1987b). After 24 h in the case of copolymer 2, most radioactivity was found in the urine (67%) and blood (17%) while with copolymer 1 radioactivity was found in the liver (52%) and in urine (40%). From the molecular weight distribution of the conjugates excreted in the urine it could not be established whether degradation of crosslinks had occurred or if renal ultrafiltration of the original conjugate having a significant polydispersity ($\bar{M}_w/\bar{M}_n$2.7-3.1) had caused a change in molecular weight distribution. After subcutaneous administration the polymers are removed very slowly from the site of injection. The applicability of the subcutaneous route to effect systemic administration is apparently limited because drug conjugates may be pinocytically captured by surrounding cells and prematurely degraded intra- or extracellularly. Oral administration of polymeric drug conjugates is probably not suitable for drug targeting due to the inefficient transfer across the intestinal wall and the premature release of drug by intestinal enzymes such as trypsin and chymotrypsin. Oral administration of the polyHPMA conjugates 1 and 2 resulted in a more rapid transfer of the galactosamide-containing conjugate 1 to the small intestine due to association with the intestinal mucosa, which offers the possibility of carbohydrate-mediated bioadhesive drug delivery. Blood and organ levels of radioactivity were found to be insignificant for both conjugates and most of the material was recovered in the urine and faeces.

The use of biodegradable crosslinks in the conjugates 1 and 2 was demonstrated to be a feasible strategy to accomplish the excretion of polyHPMA carriers in the long term. After i.p. administration, 90% of 1 and 2 was excreted in the urine and faeces within 30 days (Fig. 10).

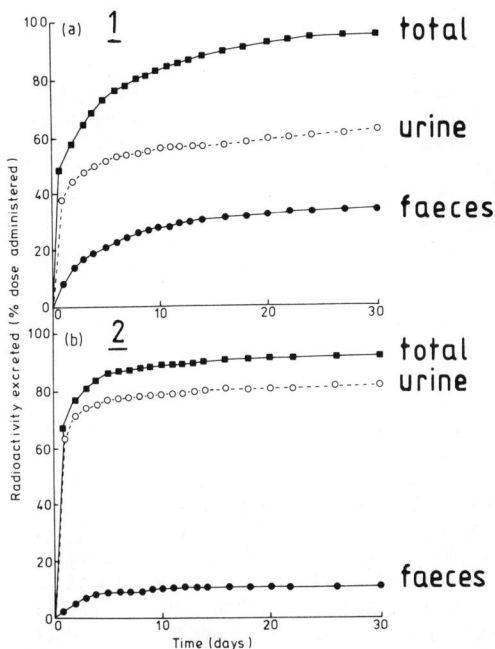

Fig. 10   Excretion of radioactivity after i.p. administration of crosslinked HPMA copolymers 1 (a) and 2 (b) (from Cartlidge et al., 1987b)

The hepatocyte-targeted conjugate 1 is largely excreted in the faeces suggesting the possibility for achieving drug targeting to the bile. Linear HPMA copolymers with molecular weights above the renal threshold value were recently shown to remain in the body after i.p. dosage in rats for extended periods. After 21 days 47-69% of the administered dose of polymer was still found in the body. The residual amount was dependent on the molecular weight ($\bar{M}_w$ 78000 - 778000) (Seymour et al., 1987b). This limits the practical usefulness of high molecular weight linear polyHPMA as a drug carrier. Linear polyHPMA with molecular weights below the renal threshold value offer a better prospect for this purpose notably in targeted drug delivery. Both the increase in molecular weight by attaching a macromolecular ligand (antibody or hormone) and the cell binding capacity of a low-molecular weight ligand (carbohydrate) will delay the excretion of polymeric carrier allowing a prolonged duration of drug action.

In vitro and **in vivo** targeting of polyHPMA conjugates of antitumor agents

The studies above deal with the in vitro and in vivo performance of polyHPMA conjugates carrying model drugs. These are currently being extended to the targeted delivery of clinically useful antitumor agents and other drugs (Kopecek et al., 1985). The results published to date for the anthracycline daunorubicin (DNR) will be taken as an example (Duncan et al., 1987; Kopecek and Duncan, 1987; Rihova and Kopecek, 1985; Rihova et al., 1986). The anthracyclines are often chosen for the evaluation of new drug delivery systems. Adriamycin (ADR) and to a lesser extent DNR are used clinically in the treatment of leukemias, lymphomas, breast cancer, sarcomas, ovarian cancer, gastric cancer and small cell lung cancer. A major clinical problem is the dose-dependent cardiotoxicity of the drugs. The cumulative dose given is therefore usually limited to 550 mg/m^2. Other side effects are vomiting, myelosuppression, tissue necrosis after extravasation and reversible alopecia. To improve the therapeutic index of ADR and DNR many types of controlled drug delivery devices have been evaluated, amongst them covalently-bound polymeric conjugates.

In polyHPMA conjugates of DNR the effects of the biodegradability of spacers and of polymer-bound galactosylamine and fucosylamine moieties were studied using mouse leukemia L1210 cells in vitro and in vivo. The fucose-receptor on the membrane of L1210 cells allows the possibility of fucosyl-mediated targeting of polymeric conjugates by enhancement of pinocytic entry. The polyHPMA-DNR conjugates as well as free DNR inhibited the growth of L1210 cells in vitro in a dose-dependent manner. The cytotoxicity of the conjugates varied with the spacer used, in the order P-Gly-Phe-Phe-Leu-DNR > P-Gly-Phe-Leu-Gly-DNR > P-Gly-Gly-DNR. This is consistent with the lysosomal degradability of the spacers which is only observed for the first two conjugates. The slight cytotoxicity found for the conjugate having a non-degradable (-Gly-Gly-) spacer was tentatively explained by a membrane-mediated toxicity of the polymer-bound drug. With all polyHPMA-DNR conjugates the total drug concentrations required for cell growth inhibition are much higher than with the free drug. The low bioavailability of drug from these polyHPMA conjugates can be improved by the additional binding of a fucosylamine moiety to the carrier whereas a bound galactosylamine is not effective. Thus, a fucosylamine carrying conjugate with a degradable spacer (-Gly-Phe-Leu-Gly-) shows strongly enhanced binding to the cells and the efficacy of the polymer-drug conjugate in cell growth inhibition increases to around 10% relative to free drug. These data support the importance of endocellular release of drug after pinocytic entry of polyHPMA-DNR conjugates in the killing of cells.

It is generally acknowledged that the anthracyclines interfere with cell division by several mechanisms (Myers, 1982). An important pathway is the intercalation of anthracylines with DNA. Another mode of cytotoxicity due to membrane binding of externally bound drug has been proposed by Tritton and coworkers. This was based on the observed in vitro cytotoxicity for L1210 cells with adriamycin bound to insoluble polymeric supports (Wingard et al., 1985).

These insoluble drug conjugates are not taken up by the cells and the cytotoxicity was explained by membrane-disturbing effects of the bound drug interacting in a high local concentration with the cell membrane. A concentration-dependent reversible binding of soluble conjugates of ADR and poly(α-L-glutamic acid) with L1210 cells in vitro has been observed. However, almost no cytotoxic effects were seen (Hoes et al., 1986).

A contribution from membrane-mediated cytotoxicity of polyHPMA-DNR conjugates as suggested by Duncan et al (1987) cannot be excluded. Clearly further studies are needed to establish whether significant changes in the membrane are occurring in the presence of polymeric conjugates of anthracyclines.

PolyHPMA conjugates of DNR were studied in vivo on L1210 cells inoculated i.p. in DBA$_2$ mice with i.p. administration of drug and drug conjugates (Kopecek and Duncan, 1987). Conjugates having nondegradable (-Gly-Gly-) and biodegradable (-Gly-Phe-Leu-Gly-) spacers were compared and the targeting potential of a fucosylamine moiety was evaluated. Three independent series of experiments were performed. Untreated animals died on day 14-17 (mean survival time, MST) and showed a weight gain of 140% relative to the starting weight due to tumor growth. Administration of a low dose (3x2 mg/kg) of free DNR produced only slight weight loss and no change in MST (15-16 days). When using a higher dose of DNR (3x5 mg/kg) significant weight loss was obtained but the survival time decreased (MST 10 days). Apparently free DNR is insufficiently cytotoxic at low doses while at a higher dose partial tumor regression occurs but also lethal toxicity develops. PolyHPMA-DNR conjugates (3x5 mg/kg) showed variable biological effects dependent on composition. The P-Gly-Gly-DNR conjugate was inactive (MST 14-18 days) whereas the conjugate P-Gly-Phe-Leu-Gly-DNR effected a definite increase in MST (19-32 days). The effect of linking fucosylamine to the latter conjugate could not be established unequivocally because different results were obtained from different experimental series. In one experiment the MST raised from 19 days without carbohydrate ligand to 50 days in the presence of ligand. In another experiment these values were 32 and 25 days respectively. Results from other dosage schedules point to a positive effect of fucosyl targeting on MST. Intraperitoneal dosage of biodegradable DNR conjugates against L1210 cells inoculated subcutaneously was found to be therapeutically effective dependent on the dose. Assuming that the conjugates are not degraded in the peritoneum this may indicate that the DNR conjugates cross compartmental barriers as observed for the carrier and reach the tumor localized in another compartment. In conclusion polyHPMA-DNR conjugates containing a biodegradable spacer effectively suppress tumor growth as indicated by the absence of weight change and a definite increase in MST. A therapeutic effect was absent when using a nondegradable drug conjugate. The data suggest that drug release by lysosomal enzymes is mandatory for pharmacological activity.

The targeted delivery of drug to T-lymphocytes has been studied with antibody-polyHPMA-DNR conjugates (Kopecek and Duncan, 1987; Rihova and Kopecek, 1985; Rihova et al., 1986). T-lymphocytes are important for all types of immune reactions and targeting of drugs might be of therapeutic value in the treatment of certain pathological states, such as transplant rejection, autoimmune disorders and T-cell leukemia. Antibodies raised in rabbits against the surface Thy 1.2 alloantigen of T-lymphocytes from mice (anti-Thy 1.2; ATS) were used as the targeting structure. These antibodies react in vivo with the T-

cells carrying the Thy 1.2 antigen. As a result the complement
system is activated effecting the lysis of these cells. ATS was
bound to polyHPMA-DNR conjugates and the cytotoxicity against T-
lymphocytes was evaluated in vitro and in vivo. Binding of ATS to
the polymeric carrier is accompanied by partial inactivation of the
antibody cytotoxicity. It is currently unknown whether this is due
to changes in the antigen binding site (F_{ab} fragment) or in the
complement activating part (F_c fragment) of the protein. DNR was
bound to ATS-polyHPMA using a nondegradable (-Gly-Gly-) and a
degradable (-Gly-Phe-Leu-Gly-) spacer. In vitro assays in the absence of complement showed that ATS conjugates having DNR bound
across a nondegradable spacer is inactive whereas the use of a
degradable spacer results in significant cell death. This argues
strongly for a drug-mediated cytotoxic effect probably after pinocytosis and endocellular drug release in accordance with the endocellular design strategy. Nevertheless quite high concentrations of
antibody and drug were needed to approach 100% killing of the cells.
In the presence of complement lower doses of antibody and drug effected cell death and the conjugate having a cleavable spacer-drug
bond was much more efficient. In the latter conjugate drug-mediated
cytotoxicity adds to antibody-mediated complement activation. Similar results were obtained in vivo by measuring the suppression of
antibody response to stimulation by sheep red blood cells (SRBC) as
a result of T-cell depletion. At a dose of 3 mg of protein/mouse
free ATS and polyHPMA-bound ATS used as controls showed 87 and 32%
suppression of anti-SRBC response, respectively. Ternary ATS-poly-
HPMA-DNR conjugates were also administered using the same dose of
protein (3 mg/mouse) and a low total dose of drug (0.4-0.6
mg/mouse). The conjugate having a lysosomally cleavable spacer
showed strong suppression of antibody response (98%) whereas the
conjugate with a noncleavable spacer was inactive. With an increased
dose of the latter conjugate (12 mg of protein/mouse and 2.5 mg of
drug/mouse) the suppression observed was equal to ATS-polyHPMA alone
with no apparent contribution from the drug. When using free drug or
mixtures of ATS and drug with the same protein doses, the suppression of anti-SRBC response could be increased to the 99% level only
by using much higher drug levels (> 30 mg/mouse). It was concluded
that DNR bound to a lysosomally cleavable bond in a ternary ATS-
polyHPMA-drug conjugate enhances the cytotoxicity of the antibody
against T-cells very effectively. The mechanism of the drug-mediated
enhancement of antibody-mediated killing of T-cells by complement
has not yet been established.

Poly(α-amino acids)

Poly(α-amino acids) mostly of the naturally occurring L-configura-
tion have been advocated by several investigators as possible carriers for cytostatic agents. Several polymers of this group are
water-soluble and the hydrophilicity/hydrophobicity of the carriers
can be adjusted by using appropriate copolymers. In this way either
water-soluble drug-carrier systems or water-insoluble depot systems
can be prepared. Poly(α-L-amino acids) are biodegradable in principle. The rate of biodegradation can be adjusted by the proper
choice of amino acids, molecular weight and the use of homopolymers

or random or block copolymers. Degradation results in the formation of peptides and amino acids. By a judicious choice of the carrier especially in the case of copolymers the formation of oligopeptides with undesirable biological effects can be avoided. It has been reported that several poly(α-amino acids) elicit an immunological response (Maurer, 1964). Apart from possible toxic effects the immunological properties must be considered in the selection of appropriate carriers. The use of poly(α-L-lysine), polyLys, poly(α-L-glutamic acid), polyGlu, poly[(N^5-(2-hydroxyethyl)-L-glutamine], polyHEG and poly(L-aspartic acid), polyAsp, will be discussed further.

Poly(α-L-lysine)

Poly(α-L-lysine), polyLys, has been investigated as a carrier for cytostatic agents by Ryser and Shen (1986) and has also been extensively studied by Arnold et al. (1983) both as an antineoplastic agent and as a tumor specific drug carrier. It was shown that poly-Lys binds tightly to cells and induces morphological changes of the membrane. This causes a rapid leakage of small nutrient molecules from the cell, perhaps promoted by phospholipase A2 activation, leading to a rapid loss of DNA, RNA and protein synthesis. Using short incubation periods HeLa cell growth is inhibited to a greater degree by polyLys than by its optical isomer poly(α-D-lysine), poly-(D-Lys). In time however cell growth is reestablished after exposure to polyLys whereas the viability of the cells is permanently suppressed by poly(D-Lys). This is explained by degradation of polyLys, whereas poly(D-Lys) remains intact. Both the lethal toxicity and the antineoplastic activity of polyLys increase strongly with increasing molecular weight as judged from HeLa cell cytotoxicity, toxicity in mice and suppression of Ehrlich ascites tumor growth in mice. By using low molecular weight polyLys as a carrier in drug conjugates the toxic effects exerted by the polymer are suppressed and drug-related cytotoxicity is more clearly delineated.
PolyLys, a strongly cationic poly(α-amino acid), is efficiently transported into cells by non-specific adsorptive pinocytosis. On this basis conjugates of methotrexate (MTX) and polyLys were proposed for effective endocellular drug delivery (Ryser and Shen, 1986). MTX was conjugated through one of its carboxyl groups to ε-amino groups of polyLys at a ratio of one MTX molecule per 29 amino groups (Ryser and Shen, 1978). Using this conjugate it was possible to transport MTX efficiently in cells which are defective in MTX transport such as a chinese hamster ovarian (CHO) cell line or a tumor line M5076 (Fig. 11).
The transport defect is efficiently circumvented by the pinocytic pathway leading to release of MTX possibly still bound to some L-lysine residues after lysosomal degradation of the MTX-polyLys conjugate. When MTX was conjugated to poly(D-Lys) no cytocidal activity against CHO cells was observed. This can be explained by the lack of degradation of the poly(D-Lys) carrier. The MTX antagonist leucovorin effectively inhibits the cytotoxic effects of MTX-poly-(Lys) conjugates. This demonstrates that the cytotoxic effects observed are due to the interaction of intracellularly released MTX or MTX-bound carrier fragments with the cytoplasmatic target enzyme di-

hydrofolate reductase (DHFR).
Ryser and Shen conducted some very interesting experiments using
poly(D-Lys) as a non-biodegradable carrier. To study the effect of
spacers on the cytocidal properties of the drug conjugates different

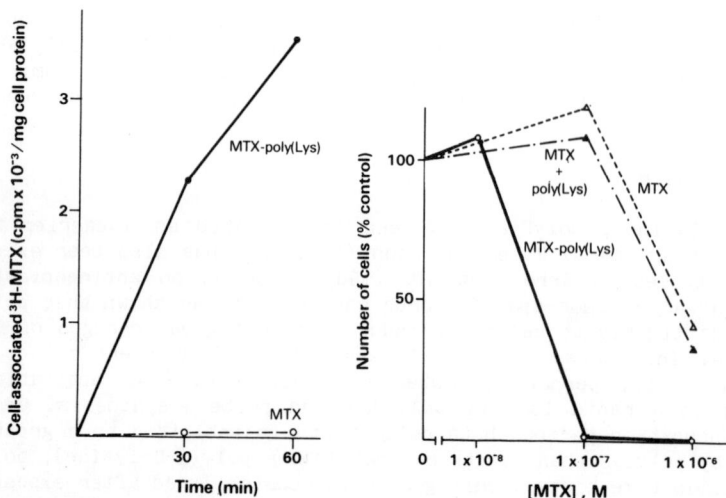

Fig. 11. Cellular uptake and cytocidal effect of MTX and MTX-polyLys
in M 5076 tumor cells. Low uptake of free MTX by M 5076
cells (left panel) is related to their 20-fold resistance
to free MTX compared to MTX-polyLys (right panel) (from
Ryser and Shen, 1986).

spacers between MTX and the carrier were evaluated. The spacers used
were either sensitive to proteases, hydrolysis at acid pH or to
chemical reduction as found inside cells.
Using protease-sensitive spacers it was found that when MTX was
first attached to spacers like Gly-Gly-Gly, Gly-Gly-Phe, Gly-Phe-Ala
or human serum albumin (HSA) and then coupled to poly(D-Lys), the
resulting conjugates were very effective against MTX-resistant CHO
cells. These results show that the influence of the type of spacer
used on the induced cytotoxic action is small. On the contrary the
selection of spacers for the coupling of anthracyclines onto proteins (Trouet, 1982) or onto polyHPMA (Duncan et al., 1987) is critical. In general a spacer with a specific length or composition is
required to obtain favourable results for a given drug-carrier
system.
Another promising approach is the use of spacers affording a hydrolyzable linkage between the drug and carrier which is readily cleaved upon exposure to an acidic pH of 4.5 to 5.5. Prelysosomal and
lysosomal cell compartments have a similar pH. Thus endocellular
drug release can be effectively accomplished by using pH sensitive
spacers. This principle was illustrated using a conjugate of daunomycin (DNR) and poly(D-Lys) coupled through a cis-aconitic acid
spacer (Fig. 12) (Shen and Ryser, 1981).

The cis-aconityl-DNR bond is acid labile due to anchimeric assistance of the cis-carboxylic acid group in the hydrolysis of the
amide bond. Incubation at acid pH causes rapid drug release, while
at neutral pH the drug-polymer bond is stable.

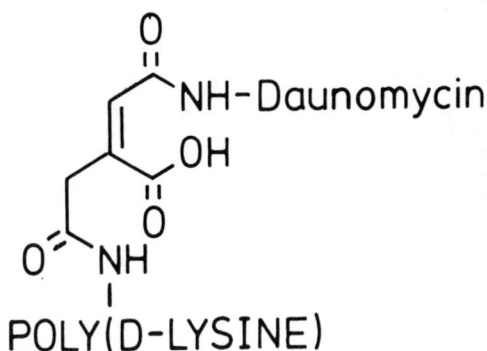

Fig. 12 Structure of DNR-poly(D-Lys) conjugate having a cis-
aconityl spacer moiety (Shen and Ryser, 1981)

The effectiveness of a disulfide spacer was investigated based on
the fact that intracellular degradation of proteins requires the
reduction of disulfide bonds. MTX was coupled to poly(D-Lys) through
a 3-(2-aminoethyl)dithiopropionic acid spacer (Shen et al., 1985).
It was found that this conjugate was effective in killing CHO cells
defective in MTX transport. The cytocidal action was not inhibited
by leupeptin, an inhibitor of lysosomal thiol proteases, nor by the
addition of NH_4Cl which increases the lysosomal pH. It was speculated that the disulfide spacer is enzymatically reduced in a prelysosomal compartment.
PolyLys conjugates of three different drugs were synthesized by
Arnold et al. (1983). The drugs used were DNR, MTX and 6-amino-
nicotinamide, (6-AN) (Fig. 13). DNR was coupled to ε-amino groups of
polyLys having a molecular weight of 35000 using dicarboxylic acid
spacers with varying chain lengths. Typical loading of DNR onto the
carrier was 1:25 w/w. The antimetabolite 6-aminonicotinamide (6-AN)
was similarly conjugated through a succinate group in a drug/carrier
ratio of 1:2.56 w/w. MTX was directly coupled to polyLys (molecular
weight 3000) mainly through its α-carboxyl group. The anthracyline-
polyLys activity on HeLa cells using a carrier molecular weight of
35000 was typically 10-15% of that of the free drug independent of
the spacer. Using a periodate oxidation coupling procedure developed
by Hurwitz et al. (1975) conjugates without any activity against
HeLa cells in culture were obtained. This procedure involves oxida-
tive opening of the daunosamine sugar between the 2' and 3' carbon
atoms generating a dialdehyde. Once the dialdehyde is formed it is
used to form Schiff bases with lysine residues which are subsequent-

ly reduced with sodium borohydride yielding stable secondary amines. It must be emphasized that the sugar ring is destroyed in the process. In vivo tests of the series of DNR-polyLys conjugates showed unexpectedly low activities due to as yet unknown factors. The MTX-

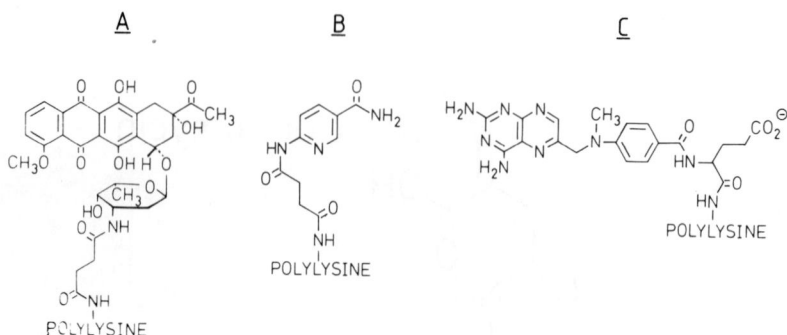

Fig. 13   Structures of conjugates of polyLys and daunomycin (A), 6-aminonicotinamide (B) and methotrexate (C) (from Arnold et al., 1983)

polyLys conjugate shows considerable activity for HeLa cells. In the presence of serum a tenfold higher activity of the conjugate was observed compared with serum-free medium. The active form of the conjugate was demonstrated to be the intact species and not the products from trypsin mediated degradation which occurs in the serum. At present the specific mechanism by which serum contributes to the enhanced cytotoxic activity of MTX-polyLys conjugates are not known. It was further observed that the activity of the conjugates is highly dependent on the cell type. Thus substantial activity was found with Ehrlich ascites cells but activity was negligible with L1210 cells cultured in serum-containing media. It was speculated that the MTX-polyLys conjugates do not interact with L1210 cells. Experiments with the (6-AN)-polyLys conjugate showed substantial cytotoxic action for HeLa cells in culture. Competition studies using nicotinamide indicated that the conjugate is catabolized by the cell to a form that retains cytotoxicity due to an antimetabolic function. The (6-AN)-polyLys conjugate also showed substantial anti-neoplastic activity on Ehrlich ascites tumor bearing white Swiss mice.

PolyLys-heparin complex

PolyLys is strongly toxic when injected intravenously to animals and its potential as a drug carrier is therefore limited. However, the complex of polyLys with heparin, an anionic biopolymer, is much less toxic, probably related to the neutralization of the positive charge of polyLys (Ryser and Shen, 1986). On addition of heparin to ^{3}H-MTX-polyLys the uptake by CHO cells decreased linearly with increasing

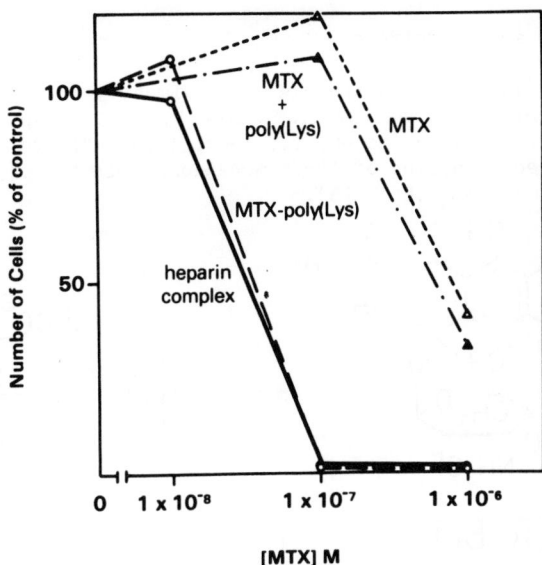

Fig. 14 Cytocidal effect of MTX plus polyLys, MTX-polyLys conjugate
and its complex with heparin on MTX-resistant M 5076 tumor
cells in culture. The complexed and uncomplexed conjugates
are equally effective in overcoming the MTX transport
defect of the tumor cells (from Shen and Ryser, 1986).

amounts of heparin. In the presence of a 10 to 100 fold heparin
excess apparently all ^{3}H-MTX-polyLys was complexed and the uptake
leveled off at 20% of the initial value. The cytotoxic action of the
complex was minimal at a heparin-polyLys ratio of 1:1 but reached a
maximum at a ratio of 100:1. Thus the decrease in uptake effected by
heparin is partly compensated by an increased cytotoxicity of the
complex. Similar effects were observed using the tumor cell line M
5076 (Fig. 14).
Various lines of evidence suggest that the uptake of polyLys-heparin
complex by cells occurs via a process resembling receptor-mediated
pinocytosis. This contrasts with the uptake of either polyLys or
heparin by non-specific adsorptive or fluid-phase pinocytosis,
respectively. The receptor involved was suggested to be a heparin
sulfate proteoglycan but was not defined in more detail. The recep-
tor-like character of the pinocytic entry of polyLys-heparin complex
may offer prospects for targeted drug delivery. It was further shown
that exposure of cells to a low pH of 4.5 greatly enhanced binding
and uptake of the complex. This was tentatively explained by a
direct translocation of the complex across the cell membrane
following a pH-induced conformational change. This type of entry of
substrates into cells which is different from pinocytosis has been
observed for diphteria toxin and certain viral proteins.
In conclusion polyLys demonstrates specific biological effects due
to its cationic charge which may be advantageously exploited in
drug-carrier systems.

Poly(L-aspartic acid), polyAsp

PolyAsp with a molecular weight of approximately 20.000 has been used as a carrier for adriamycin (ADR) by the group of Zunino (Pratesi et al., 1985; Zunino et al., 1982, 1984). The conjugate was prepared by direct coupling of 14-bromodaunorubicin to polyAsp under mildly basic conditions (Fig. 15).

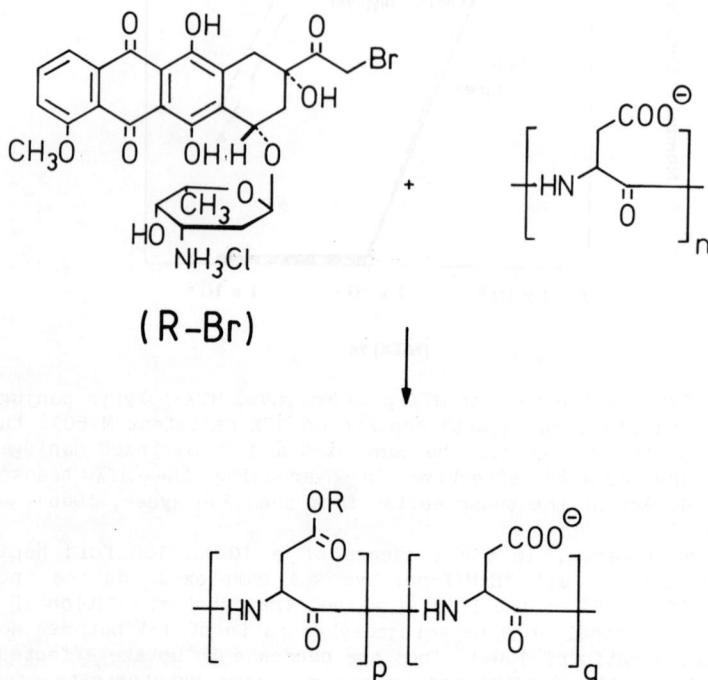

Fig. 15   Preparation of polyAsp-adriamycin conjugates (Zunino et al., 1982)

Degrees of substitution of 18-70 mol drug.mol^{-1} polyAsp were obtained. The daunomycin derivative was connected via a labile ester bond instead of using the amino group of the sugar residue to give the more stable amide bonds. The conjugate was compared with free DNR and ADR with respect to its toxicity and therapeutic efficacy. When comparing the toxicity in non-tumor bearing mice using different i.v. administration schedules the ratio between equitoxic doses of polyAsp-ADR and ADR was approximately three. This ratio also applied to the cardiotoxicity measured in mice. With respect to the ulcerogenic potential the ratio was even higher. Different tumor models in mice (early ascitic J774, i.m. M5 ascitic tumor, Lewis lung carcinoma and advanced C3H mammary carcinoma) were used to compare the conjugate with the free drug. It was shown that the conjugate provided similar or rather greater therapeutic effects than free drug at less toxic doses. This effect was more evident in the highly sensitive tumors such as the M5 model and suggests that

the therapeutic index of the conjugate is improved with respect to
the free drug. Until now no studies on cellular uptake and release
of drug from the conjugate were carried out. The conjugate was not
able to overcome the drug resistance of certain tumor cells (colon
carcinoma 26, P388 leukemia). Because the release of free drug in
the circulation was not studied, it is not known whether the anti-
tumor activity is caused by free drug released in the circulation or
by incorporation of the conjugate in the tumor cells followed by
intracellular degradation.
A new type of drug-carrier conjugate based on block copolymers was
developed by Ringsdorf and co-workers (Bader et al., 1984; Dorn et
al., 1985). This approach was extended to the use of polyAsp as the
drug-carrying block in a poly(ethylene glycol)-polyAsp block copolymer with bound ADR (Yokoyama et al., 1987). The poly(ethylene glycol) was incorporated to provide conjugates with an increased water-
solubility and stability under physiological conditions. Furthermore
the antigenicity of the conjugate eventually coupled to an immunoglobulin may be reduced. Antitumor activity and toxicity of these
novel conjugates have not been reported yet.

Poly(α-L-glutamic acid)

Poly(α-L-glutamic acid), PolyGlu, has been more extensively studied
as a carrier for cytostatics than polyAsp. Kenny (1959) evaluated
polyGlu ($\bar{M}_w$ 5.1 x 10^4 and 8 x 10^4) as a candidate for application as
plasma expander. After infusion of 40 ml of a 3% solution of polyGlu
in 0.9% aqueous sodium chloride in dogs a rise in right atrial pressure and a striking decrease in cardiac output was observed. The
excretion of polyGlu was very slow: 6-15% within 6 hours after administration and no appreciable amounts after 6 hours. At these high
concentrations polyGlu also caused the agglutination of erythrocytes. Maurer (1957) found in immunological studies that no detectable antibodies could be produced against polyGlu neither in man,
rabbit, nor guinea pig.
PolyGlu has been used by Goldberg et al (1981) to form water-insoluble salts with adriamycin (ADR). The release of ADM in saline
was 40% in 24 h at 37°C and 60% in ten days. The salt showed anti-
tumor activity in mouse mammary experiments. The toxicity of the
salt was substantially decreased, as compared to the free drug.
Various low-molecular weight prodrugs are good substrates for certain enzymes but as yet few examples in the field of polymer-drug
conjugates have been described. Conjugates of adriamycin (ADR) and
polyGlu have been prepared by us in which the drug is bound to the
γ-carboxyl group of the carrier across an amide bond (Van Heeswijk
et al., 1984). These conjugates were designed as macromolecular
prodrugs for the selective release of free drug at tumor cells
having elevated levels of γ-GT. It was hypothesized that after biodegradation of the polymeric carrier low-molecular weight γ-glutamyl
amides of ADR are formed which can be efficiently transformed to
free drug by γ-GT. Recently it has been found that enzymatically
prepared γ-glutamyl adriamycin is susceptible to γ-GT mediated
hydrolysis with formation of free drug (Stark, 1986). However, the
macromolecular γ-glutamyl prodrugs of ADR are incapable of inhibi-

ting the growth of B16 melanoma cells which are known to have membrane-bound γ-GT (Hoes et al., 1986). We explain this result by lack
of biodegradability of the polymer main chain in the conjugates on
the basis of in vitro enzymatic degradation studies. The lack of
main chain biodegradability is unexpected because only around 5% of
available carboxyl groups are substituted with drug molecules (Hoes
et al., 1985).

Hurwitz et al. (1980) prepared several conjugates of polyGlu with
daunorubicin by converting poly(γ-benzyl-L-glutamate), PBLG, with
hydrazine hydrate to the corresponding poly(glutamylhydrazide) and
subsequent coupling with daunorubicin via the carbonyl group. The
resulting conjugate contained 41% by weight of daunorubicin. Conjugates were also prepared by first reacting PBLG with 3-aminopro-
panol, followed by a reaction sequence as described earlier. Compared with free daunorubicin the polymeric derivatives were slightly
less cytotoxic to mouse Yac lymphoma cells in vitro but were equally
or more effective against the Yac lymphoma in vivo. Attachment of
daunorubicin onto carriers through a non-hydrolysable bond yielded
inactive conjugates.

Morimoto et al. (1984) coupled the alkylating agent melphalan (Fig.
16) directly to polyGlu (mol.weight 60.000 and 40.000) with a degree
of substitution of 1 molecule of melphalan per 23 glutamate residues.

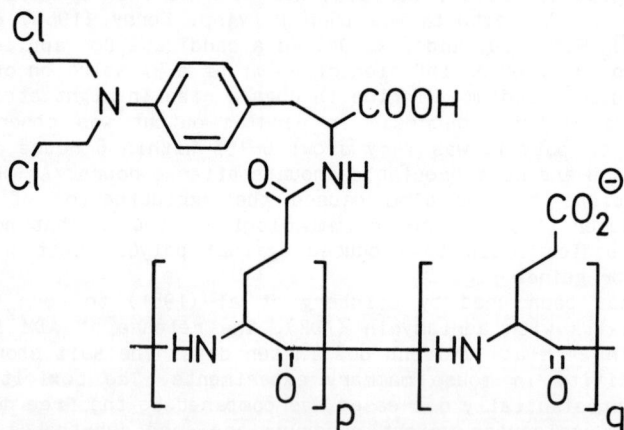

Fig. 16   Structure of polyGlu-melphalan conjugate (Morimoto et al.,
          1984)

The conjugates had 40-70% of alkylating activity in vitro as compared to free melphalan. After direct subcutaneous injection of
conjugate solutions into Yoshida sarcoma tumors in male Donryu rats
a considerable antitumor effect was observed. Histologically no
differences of the tumor tissue were seen when the conjugate or free
melphalan was injected. Model conjugates containing ^{3}H-Phe had a
tendency to be absorbed through the lymphatic routes after subcutaneous injection. The in vitro release of ^{3}H-Phe was substantially enhanced in the presence of carboxypeptidase A consistent with
enzyme-mediated carrier degradation.

High molecular weight conjugates of mitomycin C (MMC) with polyGlu were prepared using EDC by the group of Sezaki (Kato et al., 1982). Conjugates contain one MMC per eight glutamic acid units. The in vitro release of MMC in PBS (pH 7.4) showed a monoexponential curve with a half-life of 35.5 hours. The in vivo cytostatic activity was measured in BDF$_1$ mice bearing P388 leukemia or B16 melanoma in an intraperitoneal-intraperitoneal system. The MMC-polyGlu conjugate exhibited a superior effect against B16 melanoma cells in spite of a relatively low $\bar{M}_w$ of 1.4 x 10^4 (Kato et al., 1982). In a later publication (Roos et al., 1984) MMC was bound to polyGlu with a $\bar{M}_w$ of 1.1 x 10^4 or 6 x 10^4 respectively. Leukemia L1210 cells were exposed to the conjugates and to free MMC for 3 days. No major differences in growth inhibition of cells between MMC and the conjugates were observed. The effect of conjugates was much smaller after 1 hour exposure, which corresponds to the low release rate of MMC from the conjugates. The conjugates were also evaluated using P388 leukemia-bearing mice. It turned out that the polyGlu-MMC ($\bar{M}_w$ 6 x 10^4) was less effective than MMC and polyGlu-MMC ($\bar{M}_w$ 1.1 x 10^4). However, conjugates could be applied in much higher doses than free MMC. PolyGlu was used by Kato et al. (1984a) as a carrier for 1 β-D-arabinofuranosylcytosine (ara-C).

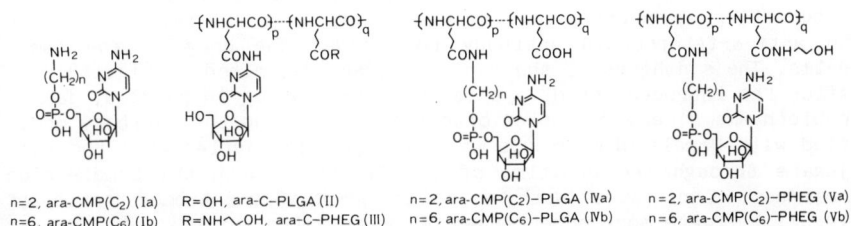

n=2, ara-CMP(C$_2$) (Ia)   R=OH, ara-C–PLGA (II)   n=2, ara-CMP(C$_2$)–PLGA (IVa)   n=2, ara-CMP(C$_2$)–PHEG (Va)
n=6, ara-CMP(C$_6$) (Ib)   R=NH~OH, ara-C–PHEG (III)   n=6, ara-CMP(C$_6$)–PLGA (IVb)   n=6, ara-CMP(C$_6$)–PHEG (Vb)

Fig. 17   Structures of conjugates of ara-C and polyGlu or its 2-hydroxyethylamide derivative (from Kato et al., 1984a)

Several conjugates were synhesized (Fig. 17) either through coupling with the 4-amino group of ara-C, or through an ω-aminoalkylphosphoryl side chain. The conjugates were also converted to the corresponding 2-hydroxyethylglutamine derivatives.
In vitro release studies indicated that the N-4 coupled conjugates released ara-C even at pH 7.0. The release was accelerated under basic and acidic conditions. Conjugates (IVa) could be cleaved by phosphodiesterase I, acid phosphatase or alkaline phosphatase respectively releasing mostly ara-C.
In vitro studies showed that the conjugates had decreased cytotoxicity against L1210 cells when compared with that of ara-C. Studies in vivo showed that all of the conjugates except ara-CMP (C2): PolyGlu (IVa, Fig. 17) had a greater antitumor activity than did ara-C in L1210 tumor bearing mice (inoculum, 1 x 10^5 cells i.p. on day 0) which were treated by a single i.p. injection of either the conjugates or the control ara-C on day 1. The largest activity (increase in life span, ILS, of 170%) was observed with a dosage of 50 mg (eq. ara-C per kg) of ara-C:PHEG (III, Fig. 17).

The authors also mention the use of a spacer (2-aminoethyl phosphoryl or 6-aminohexylphosphoryl) to connect the drug onto polyGlu. We have found that the use of peptide spacers in coupling ADR with polyGlu leads to conjugates with substantial higher activity for L1210 and B16 melanoma cells (Van Heeswijk et al., 1984; Hoes et al., 1985, 1986). This finding corresponds with the fact that ADR directly coupled onto polyGlu is very stable and is not degraded by papain.

An exciting approach is the use of conjugates which also contain antibodies against a particular tumor.

In 1975 Rowland, O'Neill and Davies (Rowland et al., 1975) reported on a conjugate of p-phenylenediamine mustard (PDM) polyGlu, and immunoglobulin from a rabbit antiserum against mouse lymphoma cells (EL4) (ratio 2:8:10 mg ml^{-1}). The conjugate was tested on mice intraperitoneally inoculated with 5 x 10^4 EL4 cells. The largest increase in survival was seen with the conjugate whereas combinations of polyGlu-PDM and free Ig gave also increased but shorter survival times than the full conjugate. The toxicity of conjugated PDM (LD$_{50}$ of ~ 200) was considerably reduced as compared with free PDM (LD$_{50}$ of ~ 5).

Tsukada et al. (1984) and Kato et al. (1984b) reported on a most interesting approach using conjugates of daunorubicin, polyGlu and an anti-α-fetoprotein antibody to deliver the drug to the tumor cells. The syntheses of the conjugates are described in Fig. 18.

After the introduction of a protected thiol group in polyGlu, daunorubicin was directly coupled onto the carrier and the antibody modified with N-maleimido groups was subsequently coupled with the conjugate through the addition of the SH-groups with the N-maleimido residues. Conjugates with a molar DM-polyGlu-antibody ratio of 10.8:1.1:1 were more potent than DM in in vitro cytotoxicity against the AFP-producing rat ascites hepatoma cell line AH 66. In therapeutic experiments the conjugates were more efficacious in prolonging the lives of AH 66 hepatoma-bearing rats than DM, antibody, a mixture of DM and antibody, or a conjugate similarly prepared with normal horse immunoglobulin.

Recent studies showed that human α-foetoprotein-producing hepatocellular carcinoma growth in nude mice could be suppressed using an daunorubicin-polyGlu-anti α-foetoprotein antibody conjugate (Tsukada et al., 1985)

It can be concluded that the use of polyGlu as a carrier for cytostatic agents is promising when spacers with optimal length and composition to promote intracellular release of the drug are used. At this moment it cannot be excluded that release of drugs near the tumor cell surface using the presence of a relative high concentration of specific enzymes near the tumor cell is also a good approach to enhance the specificity of the system. However, a more promising method is to use specific antibodies coupled to the conjugate in such a way that their function is maintained.

Natural macromolecules

Soluble natural macromolecules have been proposed as carriers for cytostatic drugs. These include DNA, dextrans, plasma proteins notably albumin, antibodies and glycoproteins (Baurain et al., 1983;

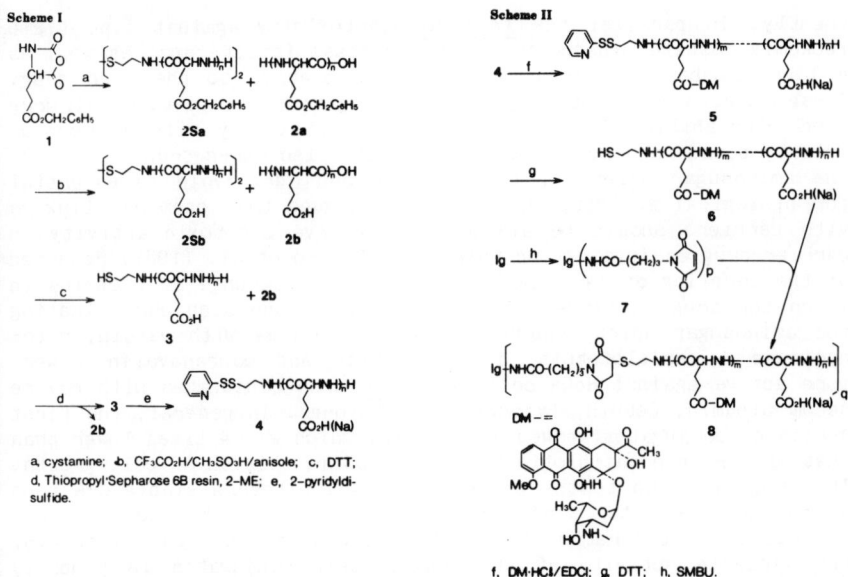

Fig. 18.   Preparation of ternary conjugates of DNR, polyGlu and
           antibody (from Kato et al., 1984b).

Poznansky and Juliano, 1984).
In 1972 Trouet and co-workers reported the use of DNA as a carrier
for the anthracyclines daunorubicin and adriamycin taking advantage
of their DNA-binding properties due to intercalation (Trouet et al.,
1972). With these noncovalent DNA-drug complexes the concept of the
lysosomotropic drug carrier was introduced. After pinocytosis of the
complex by target cells and lysosomal degradation of the carrier by
phosphodiesterases the drug is released endocellularly and effects
cell death probably by intercalation with nuclear DNA. Extensive
clinical trials with DNA-anthracycline complexes conducted on more
than 700 patients with different types of leukemia showed a high
antitumor efficacy similar to that of the free drug as well as a
greatly reduced cardiotoxicity. Other side effects were comparable
to those found for the free drug which is probably due to the limited stability of the non-covalent carrier-drug complexes in blood
(Trouet et al., 1972; Trouet and Jollés, 1984). Later, the Trouet
group turned to protein carriers with covalently linked anthracyclines to enhance the selectivity of endocellular drug release
(Baurain et al., 1983; Trouet et al., 1982). Bovine serum albumin
was succinylated to enhance its rate of pinocytosis by cells and
daunorubicin was coupled covalently with the carrier across an amide
bond either directly or by the use of a peptide spacer composed of
one to four amino acid residues. The drug release rate by the action
of tritosomes was strongly dependent on the length of the spacer. In
the directly bound conjugate or with a single amino acid spacer, no
drug release by lysosomal enzymes was demonstrated and in vivo activity against i.p. L1210 leukemia in mice was negligible. With spacer
lengths of 3 and 4 residues the enzymatic cleavability improved

greatly. In parallel the in vivo cytotoxicity against i.p. L1210
leukemia resulted in an impressive increase in life span as well as
a large number of long-term survivors relative to the free drug.
These studies were the first to demonstrate the importance of drug
bond cleavability in achieving high cytotoxicity with macromole-
cularly bound anthracyclines and possibly also other drugs.

The aminosugar moiety of daunorubicin and adriamycin is essential
for biological activity and it can be argued that chemical linkage
with carriers should be avoided to preserve cytotoxic activity in
carrier-drug conjugates. On this basis Zunino et al. (1981) reported
on the coupling of 14-bromodaunorubicin onto a range of proteins in
which the drug is linked using its methylketone side chain leaving
the aminosugar moiety unsubstituted. Conjugates with casein, ribo-
nuclease A, asialofetuin, immunoglobulin and concanavalin A were
more active against HeLa cells in vitro than conjugates with bovine
serum albumin, fetuin, lysozyme and histones. In general, the first
series of conjugates showed an activity which was 4 times lower than
that of the free antibiotic to obtain 50% inhibition of cell growth.
The drug is bound to the proteins forming either a stable C-N bond
or an unstable ester bond, but the ratio of bonds has not been reported. The molar ratio of drug to protein varied from 0.1 to 8.5.
the variable activity of the drug-protein conjugates is probably
best explained by differences in cellular uptake and degradation.

Methotrexate has been conjugated to BSA, bovine chymotrypsin and
bovine IgG by linking one of the drug carboxyl groups to the protein
lysine side chains (Chu and Whiteley, 1980). These conjugates were
as effective as the free drug in prolonging the life span of mice
bearing i.p. L1210 tumor cells. In vitro studies with L1210 cells
indicated that MTX-BSA was less effective than free drug in cell
growth inhibition. This correlates with the differences in intracellular drug concentrations found with either MTX or MTX-BSA. The
cytotoxicity of MTX is due to inhibition of the cytoplasmatic enzyme
dihydrofolate reductase (DHFR). It has been shown that MTX-BSA also
acts as an inhibitor of DHFR with an efficacy of 20% relative to
MTX. The chemotherapeutic action of MTX-BSA was suggested to result
from cellular uptake by pinocytosis and partial intracellular degradation yielding biologically active nondialyzable protein-MTX fragments. MTX-BSA was proposed for the treatment of MTX-resistant
tumors, because the defective MTX-transport mechanism in resistant
cells can be bypassed by the pinocytic entry of MTX-BSA.

The use of bovine or human serum albumin in the antibody-mediated
targeting of drugs has been advocated by several researchers (Endo
et al., 1987; Garnett et al., 1983; Ohkawa et al., 1986; Takahashi
et al., 1987). Ternary antibody-albumin-drug conjugates are currently preferentially prepared by an orthogonal coupling strategy. In
this way a high drug load can be achieved and loss of antigen-binding capacity as well as crosslinking of antibody molecules resulting in formation of insoluble products can be suppressed. Mitomycin
C was linked to HSA at the aziridine nitrogen atom through a gluta-
ric acid derived spacer. HSA-MMC was subsequently conjugated across
the single free thiol group with derivatized antibodies containing
maleimide moieties. In this way conjugates having an average molar
binding ratio of antibody/HSA/MMC of 1:1:30 were obtained. The drug-
albumin bond is slowly cleaved under physiological conditions with a

half life on the order of 3-4 days (Ohkawa et al., 1986; Takahashi et al., 1987). Methotrexate conjugates of antibody were prepared using essentially the same strategy resulting in an average molar binding ratio of antibody/HSA/MTX equal to 1:1.1:38. The drug is bound to HSA via one of its carboxyl groups resulting in a stable amide linkage (Endo et al., 1987). These conjugates mediate selective cytotoxicity to cells in vitro and in vivo superior to the free drug or antibody or a drug/antibody mixture and are discussed elsewhere in this volume.

Dextran

Dextrans are a class of polysaccharides having several advantageous properties. They have a well-defined structure, are available in many different molecular weights with a relatively narrow distribution, and are biological inert, biodegradable and possess many hydroxyl groups for chemical derivatization. This has led to the use of dextrans as plasma expanders and drug carriers.

Mitomycin C (MMC) is widely used in cancer chemotherapy due to its ability to crosslink DNA after in vivo reduction to the hydroquinone derivative (bioreductive alkylation). The drug shows severe side effects, notably bone marrow depression and gastrointestinal damage. Soluble controlled release systems of chemically linked conjugates of mitomycin C and dextrans of various molecular weights (10000, 70000 and 500000) have been evaluated by Sezaki and coworkers to decrease the toxic side effects of the free drug (Sezaki and Hashida, 1984a,b, 1985).

In the dextran conjugates (MMCD) (Fig. 19) the drug is bound at the aziridinyl nitrogen atom to an acyl spacer moiety linked to the carrier. When this amide bond remains intact the drug is inactive in bioreductive alkylation. Amide bonds are usually resistant to hydrolysis but the ring strain of the three-membered ring enhances the rate of hydrolysis significantly. As a result the drug is released by solvolysis from the carrier with a half life of only 24 h under physiological conditions (pH 7.4, 37°C). Enzymatic cleavage of the amide bond is not observed (Hashida et al, 1983). In contrast to polyHPMA conjugates discussed above, pinocytic capture by cells followed by lysosomal degradation of mitomycin C-dextran conjugates is not essential for release of free drug and is actually undesirable due to inactivation of the drug in the lysosomal compartments. Thus dextran-mitomycin C conjugates may act as sustained release devices maintaining sufficiently high free drug levels within the circulation or other body compartments for extended time periods.

To evaluate the therapeutic potential of MMCD conjugates detailed studies of the biodistribution, the pharmacokinetics of the conjugate, carrier and drug as well as the relation with in vitro and in vivo toxicity for tumor models were performed.

When administered i.v. in rats, the biodistribution pattern of three types of ^{14}C-labeled MMCD conjugates prepared from dextran of mean molecular weights of 10000, 70000 and 500000 (denoted as T-10, T-70 and T-500, respectively) showed rapid capture by primarily RES cells of the liver and spleen (Hashida et al, 1984). The plasma levels of free drug obtained after bolus injection of the MMCD conjugates were

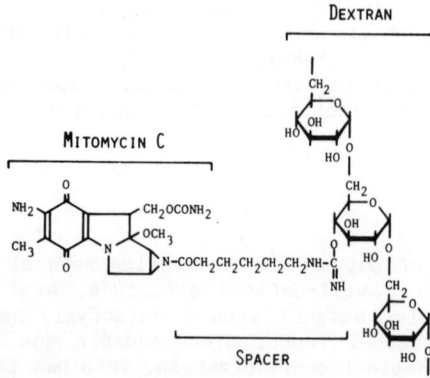

Fig. 19   Representative structure of a mitomycin C-dextran conjugate
          (from Sezaki and Hashida, 1984a)

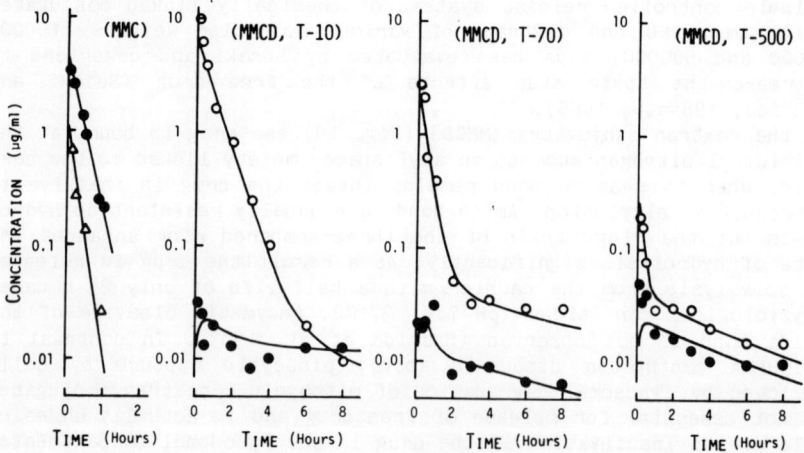

Fig. 20   The observed and simultaneously fitted (solid line) plasma
          concentration-time data for MMC after i.v. injection of MMC
          or MMCD conjugates to rats. Filled circles, free MMC after
          injection of MMC (5 mg/kg) or MMCD (5 mg equivalent
          MMC/kg); triangles, free MMC after injection of MMC (1
          mg/kg); open circles, dextran-conjugated MMC after injec-
          tion of MMCD (5 mg equivalent MMC/kg). Each point repre-
          sents the mean value of at least three rats. The curves are
          computer generated fits to the model described (from Sezaki
          and Hashida, 1984a)

approximately constant in contrast to administration of MMC (Fig. 20) illustrating that the conjugates act as a drug reservoir for a prolonged period. From pharmacokinetic analysis of the plasma levels of free and bound drug using a two-compartment model it was inferred that free drug was released from the conjugates with a conversion rate constant similar to in vitro hydrolysis with no apparent contribution from enzymes. Further the apparent distribution volumes of the MMCD conjugates were strongly dependent on the carrier size, which indicates the presence of macromolecular conjugate in tissues other than RES cells or blood. The bioavailability of free drug generated from the conjugates was calculated from the area under the plasma concentration versus time curves and was estimated to be 6-11% relative to the administration of free drug. These low values corroborate the low efficacy of MMCD conjugates after i.v. administration against tumor cells, which was previously explained as a result of rapid capture of the conjugates by RES cells.

A more promising perspective for MMCD conjugates was anticipated in local cancer chemotherapy. The prevention or cure of lymphatic metastases remaining after surgical elimination of solid tumors is of great importance. Local treatment with drugs results in absorption through the capillaries but high molecular weight compounds or emulsions of low-molecular weight drugs in oil are absorbed primarily in the lymph vessels (Takakura et al., 1984; Sezaki and Hashida, 1984a).

MMCD conjugates were studied for their possible use in lymphotropic drug delivery using i.m. administration of MMC and MMCD conjugates with various molecular weights in rats and mice. The concentrations of free and bound MMC were determined in the muscle, regional lymph nodes, thoracic lymph and plasma. The efficacy of the conjugates in curing in tumors was determined using s.c. inoculation of L1210 lymphocytic leukemia in mice as a test model; these tumor cells are known to concentrate in lymph nodes. The macromolecular conjugates disappear much more slowly from the site of injection than free MMC (Fig. 21) with a strong dependence on the molecular weight of the carrier. A remarkably high concentration of each of the conjugates was found in the regional lymph nodes for a prolonged period (Fig. 22) with a delayed and partial transfer to the thoracic lymph and with low or insignificant levels in plasma. In contrast, only small amounts of MMC appeared in the lymph nodes directly after i.m. administration of the free drug. Most of the drug was transported rapidly into the blood. Depending on the molecular weight the prodrugs accumulated primarily in the lymph nodes and subsequent transport to the thoracic lymph was observed only for the T-10 carrier. The general distribution is indicated schematically in Fig. 23.

The efficacy of the MMCD conjugates against lymph node metastases was tested using i.m. administration of free and bound MMC on day 4 in mice s.c. inoculated with L1210 leukemia on day 0. It is known that L1210 cells are concentrated in the lymph nodes. On day 7 the mice were sacrificed and the lymph nodes were weighed. The data indicate an equal or better inhibitory effect of the MMCD conjugates relative to that of the free drug against established lymph node metastasis. With all three conjugates the tumor cells were only partially eradicated. The ability of the conjugates or the free drug to prevent tumor cell implantation was measured by pretreating ani-

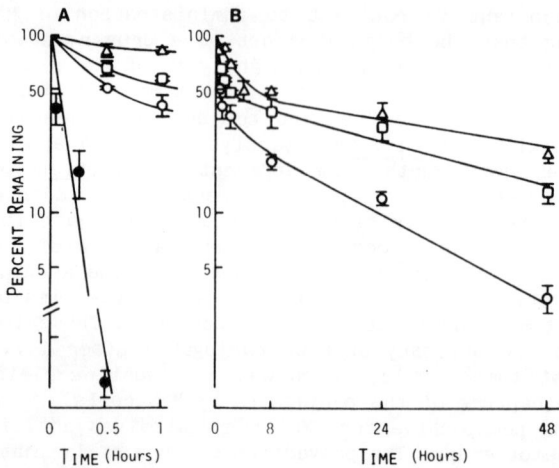

Fig. 21　Disappearance of mitomycin C and mitomycin C-dextran conjugates from the thigh muscle during the first hour (A) and
during 48 hours (B) after intramuscular injection. Filled
circles, MMC; open circles, MMCD (T-10); squares, MMCD
(T-70); triangles, MMCD (T-500). Results are expressed as
the mean ± standard error (SE) of at least four rats (from
Sezaki and Hashida, 1984a)

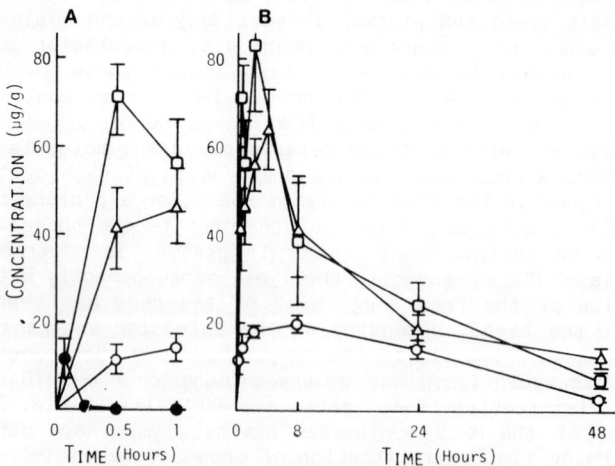

Fig. 22　Concentration of mitomycin C and mitomycin C-dextran conjugates in the regional lymph node during the first hour (A)
and during 48 hours (B) after intramuscular injection.
Filled circles, MMC; open circles, MMCD (T-10); squares,
MMCD (T-70); triangles, MMCD (T-500). Results are expressed
as the mean ± SE of at least four rats (from Sezaki and
Hashida, 1984a)

(MMC)   (MMCD, T-10)   (MMCD, T-70)   (MMCD, T-500)

THORACIC LYMPH → LYMPH NODE	THORACIC LYMPH ← LYMPH NODE	THORACIC LYMPH ← LYMPH NODE	THORACIC LYMPH ← LYMPH NODE
MUSCLE	MUSCLE	MUSCLE	MUSCLE
BLOOD	BLOOD	BLOOD	BLOOD

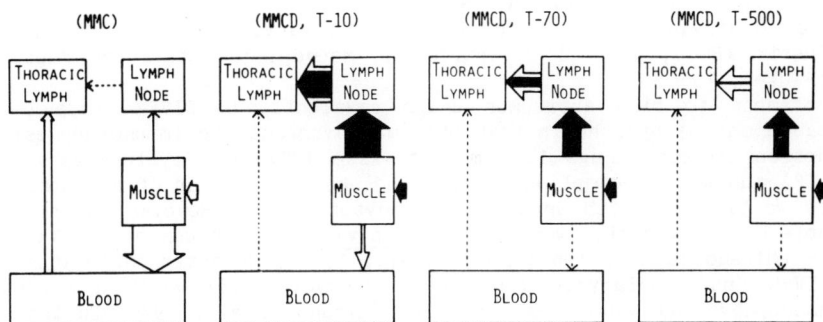

Fig. 23   Schematic representation of lymphatic transfer patterns of
mitomycin C and mitomycin C-dextran conjugates after intramuscular injection. Open arrows represent transfer of free
MMC; filled arrows represent transfer of dextran-conjugated
MMC (from Sezaki and Hashida, 1984a)

mals with MMCD 1 day before s.c. inoculation with the tumor cells.
The high molecular weight (500000) prodrug showed a very high activity while carriers of 10000 and 70000 were less effective. The in
vivo cytotoxicity correlated well with the pharmacokinetic behaviour
of the macromolecular conjugates (Fig. 22). The cytotoxic effectivity of MMCD conjugates also increased with increasing molecular
weight of the carrier when rats were treated i.p. with the conjugates after i.p. inoculation with L1210 cells (Matsumoto et al.,
1985). In addition to the better localization of macromolecular
conjugates with higher carrier molecular weights it was inferred
that the presence of positive charge in the conjugate due to imidate
bonds between spacer and carrier (Fig. 19) contributes to binding
with the tumor cells (Matsumoto et al., 1986).
From the data on MMCD conjugates it can be concluded that future
prospects exist for local chemotherapy notably lymph node metastases. Evidently the optimal balance between dose, extent of localization of macromolecular drug and rate of drug release which all
determine the effective local drug concentration is critical and has
not yet been achieved for complete eradication of tumor cells.

Copolymer of maleic anhydride and divinylether (DIVEMA)

The polymer obtained by radical copolymerization of maleic anhydride
and divinylether in a 2:1 molar ratio followed by hydrolysis of
anhydride groups (DIVEMA) is respresentative of a class of polyanions with strong biological effects. DIVEMA has immunostimulating
properties including interferon induction, antiviral activity and
tumor cell cytotoxicity as well as anticlotting activity (Butler et
al., 1985; Ottenbrite, 1985; Ottenbrite and Kaplan, 1985). The anti-

tumor activity of DIVEMA and related polymers is probably mediated
by activation of macrophages which recognize and destroy tumor cells
in a specific way by as yet unknown mechanisms. In this way a highly
selective chemotherapy may be possible as was described for other
immunoadjuvants such as the bacteria C. Parvum, Bacillus Calmuette
Guerin (BCG), muramyl dipeptide and macrophage activation factors
(Fidler, 1986).
Attempts to bind antitumor drugs covalently to DIVEMA have been
based on the hypothesis that this may overcome the immunosuppresive
action of some drugs like methotrexate (MTX) and cyclophosphamide
(CP) and will accomplish carrier-mediated activation of macrophages
in combination with drug-mediated cytotoxicity. However unexplain-
able high toxicities were observed in vivo for conjugates of MTX and
CP (Hirano et al., 1980; Dorn et al., 1985). Recently conjugates of
DIVEMA and adriamycin (ADR) or daunomycin (DM) which are known
macrophage activators were studied (Hirano et al., 1986; Zunino et
al., 1987).
Compared with free drug DIVEMA-ADR conjugates with a solvolytically
labile ester bond showed almost equal or slightly less inhibition of
growth in cell cultures (Zunino et al., 1987) whereas amide-bound
conjugates of ADR or DM and DIVEMA were only slightly cytotoxic
under in vitro conditions (Hirano et al., 1986). These results
parallel the rate of release of free drug which is faster for the
ester-bound drug. In contrast both types of DIVEMA conjugates resulted in significant increase in the life span of treated animals
compared with free drug, carrier alone, or a mixture of carrier and
drug. This argues for a synergetic contribution from the carrier and
the drug to the cytotoxicity of the conjugates under natural biological conditions. The elucidation of this synergism will require
studies on the body distribution, pharmacokinetics, drug release
rate and immunostimulating effects of the conjugate.

Other carrier-drug conjugates and new developments
Many conjugates of polymeric carriers and drugs including antitumor
agents to effect a sustained release of free drug for prolonged
periods have been proposed and studied in addition to those discussed above. A review on these conjugates is beyond the scope of this
chapter and can be found elsewhere (Dorn et al., 1985; Duncan and
Kopecek, 1984; Ferruti and Tanzi, 1986; Ghosh and Maiti, 1985; Gros
et al., 1981; Ringsdorf, 1975). Most conjugates have been prepared
from polyvinyl polymers often obtained by copolymerization of
acryloylated drugs with different solubilizing comonomers (see reviews noted above; Ouchi et al., 1986a, 1987; Pato et al., 1982).
Less frequently polysaccharides, poly(amino acids) (Drobnik et al.,
1979; Roos et al., 1984) and polyethers (Ouchi et al., 1986b) have
been explored. Antitumor drugs include adriamycin and daunomycin,
chlorambucil, nitrogen mustard derivatives, mitomycin C, 5-fluoro-
uracil, methotrexate and muramyl dipeptide.
A recent development in drug carriers is the use of lipoproteins
(Van Berkel et al., 1986) and synthetic polymers which mimic these
natural carriers (Dorn et al.; Illum et al., 1986). Four classes of
differently sized lipoproteins can be distinguished, including high
density lipoprotein (HDL, 10 nm) low density lipoprotein (LDL, 23
nm), very low density lipoprotein (VLDL, 30-90 nm) and chylomicrons

(10-100 nm). The lipoproteins are composed of a hydrophobic core
with a hydrophilic shell; they transport cholesterol, triacylgly-
cerols and phospholipids in blood by insertion of these molecules or
their hydrophobic counterparts in the hydrophobic core (Van Berkel
et al., 1986). Recently, amphiphilic synthetic polymers have been
designed which spontaneously form micelles in aqueous environments
(Bader et al., 1984; Dorn et al., 1985) or adopt a globular conformation with a diameter of approximately 10 nm (Illum et al., 1986)
with a hydrophobic core stabilized by a hydrophilic shell. Lipophilic drugs can be trapped in the hydrophobic core either nonco-
valently (Illum et al., 1986) or covalently (Bader et al., 1984;
Dorn et al., 1985). The amphiphilic carrier-drug conjugates proposed
by Ringsdorf and his coworkers (Bader et al., 1984; Dorn et al.,
1985) are composed of a hydrophilic poly(ethylene oxide) A-block and
a hydrophobic B-block (Fig. 24). The B-block was prepared from biodegradable poly(α-L-lysine) which was partially substituted with
cyclophosphamide (CP)-spacer moieties and varying amounts of palmitic acid residues which regulated the hydrophobicity of the microenvironment of the drug-carrier bond in the final micelle. Experimentally, the rate of drug release which occurs by chemical solvolysis could be widely varied dependent on the amount of palmitic acid
residues and a significant inhibition of L1210 cell growth due to
CP-mediated DNA-crosslinking could be obtained. In principle, these
colloidal micelle-forming polymeric drug carriers enable a high
loading with hydrophobic drugs and can advantageously protect the
drug against inactivation and interaction with cells and plasma
proteins during transport. Therefore they might serve as both particulate carriers such as liposomes or nanocapsules and soluble
carrier-drug conjugates. The evaluation of the potential of colloidal drug carriers awaits the results of studies on their stability,
size and biocompatibility as well as the drug bioavailability.

CONCLUSIONS AND PERSPECTIVES

Covalently-bound polymer-drug conjugates are currently under
development and have not yet reached the stage of clinical application. One can ask the question whether the selectivity problem in
cancer chemotherapy or at least the less ambitious goal of improving
the therapeutic index with these conjugates can be solved in the
future. It has been demonstrated that soluble polymer-drug conjugates can be designed to act as devices for the extracellular
release of drugs including cytostatics. Drug levels can thus be
maintained at a low yet therapeutically effective value for extended
periods in the circulation or within body compartments. At present
only partial remissions of tumor models under _in vivo_ conditions
have been reported using this approach.
Improvement of the antitumor efficacy of extracellular polymer-drug
conjugates may be feasible within the next few years. To this end
further knowledge on the pharmacokinetics of polymer-drug conjugates
is indispensable. In general polymer-drug conjugates are amphiphilic
because a hydrophilic carrier is combined with a hydrophobic drug.
The effect of the bound drug on the conformational properties of the
polymer, the formation of micelles and the binding with plasma proteins or cells are areas largely unexplored. Yet these aspects are

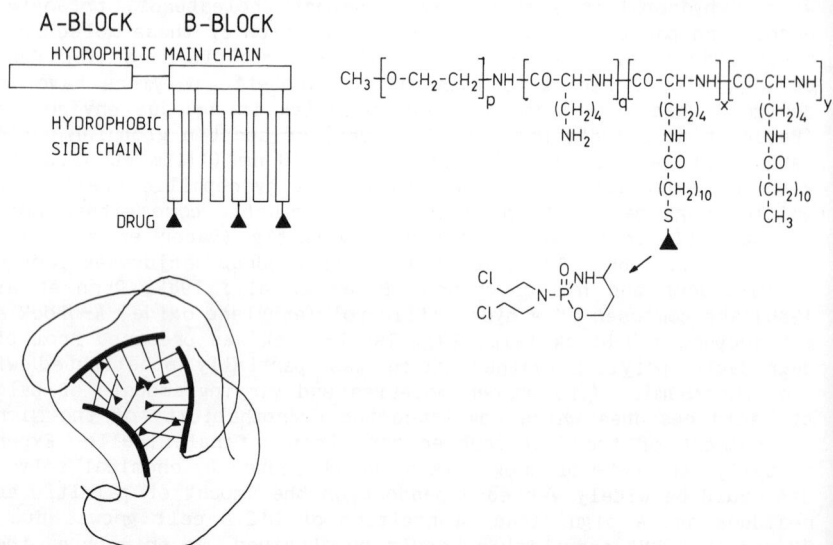

Fig. 24   Structure of amphilic block copolymer-drug conjugates (from
          Bader et al., 1984)

of great importance in the transport and biodistribution of conjugates and the release of drug. Fine-tuning of the drug release rate
for a given carrier-drug combination may achieve a better cytotoxic
effect. The cleavability of the drug-polymer bond is often less than
optimal. The rate of hydrolytic cleavage may be affected by structural variations in the vicinity of the drug-polymer bond. By using
such an approach it may be possible to increase the drug bioavailability.

The potential of carriers which are biologically active for use as
polymeric drug delivery devices is less certain at present. The
mechanisms by which such conjugates act under in vivo conditions
and the relative contributions from the carrier and the drug are far
more complicated and only partially known. The elucidation of these
mechanisms is probably imperative for a prospect of clinical accep-
tance for this type of drug conjugate.

The use of biodegradable carriers in polymer-drug conjugates both
synthetic polymers such as poly(α-L-amino acids) and proteins gains
increasing interest. The pharmacokinetics, biodistribution and in
vivo degradation of drug conjugates especially those of poly(α-L-
amino acids) are however not well elucidated and should be studied
more extensively.

Promising prospects are beginning to emerge in the field of targeted
delivery of cytostatic agents, especially by the use of antibody-
polymer-drug conjugates. The main advantage of using an intermediate
polymeric carrier is to increase the amount of drug carried by the

antibody while retaining its activity. To achieve this, crosslinking of antibodies by polymer-drug chains should be avoided because the drug load decreases and insoluble or inactive antibody aggregates may be formed. Thus in the preparation of antibody-polymer-drug conjugates, polymers having a single functional group for antibody coupling and a different type of functional group for drug coupling are clearly advantageous. Albumin and polyGlu have been used in such an orthogonal coupling strategy. With targeted drug delivery systems extracellular drug release is feasible, but greater selectivity of drug release is expected by the use of the endocellular design strategy. In the latter case drug release is induced after antigen-antibody binding which is a fast process. The subsequent delivery of drug may occur relatively rapidly and may yield high endocellular drug concentrations. The actual mechanism of action of antibody-drug conjugates may involve antibody-mediated activation of the humoral and cellular defense mechanisms. Drug release from antibody conjugates may take place in cells acting in the immunological process and not only in the target cells. Antibody-mediated drug delivery also has potential limitations which will be discussed in more detail elsewhere in this volume. A few drug-related potential obstacles should also be mentioned. The limited half-life of drug-carrying antibodies in the circulation due to degradation will result in loss of selectivity in drug targeting. Further, the capacity of antibodies and their conjugates to extravasate is subject to diffusional limitations. Thus tumors which are rapidly accessible in the circulation or in body compartments offer better prognosis for chemotherapy than solid tumors. Also the acquired resistance of cells due to chronic drug exposure remains a potential hazard. Nevertheless in the near future the potential of antibody-mediated targeting of drugs including the use of polymeric carriers will be revealed.

ACKNOWLEDGMENT

This work has been supported by the Dutch Foundation for Medical and Health Research (MEDIGON) (Grant no. 900-535-050).
We thank miss L. Rotman for typing the manuscript.

REFERENCES

Anderson, J.H. and Kim, S.W. (eds) (1986) "Advances in drug delivery systems", Elsevier Science Publ., Amsterdam, The Netherlands. Also published as J. Controlled Release 2 (1985).

Arnold, L.J., Dagan, A. and Kaplan, N.O. (1983) "Poly(L-lysine) as an antineoplastic agent and tumor-specific drug carrier" in Targeted drugs (Goldberg, E.P., ed.), Wiley and Sons, New York.

Bader, H., Ringsdorf, H. and Schmidt, B. (1984) "Watersoluble polymers in medicine", Angew. Makromol. Chem. 123/124, 457-485.

Bailey, W.J. and Gapud, B. (1985) "Synthesis of biodegradable addition polymers", Annals N.Y. Acad. Sci., __446__, 42-50.

Baker, R. (1987) "Controlled release of biologically active agents", Wiley, New York, USA.

Barondes, S.H. (1986) "Vertebrate lectins: properties and functions" in The Lectins, properties, functions and applications in biology and medicine (Liener, I.E., Sharon, N. and Goldstein, I.J., eds.), pp. 437-466, Academic Press, New York, USA

Baurain, R., Masquelier, M., Deprez-De Campeneere, D. and A. Trouet (1983) "Targeting of daunorubicin by covalent and reversible linkage to carriers. Lysosomal hydrolysis and antitumoral activity of conjugates prepared with peptidic spacer arms" Drugs Exptl. Clin. Res. IX, 303-311.

Butler, G.B., Xing, Y., Gifford, G.E. and Flick, D.A. (1985) "Physical and biological properties of cyclopolymers related to DIVEMA ("Pyran copolymer")", Annals N.Y. Acad. Sci. __446__, 149-159.

Cartlidge, S.A., Duncan, R., Lloyd, J.B., Rejmanova, P. and Kopecek, J. (1986) "Soluble crosslinked N-(2-hydroxypropyl)methacrylamide copolymers as potential drug carriers. 1. Pinocytosis by rat visceral yolk sacs and rat intestine cultured in vitro. Effect of molecular weight on uptake and intracellular degradation", J. Controlled Rel. __3__, 55-66

Cartlidge, S.A., Duncan, R., Lloyd, J.B., Kopeckova-Rejmanova, P. and Kopecek, J. (1987a) "Soluble crosslinked N-(2-hydroxypropyl)-methacrylamide copolymers as potential drug carriers. 2. Effect of molecular weight on blood clearance and body distribution in the rat after intravenous administration. Distribution of unfractionated copolymer after intraperitonal, subcutaneous or oral administration", J. Controlled Release __4__, 253-264.

Cartlidge, S.A., Duncan, R., Lloyd, J.B., Kopeckova-Rejmanova, P. and Kopecek, J. (1987b) "Soluble crosslinked N-(2-hydroxypropyl)-methacrylamide copolymers as potential drug carriers. 3. Targeting by incorporation of galactosamine residues. Effect of route of administration", J. Controlled Rel. __4__, 265-278.

Chakravarty, P.K., Carl, P.L., Weber, M.J. and Katzenellenbogen (1983a) "Plasmin-activated prodrugs for cancer chemotherapy. 1. Synthesis and biological activity of peptidylacivicin and peptidylphenylene diamine mustard" J. Med. Chem. __26__, 633-638.

Chakravarty, P.K., Carl, P.L., Weber, M.J. and Katzenellenbogen, J.A. (1983b) "Plasmin-activated prodrugs for cancer chemotherapy. 2. Synthesis and biological activity of peptidyl derivatives of doxorubicin", J. Med. Chem. __26__, 638-644.

Chu, B.C.F. and Whiteley, J.M. (1980) "The interaction of carrier-bound methotrexate with L1210 cells", Mol. Pharmacol. __17__, 382-387.

Chytry, V., Kopecek, J., Leibnita, E., O'Hare, K., Scarlett, L. and Duncan, R. (1987) "Copolymers of 6-O-methacryloyl-D-galactose and N-(2-hydroxypropyl)methacrylamide: targeting to liver after intravenous administration to rats" in New Polymeric Materials, in press.

Davis, S.S. and Illum, L. (1986) "Colloidal delivery systems - opportunities and challenges" in Site-specific drug delivery (Tomlinson, E. and Davis, S.S., eds), pp 93-110, Wiley and Sons, Chichester.

Dorn, K., Hoerpel, G. and Ringsdorf, H. (1985) "Polymeric antitumor agents on a molecular and cellular level" in Bioactive polymeric systems (Gebelein, C.G. and Carrahar, C.E., eds.) pp. 531-585, Plenum Press, New York.

Drobnik, J., Saudek, V., Vlasak, J. and Kalal, J. (1979) "Polyaspartamide- a potential drug carrier", J. Polym. Sci., Polym. Symp. 66, 65-74.

Drobnik, J. and Rypacek, J. (1984) "Soluble synthetic polymers in biological systems", Adv. Polym. Sci. 57, 1-50.

Duncan, R., Rejmanova, P., Kopecek, J. and Lloyd, J.B. (1981) "Pinocytic uptake and intracellular degradation of N-(2-hydroxypropyl)-methacrylamide copolymers. A potential drug delivery system", Biochim. Biophys. Acta 678, 143-150.

Duncan, R., Cable, H.C., Lloyd, J.B., Rejmanova, P. and Kopecek, J. (1983a) "Polymers containing enzymatically degradable bonds, 7. Design of oligopeptide side chains in Poly [N-(2-hydroxypropyl)-methacrylamide] copolymers to promote efficient degradation by lysosomal enzymes", Makromol. Chem. 184, 1997-2008.

Duncan, R., Kopecek, J., Rejmanova, P. and Lloyd, J.B. (1983b) "Targeting of N-(2-hydroxypropyl)methacrylamide copolymers to liver by incorporation of galactose residues", Biochim. Biophys. Acta 755, 518-521.

Duncan, R., Cable, H.C., Rejmanova, P., Kopecek, J. and Lloyd, J.B. (1984) "Tyrosinamide residues enhance pinocytic capture of N-(2-hydroxypropyl)methacrylamide copolymers", Biochim. Biophys. Acta 799, 1-8.

Duncan, R., Seymour, L.C.W., Scarlett, L., Lloyd, J.B., Rejmanova, P. and Kopecek, J. (1986) "Fate of N-(2-hydroxypropyl)methacrylamide copolymers with pendent galactosamine residues after intravenous administration to rats", Biochim. Biophys. Acta 880, 62-71.

Duncan, R., Kopeckova-Rejmanova, P., Strohalm, J., Hume, I., Cable, H.C., Pohl, J., Lloyd, J.B. and Kopecek, J. (1987) "Anticancer agents coupled to N-(2-hydroxypropyl)methacrylamide copolymers. I. Evaluation of daunomycin and puromycin conjugates in vitro", Br. J. Cancer 55, 165-174.

Duncan, R. and Kopecek, J. (1984) "Soluble synthetic polymers as potential drug carriers", Adv. in Polymer Sci. 57, 51-101.

Endo, N., Kato, Y., Takeda, Y., Saito, M., Umemoto, N., Kishida, K. and Hara, T. (1987) "In vitro cytotoxicity of a human serum albumin-mediated conjugates of methotrexate with anti-MM46 monoclonal antibody", Cancer Res. 47, 1076-1080.

Ferruti, P. and Tanzi, M.C. (1986) "New polymeric and oligomeric matrices as drug carriers", CRC Critical Reviews in therapeutic drug carrier systems, Vol. 2(2), pp. 175-244.

Fidler, I.J. (1986) "Immunomodulation of macrophages for cancer and antiviral therapy" in Site-specific drug delivery (Tomlinson, E. and Davis, S.S., eds.) pp. 111-134, Wiley, Chichester.

Garnett, M.C., Embleton, M.J., Jacobs, E. and Baldwin, R.W. (1983) "Preparation and properties of a drug-carrier-antibody conjugate showing selective antibody-directed cytotoxicity in vitro", Int. J. Cancer 31, 661-670.

Goldberg, E.P., Terry, R.N. and Moshe, L. (1981) "Polymeric drugs with tissue binding properties for localized chemotherapy", Org. Coat. Plast. Chem. 44, 132-136

Ghosh, M. and Maiti, S. (1985) "Polymeric anticancer agents - an overview" in Polymeric materials in medication (Gebelein, C.G. and Carrahar, C.E., eds.) pp. 103-114, Plenum Press, New York.

Gros, L., Ringsdorf, H. and Schupp, H. (1981) "Polymere Antitumor-mittel auf molekularer and zellulärer Basis?" Angew. Chem. 93, 311-332.

Hashida, M., Takakura, Y., Matsumoto, S., Sasaki, H., Kato, a., Kojima, T., Muranishi, S. and Sezaki, H. (1983) "Regeneration characteristics of mitomycin C-dextran conjugate in relation to its activity", Chem. Pharm. Bull. 31(6) 2055-2063.

Hashida, M., Kato, A., Takakura, Y. and Sezaki, H. (1984) "Deposition and pharmacokinetics of a polymeric prodrug of mitomycin C, mitomycin C-dextran conjugate, in the rat", Drug Metabolism and Disposition 1, 492-499.

Hirano, T., Ringsdorf, H. and Zaharko, D.S. (1980) "Antitumor activity of monomeric and polymeric cyclophosphamide derivatives compared with in vitro hydrolysis" Cancer Res. 40, 2263-2267.

Hirano, T., Ohashi, S., Morimoto, S., Tsuda, K., Kobayashi, T. and Tsukagoshi, S. (1986) "Synthesis of antitumor-active conjugates of adriamycin or daunomycin with the copolymer of divinyl-ether and maleic anhydride" Makromol. Chem., 187, 2815-2824.

Hoes, C.J.T., Potman, W., Van Heeswijk, W.A.R., Mud, J., De Grooth,

B.G., Greve, J. and Feijen, J. (1985) "Optimization of macromolecular prodrugs of the antitumor antibiotic adriamycin", J. Controlled Rel. 2, 205-213.

Hoes, C.J.T., Potman, W., de Grooth, B.G., Greve, J. and Feijen, J. (1986) "Chemical control of drug delivery" in Innovative Approaches in Drug Research (Harms, A.F., ed.), pp. 267-283, Elsevier Science Publ., Amsterdam, The Netherlands.

Hurwitz, E., Levy, R., Haron, R., Wilchek, M., Arnon, R. and Sela, M. (1975) "The covalent binding of daunomycin and adriamycin to antibodies, with retention of both drug and antibody activities", Cancer Res. 35, 1175-1181

Hurwitz, E., Wilchek, M. and Pitha, J. (1980) "Soluble macromolecules as carriers for daunorubicin", J. Appl. Biochem. 2, 25-35

Illum, L., Huguet, J., Vert, M. and Davis, S.S. (1986) "A sustained delivery system for intramuscular administration of lipophilic drugs using globular partially quaternized poly[thio-1-(N,N-diethyl aminomethyl)-1-ethylene]", J. Controlled Rel. 3, 77-85.

Kato, A., Takakura, Y., Hashida, M., Kimura, T. and Sezaki, H. (1982) "Physico-chemical and antitumor characteristics of high molecular weight prodrugs of mitomycin C", Chem. Pharm. Bull. 30, 2951-2957

Kato, Y., Saito, M., Fukushima, H., Takeda, Y. and Hara, T. (1984a) "Antitumor activity of 1-β-D-arabinofuranosylcytosine conjugated with polyglutamic acid and its derivative", Cancer Res. 44, 25-30

Kato, Y., Umemoto, N., Kayama, Y., Fukushima, H., Takeda, Y., Hara, T. and Tsukada, Y (1984b) "A novel method of conjugation of daunomycin with antibody with a poly-L-glutamic acid derivative as intermediate drug carrier. An anti-α-fetoprotein antibody-daunomycin conjugate", J. Med. Chem. 27, 1602-1607

Kenny, A.D. (1959) "Evaluation of sodium poly-α,L-glutamate as a plasma expander", Proc. Soc. Exp. Biol. Med. 100, 778-780

Kim, S.W., Petersen, R.V. and Feijen, J. (1980) "Polymeric drug delivery systems" in: Medicinal Chemistry (Ariens, E.J., ed.), Drug Design, vol. X, pp. 193-250, Academic Press, New York.

Kopecek, J. (1984a) "Controlled biodegradability of polymers - a key to drug delivery systems", Biomaterials 5, 19-24.

Kopecek, J. (1984b) "Synthesis of tailor-made soluble polymer drug carriers" in Recent Advances in Drug Delivery Systems (Anderson, J.M. and Kim, S.W., eds.) pp. 41-62, Plenum Publ. Co., New York.

Kopecek, J., Sprincl, L. and Lim, J. (1973) "New types of synthetic infusion solutions, I. Investigation of the effect of solutions of some hydrophilic polymers on blood", J. Biomed. Mater. Res. 7, 179-

191.

Kopecek, J., Rejmanova, P. and Chytry, V. (1981) "Polymers containing enzymatically degradable bonds, 1. Chymotrypsin catalyzed hydrolysis of p-nitroanilides of phenyl alanine and tyrosine attached to side chains of copolymers of N-(2-hydroxypropyl)methacrylamide", Makromol. Chem. 182, 799-809.

Kopecek, J., Rejmanova, P., Strohalm, J., Ulbrich, K., Rihova, B., Chytry, V., Lloyd, J.B. and Duncan, R. (1985) "Synthetic polymeric drugs", Eur. Patent application, no. 0 187547.

Kopecek, J. and Duncan, R. (1987) "Targetable polymeric prodrugs", J. Controlled Rel., in press.

Lloyd, J.B. (1986) "Endocytosis and lysosomes: recent progress in intracellular traffic" in Targeting of drugs with synthetic systems (Gregoriadis, G., Senior, J. and Poste, G., eds) pp. 57-63, Plenum Publ. Co., New York.

Matsumoto, S., Arase, Y., Takakura, Y., Hashida, M. and Sezaki, H. (1985) "Plasma disposition and in vivo and in vitro antitumor activities of mitomycin C-dextran conjugate in relation to the mode of action" Chem. Pharm. Bull. 33 2941-2947.

Matsumoto, S., Yamamoto, A., Takakura, Y., Hashida, M., Tanigawa, N. and Sezaki, H. (1986) "Cellular interaction and in vitro antitumor activity of mitomycin C-dextran conjugate" Cancer Res. 46, 4463-4468

Maurer, P.H. (1957) "Attempts to produce antibodies to a preparation of polyglutamic acid", Proc. Soc. Exp. Biol. 96, 394-396

Maurer, P.H. (1964) "Use of synthetic polymers of amino acids to study the basis of antigenicity", Progr. Allergy 8, 1-40.

McCormick, L.A., Seymour, L.C.W., Duncan, R. and Kopecek, J. (1986) "Interaction of a cationic N-(2-hydroxypropyl)methacrylamide copolymer with rat visceral yolk sacs cultured in vitro and rat liver in vivo", J. Bioact. Comp. Polymers 1, 4-19.

Morimoto, Y., Sugibayashi, K., Sugihara, S., Hosoya, K., Nozaki, S. and Ogawa, Y. (1984) "Antitumor agent poly(amino acid) conjugates as a drug carrier in cancer chemotherapy", J. Pharm. Dyn. 7, 688-698

Myers, C.E. (1982) "Anthracyclines" in Pharmacologic principles of cancer treatment (Chabner, B., ed.), pp. 416-434, Saunders, Philadelphia, USA

Ohkawa, K., Tsukada, Y., Hibi, N., Unemoto, N. and Hara, T. (1986) "Selective in vitro and in vivo growth inhibition against human yolk sac tumor cell lines by purified antibody against human α-fetoprotein conjugated with mitomycin C via human serum albumin", Cancer Immunol. Immunother. 23, 81-86.

Ottenbrite, R.H. (1985) "Bioactive carboxylic acid polyanions" in Bioactive polymeric systems (Gebelein, C.G. and Carrahar, C.E., eds.) pp. 513-529, Plenum Press, New York.

Ottenbrite, R.M. and Kaplan, A.M. (1985) "Some biologically active copolymers of maleic anhydride", Annal N.Y. Acad. Sci. 446, 160-168.

Ouchi, T., Fujie, H., Jokei, S., Sakamoto, Y., Chihashita, H., Inoi, T. and Vogl, O. (1986a) "Synthesis of acryloyl-type polymer fixing 5-fluorouracil residues through D-glucofuranoses and its antitumor activity", J. Polym. Sci. Polym. chem. 23, 2059-2074.

Ouchi, T., Yuyama, H., Inui, T., Murakami, H., Fujie, H. and Vogl, O. (1986b) "Synthesis of polyether-bound 3-(5-fluorouracil-1-yl)propanoic acid and its hydrolysis reactivity" Eur. Polym. J. 22, 537-54).

Ouchi, T., Hagita, K., Kawashima, M., Inoi, T. and Tahiro, T. (1987) "Synthesis of vinyl polymer fixing 5-fluorouracils through organosilicon groups via carbamoyl bonds and its antitumor activity" Macromolecules, submitted.

Pastan, I. and Willingham, M.C. (eds.), (1985) "Endocytosis", Plenum Press, New York.

Pato, J., Azori, M. and Tudos, F. (1982) "Polymeric prodrugs, 1. Synthesis by direct coupling of drugs" Makromol. Chem. Rapid Comm. 3, 643-647.

Pato, J., Azori, M., Ulbrich, K. and Kopecek, J. (1984) "Polymers containing enzymatically degradable bonds, 9. Chymotrypsin catalyzed hydrolysis of a p-nitroanilide drug model bound via oligopeptides onto poly(vinylpyrrolidone-co-maleic anhydride), Makromol. Chem. 185, 231-237.

Poznansky, M.J. and Juliano, R.L. (1984) "Biological approaches to the controlled delivery of drugs : a critical review", Pharmacol. Revs. 36, 277-336.

Pratesi, G., Savi, G., Pezzoni, G., Bellini, O., Penco, S., Tinelli, S. and Zunino, F. (1985) "Poly-L-aspartic acid as a carrier for doxorubicin: a comparative in vivo study of free and polymer-bound drug", Br. J. Cancer 52, 841-848

Quigley, J.P. (1979) "Proteolytic enzymes of normal and malignant cells" in Surfaces of normal and malignant cells (Hynes, R.O., ed.) pp. 247-286, Wiley and Sons, New York.

Renard, C., Michel, A. and Tulkens, P.M. (1986) "Hydrolysis of Pro--Ala dipeptides by lysosomal hydrolases. Models for the study of lysosomotropic amino acid prodrugs of penicillins", J. Med. Chem. 29, 1291-1293.

Rejmanova, P., Kopecek, J., Pohl, J., Baudys, M. and Kostka, V. (1983) "Polymers containing enzymatically degradable bonds. 8. Degradation of oligopeptide sequences in N-(2-hydroxypropyl)methacrylamide copolymers by bovine spleen cathypsin B", Makromol. Chem. 184, 2009-2020.

Rejmanova, P., Kopecek, J., Duncan,. R. and Lloyd, J.B. (1985) "Stability in rat plasma and serum of lysosomally degradable oligopeptide sequences in N-(2-hydroxypropyl)methacrylamide copolymers", Biomaterials 6, 45-48.

Rihova, B., Kopecek, J., Kopeckova-Rejmanova, J., Strohalm, J., Plocova, D. and Semoradova, H. (1986) "Bioaffinity therapy with antibodies and drugs bound to soluble synthetic polymers", J. Chromat. 376, 221-233.

Rihova, B. and Kopecek, J. (1985) "Biological properties of targetable poly [N-(2-hydroxypropyl)methacrylamide] - antibody conjugates" J. Controlled Rel. 2, 289-310

Ringsdorf, H. (1975) "Structure and properties of pharmacologically active polymers", J. Polymer Sci. Symp. 51, 135-153.

Robinson, J.R. and Lee, V.H.L. (eds.) (1987) "Controlled drug delivery: fundamentals and applications", Marcel Dekker Inc., New York.

Roos, C.F., Matsumoto, S., Takakura, Y., Hashida, M. and Sezaki, H. (1984) "Physicochemical and antitumor characteristics of some polyamino acid prodrugs of mitomycin C", Int. J. Pharm. 22, 75-87.

Rowland, G.F., O'Neill, G.J. and Davies, D.A.L. (1975) "Suppression of tumor growth in mice by a drug-antibody conjugate using a novel approach to linkage", Nature 255, 487-488

Ryser, H.J.-P. and Shen, W.-C. (1978) "Conjugation of methotrexate to poly(L-lysine) increases drug transport and overcomes drug resistance in cultured cells", Proc. Natl. Acad. Sci. USA 75, 3867-3870.

Ryser, H.J.-P. and Shen, W.-C. (1986) "Drug-poly(lysine) conjugates: their potential for chemotherapy and for the study of endocytosis" in Targeting of drugs with synthetic systems (Gregoriadis, G., Senior, J. and Poste, G., eds), pp. 103-121, Plenum Publ. Co., New York.

Schneider, Y.-J., Abarca, J., Aboud-Pirak, E., Baurain, R., Ceulemans, F., Deprez-De Campeneere, D., Lesur, B., Masquelier, M., Otte-Slachmuylder, C., Rolin-Van Swieten, D. and Trouet, A. (1983) "Drug targeting in human cancer chemotherapy" in Receptor-mediated targeting of drugs (Gregoriadis, G., Poste, G., Senior, J. and Trouet, A., eds) pp. 1-26, Plenum Press, New York.

Seymour, L.W., Duncan R., Kopeckova-Rejmanova, P. and Kopecek, J. (1987a) "Potential of sugar residues attached to N-(2-hydroxypropyl)methacrylamide copolymers as targeting groups for the selective delivery of drugs", J. Bioact. Comp. Polymers, in press.

Seymour, L.W., Duncan, R., Strohalm, J. and Kopecek, J. (1987b) "Effect of molecular weight ($\bar{M}_w$) of N-(2-hydroxypropyl) methacrylamide copolymers on body distribution and rate of excretion after subcutaneous, intraperitoneal and intravenous administration to rats", J. Biomed. Mater. Res., 21, 1341-1358

Sezaki, H. and Hashida, M. (1984a) "Cancer drug delivery systems: macromolecular prodrugs of mitomycin C", in Pharmacokinetics (Benet, L.Z., Levy, G. and Ferraiolo, B.L., eds.), pp. 345-358, Plenum Publ. Co.

Sezaki, H. and Hashida, M. (1984b) "Macromolecule-drug conjugates in targeted cancer chemotherapy" in CRC Critical Reviews in therapeutic drug carrier systems, 1, 1-38.

Sezaki, H. and Hashida, M. (1985) "Macromolecules as drug delivery systems" in Directed drug delivery (Borchardt, R.T., Repta, A.J. and Stella, V., eds.) pp. 189-208, Humana Press.

Shen, W.-C. and Ryser, H.J.-P. (1981) "Cis-aconityl spacer between daunomycin and macromolecular carriers: a model of pH-sensitive linkage releasing drug from a lysosomotropic conjugate", Biochem. Biophys. Res. Comm. 102, 1048-1054 (1981).

Shen, W.-C., Ryser, H.J.-P. and LaManna, L. (1985) "Disulfide spacer between methotrexate and poly(D-lysine)", J. Biol. Chem. 260, 10905-10908.

Stark, A. (1986) "Anti-cancer N-γ-glutamyl derivatives of daunomycin and adriamycin, their preparation and pharmaceutical compositions containing them", Israel patent no. 68449/3.

Stella, V.J. and Himmelstein, K.J. (1985) "Prodrugs: a chemical appraoch to targeted drug delivery" in Directed drug delivery (Borchardt, R.T., Repta, A.J. and Stella, V.J., eds) pp. 247-268, Humane Press, Clifton N.Y.

Subr, V., Kopecek, J. and Duncan, R. (1986) "Degradation of oligopeptide sequences connecting poly [N-(2-hydroxypropyl)methacrylamide] chains by lysosomal cysteine proteinases", J. Bioact. Comp. Polymers 1, 133-146.

Takahashi, Y., Mai, M., Umemoto, N., Kato, Y., Hara, T. and Tsukada, Y. (1987) "Conjugates of mitomycin C with anti-alpha-fetoprotein antibody : antitumor effect against alpha-fetoprotein-producing human gastric carcinoma implanted in nude mice", NCI Monogr. 3, 101-105.

Takakura, Y., Matsumoto, S., Hasida, M. and Sezaki, H. (1984) "Enhanced lymphatic delivery of mitomycin C conjugated with dextran"

Cancer Res. 44 2505-2510.

Trouet, A., Deprez-De Campeneere, D. and De Duve, C. (1972) "Chemotherapy through lysosomes with a DNA-daunorubicin complex", Nature New Biology 239, 110-112.

Trouet, A., Masquelier, M., Baurain, R. and Deprez-De Campeneerde, D. (1982) "A covalent linkage between daunorubicin and proteins that is stable in serum and reversible by lysosomal hydrolases, as is required for a lysomotropic drug-carrier conjugate: in vitro and in vivo studies" Proc. Natl. Acad. Sci. USA 79, 626-629.

Trouet, A. and Jollés, A. (1984) "Targeting of daunorubicin by association with DNA or proteins : a review", Seminars in Oncology 11, 64-72.

Tsukada, Y., Kato, Y., Umemoto, N., Takeda, Y., Hara, T. and Hirai, H. (1984) "an anti-α-fetoprotein antibody-daunomycin conjugate with a novel poly-L-glutamic acid derivative as intermediate drug carrier", J. Natl. Cancer Institute 73, 721-729

Tsukada, Y., Ohkawa, K. and Hibi, N. (1985) "Suppression of human α-foetoprotein-producing hepatocellular carcinoma growth in nude mice by an anti α-foetoprotein antibody-daunorubicin conjugate with a poly-L-glutamic acid derivative as intermediate drug carrier", Br. J. Cancer 52, 111-116

Ulbrich, K., Konak, C., Tuzar, Z. and Kopecek, J. (1987) "Solution properties of drug carriers based on poly[N-(2-hydroxypropyl)methacrylamide] containing biodegradable bonds", Makromol. Chem. 188, 1261-1272.

Van Berkel, T.J.C., Kar Kruyt, J., Harkes, L., Nagelkerke, J.F., Spanjer, H. and Kempen, H.J.M. (1986) "Receptor-dependent targeting of native and modified lipoproteins to liver cells" in Site-specific drug delivery (Tomlinson, E. and Davis, S.S., eds.), pp. 49-68.

Van Heeswijk, W.A.R., Stoffer, T., Eenink, M.J.D., Potman, W., Van der Vijgh, W.J.F., Van der Poort, J., Pinedo, H.M., Lelieveld, P. and Feijen, J. (1984) "Synthesis, characterization and antitumor activity of macromolecular prodrugs of adriamycin" in Recent advances in drug delivery systems (Anderson, J.M. and Kim, S.W., eds.) pp. 77-100, Plenum Press, New York.

Wingard, L.B., Tritton, T.R. and Egler, K.A. (1985) "Cell surface effects of adriamycin and carminomycin immobilized on crosslinked polyvinylalcohol", Cancer Res. 45, 3529-3536.

Yokoyama, M., Inoue, S., Kataoka, K., Yui, N. and Sakurai, Y. (1987) "Preparation of adriamycin-conjugated poly(ethylene glycol)- poly-(aspartic acid) block copolymer", Makromol. Chem., Rapid Commun. 8, 431-435

Zunino, F., Gambetta, R., Vigevani, A., Penco, S., Geroni, C. and Di

Marco, A. (1981) "Biological activity of daunorubicin linked to proteins via the methylketone side chain", Tumori 67, 521-524.

Zunino, F., Giuliani, F., Savi, G., Dasdia, T. and Gambetta, R. (1982) "Anti-tumor activity of daunorubicin linked to poly-L-aspartic acid", Int. J. Cancer 30, 465-470

Zunino, F., Savi, G., Giuliani, F., Gambetta, R., Supino, R., Tinelli, S. and Pezzoni, G. (1984) "Comparison of antitumor effects of daunorubicin covalently linked to poly-L-amino acid carriers", Eur. J. Cancer Clin. Oncol. 20, 421-425

Zunino, F., Pratesi, G. and Pezzoni, G. (1987) "Increased therapeutic efficacy and reduced toxicity of doxorubicin linked to pyran copolymer via the side chain of the drug" Cancer Treatment Reports, in press.

Drug Carrier Systems
Edited by F.H.D. Roerdink and A.M. Kroon
© 1989 John Wiley & Sons Ltd.

BIODEGRADABLE POLYMERS FOR CONTROLLED RELEASE OF
PEPTIDES AND PROTEINS.

F G Hutchinson and B J A Furr, Imperial Chemical Industries
PLC, Pharmaceuticals Division, Mereside, Alderley Park,
Macclesfield, Cheshire SK10 4TG

1. Introduction

As a result of advances in molecular genetics over the last
ten years or so peptides and proteins have emerged as a major class
of therapeutic agents and recombinant DNA technology has allowed
the production of many of these macromolecular agents which have
interesting and useful pharmacological activity. At the same time
there have been significant improvements in synthetic techniques
for the total chemical synthesis of lower molecular peptide
hormones such as 'Zoladex'* [ICI 118630, D-Ser (Bu^t)^6-Azgly^{10}-LH
RH, Fig 1] which is a highly potent synthetic analogue of
luteinising hormone releasing hormone.

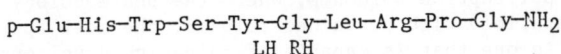

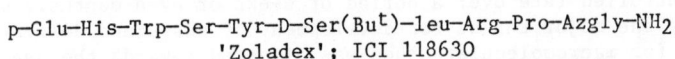

$$\text{p--Glu--His--Trp--Ser--Tyr--Gly--Leu--Arg--Pro--Gly--NH}_2$$
LH RH

$$\text{p--Glu--His--Trp--Ser--Tyr--D--Ser(Bu}^t\text{)--leu--Arg--Pro--Azgly--NH}_2$$
'Zoladex'; ICI 118630

Figure 1   Structures of LHRH and 'Zoladex'

These products of biotechnology, or total chemical
synthesis, present enormous challenges to the pharmaceutical
scientist. There still remains the major problem of effective
delivery of these agents to the body. In order to realise the full
therapeutic and commercial potential of this class of drugs the
pharmaceutical scientist has to identify and manufacture practical
and effective formulations. Although polypeptides are an example
of a drug class whose full potential is not realised because of our
present inability to direct selectively the agent to the site of
action or to control, temporally, systemic or tissue concentrations
in relation to biological need, the very high potency of many of
these agents, e.g. LH-RH analogues, renders them eminently suitable
for incorporation into, and release from, sustained delivery dosage
forms.

* 'Zoladex' is a trademark, the property of Imperial
Chemical Industries PLC

These macromolecular agents are usually ineffective by the oral route as they are rapidly degraded and deactivated by proteolytic enzymes in the gastrointestinal tract. Even if stable to enzymatic digestion, their molecular weights are too high for absorption through the intestinal wall to occur. Obvious benefits would accrue if oral formulations could be defined which offer both an acceptable bioavailability of compound, and the convenience of the oral route. Consequently this technical target will continue to provide impetus for expanding programmes of research opposite oral delivery. However, in the authors view the inherent properties of peptides and proteins, physicochemical and biological, will inevitably prejudice this route as a general solution which can be applied to all macromolecular agents. Such programmes are likely to meet with only a limited success and the same can be said of other routes of administration including intranasal (Anik et al, 1984; Petri et al, 1984), buccal (Anders et al, 1983), intravaginal (Okada et al, 1982; Okada et al, 1983; Okada et al, 1984), rectal (Yoshikawa et al, 1985) and percutaneous (Uda and Yamada, 1984). All of these alternative routes are associated with a low and variable bioavailability. Consequently polypeptides and proteins are normally administered parenterally (subcutaneous, intramuscular and intravenous injection) but even by this route major problems are encountered. These macromolecular agents have very short elimination half—lives and as a consequence frequent injections are required to produce an effective therapy.

For polypeptide hormones, where the pharmacology of the agent is compatible with sustained release, the most appropriate dosage form is one that is capable of releasing drug continuously at a controlled rate over a period of weeks or even months. Until recently the major thrust for development of sustained delivery systems for macromolecular drugs was directed towards the use of hydrolytically and enzymatically stable polymers. Much of this work concentrated on the use of synthetic high molecular weight materials as the rate controlling barrier or matrix and the various carriers are summarised in Table 1.

TABLE 1   Non—erodible carriers for sustained release of
            polypeptides

Hydrogels

Cross—linked polyacrylamide (Davis, 1972; Davis, 1974)
Cross—linked polyvinyl alcohol (Langer and Folkman, 1976)
Cross—linked poly (hydroxyethyl methacrylate) (Langer and Folkman, 1976)

Hydrophobic polymers

Ethylene/vinyl acetate copolymer (Langer, 1984; Cohen et al, 1984; Hsu and Langer, 1985)
Silicone elastomers (Lotz and Syllwasschy, 1979; Hsieh et al, 1985)
Microporous polypropylene (Kruisbrink and Boer, 1984)
Cross—linked (meth)acrylates (Yoshida et al, 1985)

However carrier systems based on these polymers have the major disadvantage that surgical retrieval of the expired delivery system is required.

It is preferred, therefore, that the carrier system should be biodegradable and so would ultimately disappear from the site of administration. Currently, a number of biodegradable polymers are being evaluated as carriers for the sustained release of low molecular weight drugs (Table 2).

TABLE 2   BIODEGRADABLE POLYMERS USED IN DRUG DELIVERY
Polylactic acid (Polylactide)
Polyglycolic acid (Polyglycolide)
Poly (lactic acid—co—glycolic acid)
Poly (ε—caprolactone)
Poly (hydroxybutyric acid)
Poly ortho—esters
Poly acetals
Poly dihydropyrans
Poly cyanoacrylates
Synthetic polypeptides
Cross—linked polypeptides

Long experience over many years with homo— and co—polymers of lactic and glycolic acids has shown that these materials are inert and biocompatible in the physiological environment and degrade to toxicologically acceptable products (Wise et al, 1979). Consequently, these polymers are invariably the materials of choice in the initial design of parenteral sustained delivery systems using a biodegradable carrier, particularly when release over many weeks are required. We have adopted this approach and were the first group to identify and characterise the mechanisms of transport which allow movement of these polypeptide drugs from biodegradable formulations based on these polyesters (Hutchinson, 1982). Formulations may be developed as solid injectable depots or as injectable suspensions of microparticles.

The utility of these particular types of biodegradable formulations is well illustrated using the luteinising hormone releasing hormone analogue, 'Zoladex'. Solid, injectable depot formulations of this analogue, approximately 1 cm long by 1 mm diameter, which release 3.6 mg of drug over 28 days have been developed and used in the treatment of hormone responsive prostate and mammary tumours in animals and the human. Currently marketing approval for this formulation for the treatment of prostate cancer is at an advanced stage in a number of territories and the product has already been launched in the United Kingdom.

It has been further demonstrated that the technology can be applied successfully to other polypeptide drugs.

2. Rationale

The successful development of sustained—release biodegradable delivery systems for peptide drugs such as 'Zoladex' requires recognition and resolution of a number of major problems

posed by these macromolecular agents. Firstly, the mechanism most
commonly used to achieve sustained release, namely controlled
diffusion through a matrix or membrane, may not be appropriate for
a high molecular weight polypeptide. Design of a sustained release
dosage form must take into account both the properties of the rate
controlling polymer and drug. For diffusion of the drug through
the polymer to occur it must have some solubility in the high
molecular weight carrier; this is often the case for low molecular
weight drugs. In contrast, it is well established that, in the
absence of specific chemical interactions, polypeptides will either
be insoluble in, or incompatible with, any polymer such as a
polyester, which has a totally dissimilar structure, because of
entropic and enthalpic factors (Bohn, 1975). Consequently, low or
negligible solubility of the macromolecular drug in a polymer, such
as polyester, will prevent diffusional transport of the agent
through the polymer phase. With regard to the properties of the
drug the most important of these are its size, shape and solubility
(Baker and Lonsdale, 1974). There is an approximate log—log
correlation between molecular weight (M) and diffusion coefficient
(D) where

$\log D = a - b \log M$ (where a and b are abitrary constants)
such that D decreases as molecular weight increases. For
polypeptides M is large and the diffusion coefficient becomes
vanishingly small because the diffusant cannot be accommodated by
the free volume of polymer arising from rotational and
translational segmental mobility.

Consequently, for reasons of incompatibility and molecular
size, polymers such as polyesters are not likely to allow
partition—dependent diffusion of polypeptides through the polyester
phase.

Secondly, polypeptides are biologically labile and can be
readily degraded by tissue enzymes. They must, therefore, be
effectively protected at the depot site if active drug is to be
released continuously. The difficulty of achieving this is
emphasised by the fact that synthetic polypeptides have actually
been used as biodegradable carriers for drugs such as steroids and
narcotic antagonists (Mitra et al, 1979; Sidman et al, 1981).

Thirdly, excipients used to achieve sustained release of
macromolecular agents might provoke an adjuvant induced
immunological response, which may be related to the nature of the
excipient, the delivery rate, or profile of release. There is some
evidence that sustained release of large proteins may be an
effective means of raising antibodies to them (Langer, 1981).

Finally, long—lasting depots might become encapsulated by
fibrous tissue, thus inhibiting further release of drug. This is
certainly the case for non—degradable silicone elastomer implants
(Anderson et al, 1981).

These imposing problems opposite sustained polypeptide
delivery have been resolved by the design of biodegradable delivery
systems based on polyesters such as poly (d,1—lactide) and poly

(d,1-lactide-co-glycolide) to give formulations which allow release
of polypeptides over an extended period of time. These studies
have encompassed polymer synthesis to give polyesters of a defined
structure and molecular weight, characterisation of the degradation
processes leading to erosion and ultimate disappearance of the
biodegradable carrier, morphological studies on the degrading
polymer and finally drug release studies and biological evaluation
of defined dosage forms.

2. Biodegradable polyesters derived from lactic and glycolic acids

These simple biodegradable homo- and co-polymers were
prepared at elevated temperature by the ring-opening polymerisation
of dry, freshly prepared acid dimers, d,1-lactide and glycolide,
by using organo-tin compounds as catalysts. Control of molecular
weight was achieved by using a chain transfer agent such as d,1-
lactic acid. In this way polymers (Fig 2) of variable composition,
having intrinsic viscosities from <0.1 to >1, can be prepared. The
polymers can be further characterised by size exclusion
chromatography relative to polystyrene standards to define number
average molecular weight (Mn) and weight average molecular weight
(Mw) and polydispersity (P = Mw/Mn). Number average molecular
weight is defined as

$$Mn = \frac{\Sigma\, n_i\, M_i}{\Sigma\, n_i} \quad \text{where } n_i \text{ is the number of molecules of polymer having molecular weight } M_i$$

and weight average molecular weight is defined as

$$Mw = \frac{\Sigma\, W_i\, M_i}{\Sigma\, W_i} = \frac{\Sigma\, n_i\, M_i^2}{\Sigma\, n_i\, M_i} \quad \text{where } W_i \text{ is the weight of the polymer molecules having molecular weight } M_i$$

Additionally, the polymers can be characterised by [13]C
nuclear magnetic resonance spectroscopy to define the distribution
of co-monomers and polymer structure (that is, the average values
for n and m of co-polymers shown in Fig 2).

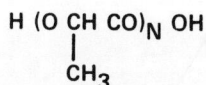

H (O CH CO)$_N$ OH   **Polylactic acid/Polylactide**
     |
     CH$_3$

H (O CH$_2$ CO)$_M$ OH   **Polyglycolic acid/Polyglycolide**

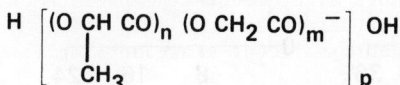

H $\left[(O\ CH\ CO)_n\ (O\ CH_2\ CO)_m^{\ -} \right]_p$ OH   **Poly (lactide-co-glycolide)**
         |
         CH$_3$

Figure 2   Polymers and co-polymers of lactic and glycolic
                  acids

3. Degradation studies on poly (d,1—lactide—co—glycolide) and poly
(d,1—lactide)

As polypeptides have high molecular weight and are water
soluble their release from these biodegradable polyesters, by
classical partition—dependent diffusion of the macromolecular drug
through the polyester, is unlikely to occur.

Consequently, degradation of the poly (d,1—lactide) or poly
(d,1—lactide—co—glycolide) will be a critical factor in determining
transport of the high molecular weight polypeptide from the dosage
form. The degradation of these polymers in the absence of drug has
been characterised in terms of molecular weight and distribution,
weight loss, water uptake and morphology of the hydrated and
degraded polymer.

Degradation of the polymers in vitro in buffer at pH 7.4
results in progressive changes in molecular weight and molecular
weight distribution. Under these conditions degradation is not
enzyme mediated and must occur by simple hydrolytic cleavage of
ester groups. The profile of weight loss and change in molecular
weight, for polymers having the most probable distribution (P 2),
are consistent with this. High molecular weight polymers degrade
to lower molecular weights, as measured by viscosity, yet retain
their water insolubility (Fig 3). Only after an extended time of
degradation does any weight loss occur. In contrast very low
molecular weight polymers can degrade with weight loss
immediately.

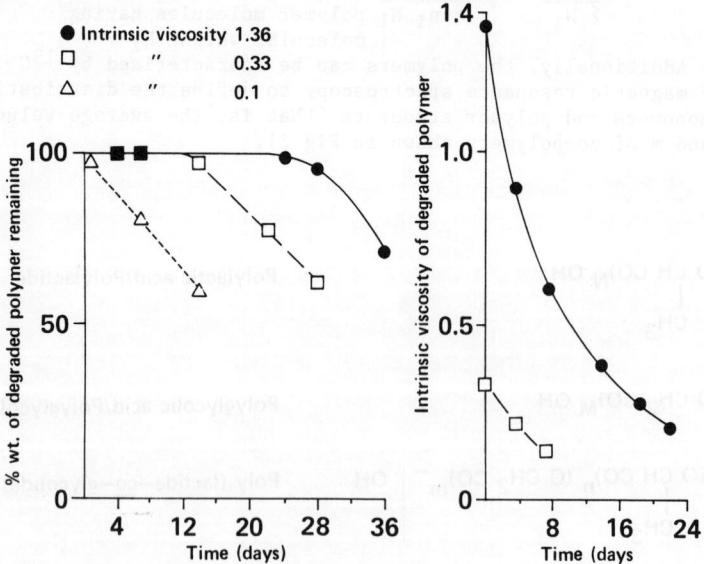

Figure 3   In vitro degradation of 50/50 molar poly (d,1—
lactide—co—glycolide) at 37°C in buffer at pH 7.4

Similar results are obtained with high lactide containing polymers except that the time scale of events is more extended for these more hydrolytically stable polymers. These results are consistent with bulk hydrolysis in the _in vitro_ condition and this correlates broadly with degradation of these polymers _in vivo_ suggesting that even in subcutaneous tissue, enzyme mediated degradation is significantly less important than simple hydrolysis. In this event, polylactides could effectively protect polypeptides at the depot site from the influence of degradative enzymes.

For these degradation experiments, if the logarithm of the number average molecular weight is plotted as a function of time, then for high molecular weight polymers an essentially linear relationship is seen to hold except at extended times of degradation where a discontinuity arises (Fig 4).

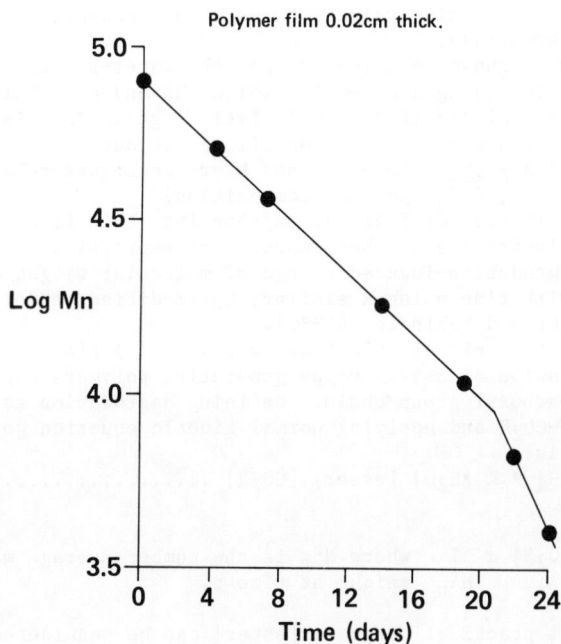

Polymer film 0.02cm thick.

Figure 4   In vitro degradation of poly (d,1–lactide–co–
             glycolide) at 37°C in buffer at pH 7.4 and
             dependence of number average molecular weight on
             time of degradation

Pitt and Schindler (1980) studying poly (d,1–lactide) have seen a similar behaviour but have ignored the nature of, and reasons for, the discontinuity. In fact, this arises because of water—uptake by the degrading polymer. For an amorphous polyester, water—uptake will be governed, empirically by the intrinsic hydrophilicity of the repeat units and by end group effects. For these polyesters the end groups are alkoxylic and carboxylic and these increase as molecular weight falls. That is, as degradation proceeds the essentially hydrophobic polymer becomes more hydrophilic. The profile of water uptake at 37°C in buffer at pH 7.4, for polymers which have been dried rigorously, has been studied as a function of time. For these, water—uptake is determined by two events. The first is simple diffusional ingress into the dried material and in the absence of degradation this would occur to a level that would be characteristic of the equilibrium swelling of this kind of polyester. However, these polymers are hydrolytically unstable and following, or even during, this initial diffusional phase, the polymer can degrade and so take up more water. For high molecular weight polymers, having a normal distribution (P 2), these two phases of water—uptake are separated by an interval during which water—uptake increases hardly at all. In contrast, low molecular weight polymers have essentially a continuous water—uptake.

It can be shown empirically that the water—uptake for a thin polymer film having a molecular weight Mn and a polydispersity P, in the absence of significant hydrolytic degradation, is described approximately by the hyperbolic function:

$$[H_2O] = a + \frac{b}{P\,M_n}$$ (where a and b are constants related to polymer composition)

If the initial diffusional ingress into thin films is assumed to be instantaneous then approximate expressions can be derived for degradation—induced change of molecular weight and water—uptake with time using a similar, but modified, model proposed by Pitt and Schindler (1980).

Degradation of poly (lactide—co—glycolide) proceeds by hydrolytic scission of ester groups generating polymers containing one terminal carboxyl group/chain. Defining degradation as appearance of —CO_2H and applying normal kinetic equation governing ester hydrolysis:

$$\frac{d\,[CO_2H]}{dt} = K\,[H_2O]\,[ester]\,[CO_2H] \quad \dotsfill \quad 1$$

and $[CO_2H] \propto \dfrac{1}{M^t_n}$ where M^t_n is the number average molecular weight at time t.

For all practical purposes [ester] can be considered a constant and as

$$[H_2O] = a + \frac{b}{P\,M_n}$$ equation 1 reduces to

$$d\,[1/M^t_n] = k.\ a + \frac{b}{P\,M^t_n}\ .\ \frac{1}{M^t_n} \quad \dotsfill \quad 2$$

Where k = K [ester]

At t = 0, M^t_n = M^o_n where M^o_n is initial number average molecular weight and the solution to equation 2 is:

$$M^t_n = M^o_n \ e^{-akt} + \frac{b(e^{-akt}-1)}{aP} \quad \dots\dots\dots\dots\dots\dots\dots\dots \ 3$$

$$\text{and } [H_2O]_t = a \left[1 + \frac{b}{a \ P \ M^o_n \ e^{-akt} + b \ (e^{-akt}-1)} \right] \dots\dots \ 4$$

It should be noted that these derived expressions relate to the condition where the polymers have initially a normal distribution (i.e. P 2) and hydrolysis of the polymer chains is essentially a random process. The equation also shows clearly that degradation of the polymer is dependent not only on molecular weight and composition but is also dependent on the molecular weight distribution (and indeed the type of distribution).

Polymer film 0.02cm thick.

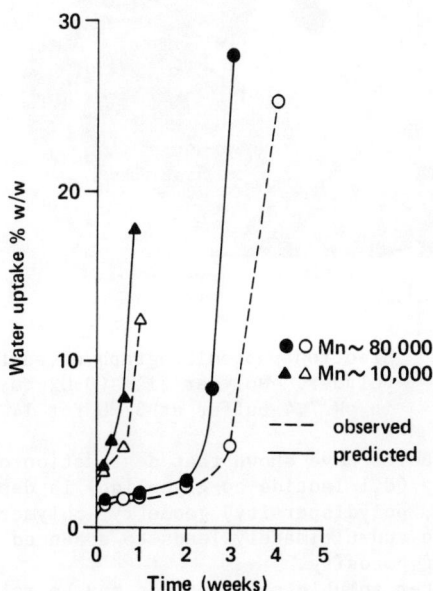

Figure 5   In vitro degradation of poly (d,1—lactide—co glycolide) at 37°C in buffer at pH 7.4. Effect of degradation on water—uptake by polymer

It can be seen from Fig 5 that the derived equation for water—uptake correlates broadly with experimentally determined events. Thus, hydrolytic degradation is characterised by reduction in molecular weight, enhanced water—uptake and ultimately weight loss of polymer. All these events occur at a temperature which is below or near to the glass transition temperature of the polyester. This in turn implies that morphological changes are likely to occur within the polymer whilst hydrolysis is occurring. This is confirmed by scanning electronic—microscopy of the degraded products which shows the development of porosity within the degrading polyester (Fig 6).

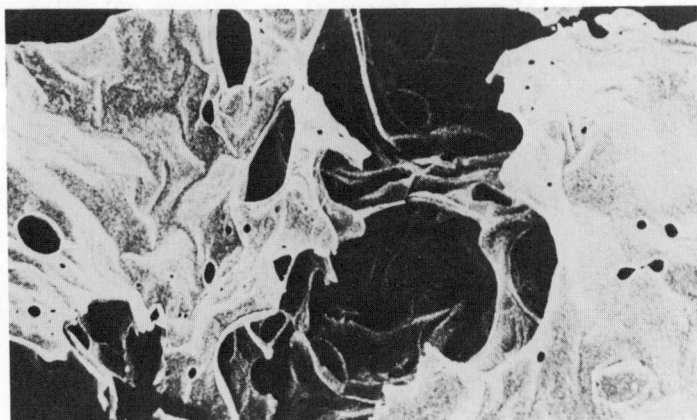

Figure 6   Electron photomicrograph of surface of degraded
            polymer. Polymer film (0.02 cm thick) incubated
            in pH 7.4 buffer at 37°C for 14 days

These studies have shown that degradation of poly (d,1—lactide) and poly (d,1—lactide—co—glycolide) is dependent on molecular weight, polydispersity, geometry, polymer composition and polymer structure and ultimately leads to enhanced water—uptake and the generation of porosity.

Thus, water soluble polypeptides may be released from these biodegradable polyesters since enhanced water uptake and the generation of porosity should facilitate transport of polypeptide from the dosage form. This is likely to involve diffusion through aqueous pores generated in the drug/polymer matrix. In this event, the release of polypeptide will differ mechanistically from the processes thought to occur during release of steroids, narcotic antagonists and antimalarials from poly (d,1–lactide) and poly (d,1–lactide–cl–glycolide) (Wise et al, 1979). Whereas these low molecular weight drugs will diffuse, by a simple partition–dependent process, through intact polymer membranes in diffusion cell experiments, these same polymer membranes are totally impermeable to polypeptides.

4. Release studies with 'Zoladex'

Chronic administration of LH–RH analogues, such as 'Zoladex' has been shown to cause a reversible chemical castration which leads to regression of hormone–responsive animal and human mammary and prostate tumours (Furr and Hutchinson, 1985).

Because of low oral potency, the drugs have usually been administered parenterally once or more times daily. A biodegradable formulation, based on poly (lactide–co–glycolide), either as a subdermal depot or an injectable suspension, which will deliver the drug over a period of 28 days or even longer, would be more clinically acceptable. Research was focussed on 'Zoladex' (molecular weight 1269) from solid depots because this was thought more likely to afford a clearer understanding of the physicochemical parameters which allow transport of drug from the dosage form.

Continuous release of the polypeptide _in vivo_ can be measured qualitatively by the biological effect elicited in regularly cycling adult female rats. Normally, these rats have an oestrous cycle of 4 days and the occurrence of oestrus is indicated by the presence of cornified cells in vaginal smears. In rats given subdermal depots of 'Zoladex' release of drug at an effective rate will cause a fall in circulating oestrogens, which in turn leads to a suppression of oestrus and absence of cornified smears. Rats therefore show an extended period of dioestrus.

With respect to polymer composition our studies have shown that for amorphous homo– and co–polymers of approximately the same high molecular weight increasing lactide content results in slower degradation. This is reflected in the biological effect elicited in rats using subdermal implants containing small amounts of drug in high molecular weight carriers (Fig 7). When administered to female rats some drug is released initially as judged by biological response but this immediate release soon ceases as biological effect is not maintained. There then follows a period during which either drug is not released at all or is released at an ineffective rate. At some later time point release recommences and continues until the depot has been fully depleted of drug. For polymers of similar molecular weight and distribution it can be seen that the interval between the two phases of release is shortest for the most rapidly degradable polymer.

On the basis of degradation studies, transport of drug from these depots is likely to be governed by various properties of the rate controlling polyester. These properties include polymer composition, molecular weight and distribution, level of drug incorporation, morphology of the drug/polymer mixture, degradation characteristics of the polymer and geometry. It can be shown that

release of polypeptide from these biodegradable polyesters occurs
by diffusion through aqueous pores generated in the dosage form
(Fig 8). These aqueous channels, which facilitate drug release are
generated by two distinct and separate mechanisms. The first
involves leaching of drug from polypeptide domains at or near the
surface of the delivery system and essentially is a dissolution/
diffusion controlled event. However drug within the body of the
depot, existing in isolated domains not continuous or contiguous to
the surface cannot be released until the second mechanism becomes
operative. This second mechanism involves degradation of the
polyester and is associated with the generation of microporosity
in, and enhanced water-uptake by, the degrading polymer.

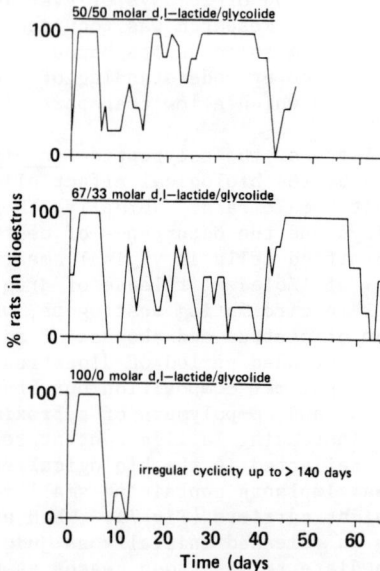

Figure 7   Regularly cycling female rats treated with
            100 µg 'Zoladex' using subcutaneous implants
            containing 3% w/w drug

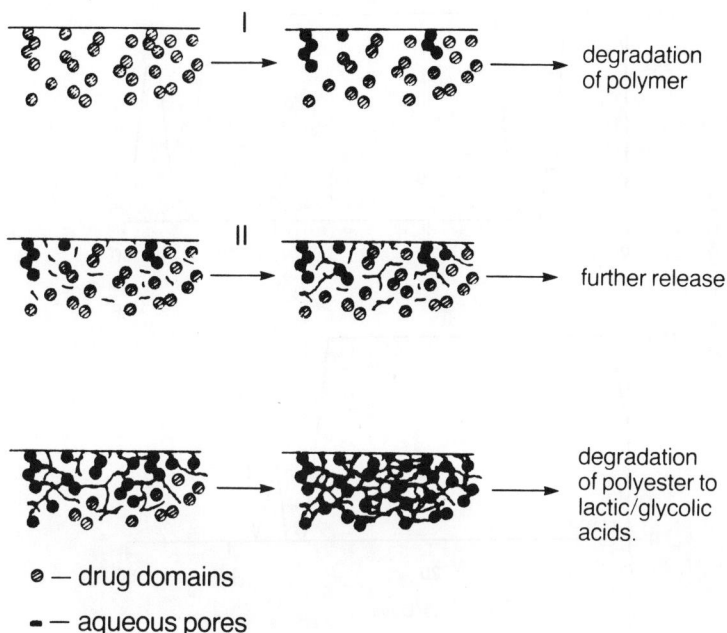

I. initial diffusional release
II. degradation induced release

Figure 8   Mechanisms of release of polypeptides from
            formulations based on poly (d,1—lactide) and
            poly (d,1—lactide—co—glycolide)

Typical parameters controlling the initial phase of release
are, for example, drug loading, morphology and geometry whereas the
second phase is intrinsically related to the degradation properties
of the polyester. When these two phases of release do not overlap
discontinuous release is observed [Fig 9(a)].

However, by controlling the properties of the polymer the
initial phase of release can be made to overlap with the second
phase and depots can be defined which give continuous release over
28 days both in vitro and in vivo [Fig 9(b)]. These depots have
been used to induce a castration like effect in rats and thereby to
inhibit growth of mammary and prostate tumours (Furr and
Hutchinson, 1985).

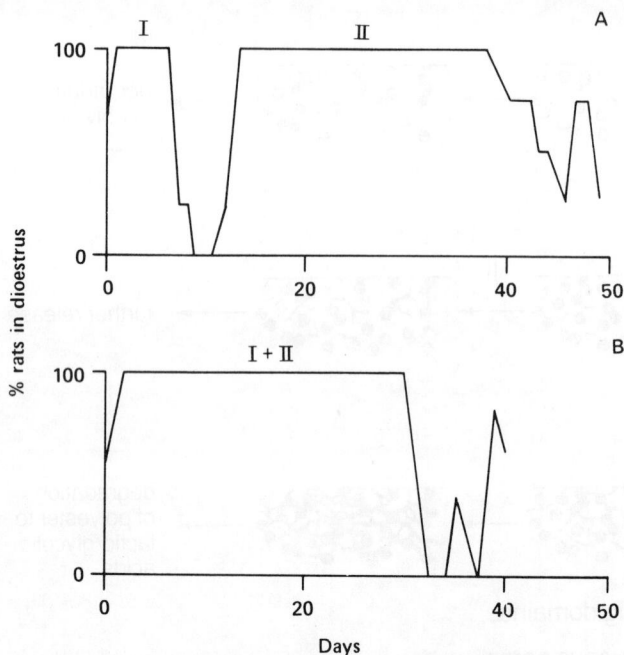

Figure 9   Effect of subdermal depots containing 300 µg
            'Zoladex' administered to regularly cycling
            adult female rats
            (a) Depots containing 3% (w/w) 'Zoladex' in high
                molecular weight polymer
            (b) Depots containing 20% (w/w) 'Zoladex' in low
                molecular weight polymer
            I — Initial release due to leaching from surface
            II— Degradation induced release

5. Use of 'Zoladex' depots in animal models for hormone—responsive
    prostate and mammary tumours
        The effect of a single subcutaneous depot containing 500 µg
'Zoladex' on the growth of rat dimethylbenzanthracene—induced
mammary tumours is shown in Fig 10. This experiment is a model for
advanced mammary cancer and in control animals given placebo depots
tumours have doubled in size in 4 weeks. In contrast tumours
regress markedly in rats given the single depot.
        If rats are given depots at the start of the experiment and
then at weeks 4 and 8 a far more profound regression occurs (Fig
11). By week 11 none of the tumours present at the start of the
experiment is palpable. As the effect of the final depot becomes
exhausted, around week 16, recovery in growth of the tumours
occurs.

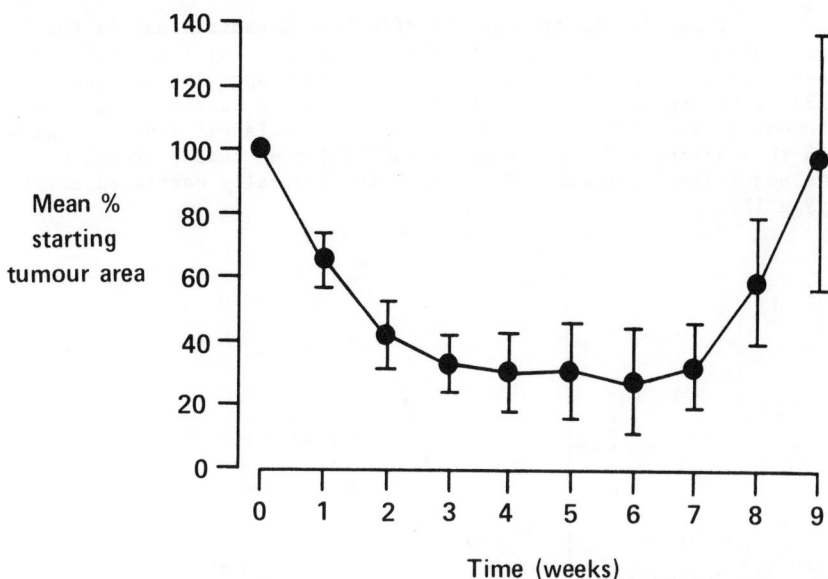

Figure 10   Effect of a single subcutaneous depot
containing 500 µg of 'Zoladex' on the growth of
dimethyl benzanthracene-induced mammary
tumours

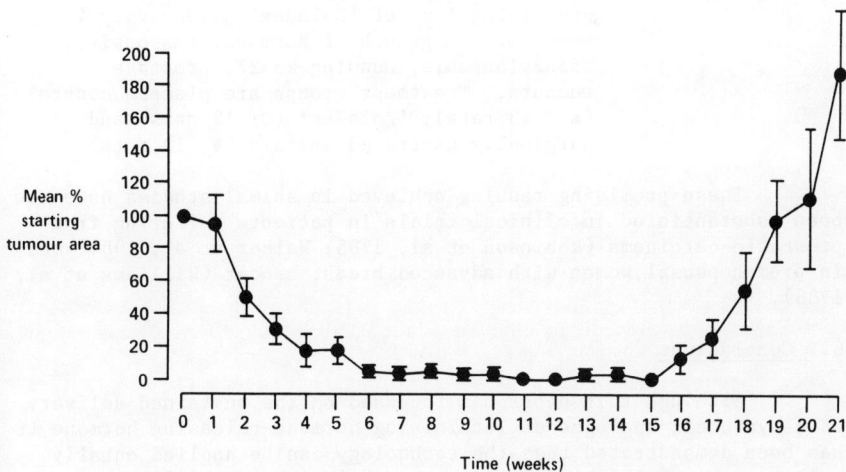

Figure 11   Effect of a single subcutaneous depot
containing 500 µg 'Zoladex' given at weeks 0, 4
and 8 on the growth of rat dimethyl
benzanthracene-induced mammary tumours

These depots are equally effective in animal models for prostate carcinoma. Single depots containing 1 mg 'Zoladex' given every 28 days to male rats bearing hormone—responsive Dunning R3327, transplantable prostate tumours cause a marked inhibition of tumour growth. Chemical castration using 'Zoladex' depots is shown in this animal model to cause an inhibition of tumour growth to values indistinguishable from those in surgically castrated animals (Fig 12).

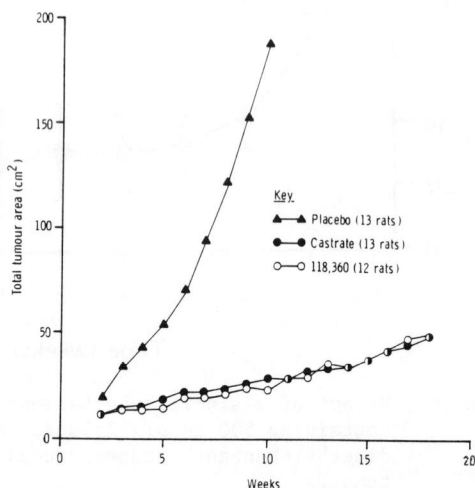

Figure 12   Effect of a single subcutaneous depot containing 1 mg of 'Zoladex' given every 4 weeks on the growth of hormonal—responsive, transplantable, Dunning R3327, prostate tumours. Treatment groups are placebo controls (▲ ; 13 rats); 'Zoladex' (O; 12 rats) and surgically castrated animals (●; 13 rats)

These promising results achieved in animal studies have now been substantiated in clinical trials in patients suffering from prostatic carcinoma (Robinson et al, 1985; Walker et al, 1984) and in premenopausal women with advanced breast cancer (Williams et al, 1986).

6. Conclusions

Although this paper has focussed on the sustained delivery of a synthetic analogue of luteinising hormone—releasing hormone it has been demonstrated that the technology can be applied equally well to other polypeptides and proteins (Hutchinson, 1982). Thus using acceptable biodegradable polymers based on lactic and glycolic acids it is possible to design a diversity of polymer types which allow continuous release of polypeptides, an objective that some 5—6 years ago was thought impossible.

REFERENCES

Anders, R., Merkle, H.P., Schurr, W. and Ziegler, R. (1983). 'Buccal absorption of protirelin: an effective way to stimulate thyrotropin and prolactin', J. Pharm. Sci., 72, 1481–1483.

Anderson, J.M., Niven, H. Pellagalli, J., Olanoff, L.S. and Jones, R.D. (1981). 'The role of the fibrous capsule in the function of implanted drug–polymer sustained release systems', J. Biomed. Mater. Res., 15, 889–902.

Anik, S.T., Sanders, L.M., Chaplin, M.D., Kushinsky, S. and Nerenberg, C. (1984). 'Delivery systems of LHRH and its analogs', in LHRH and its Analogs: Contraceptive and Therapeutic Applications (Eds. B.H. Vickery, J.J. Nestor Jr. and E.S.E. Hafez), pp. 421–435, MTP Press Limited, Boston, USA.

Baker, R.W. and Lonsdale, H.R. (1974). 'Controlled release: mechanisms and rates' in Controlled Release of Biologically Active Agents Volume 47 in Advances in Experimental Medicine and Biology (Eds. A.C. Tanquary and R.E. Lacey), pp 15–71, Plenum Press, New York.

Bohn, L. (1975). 'Compatible Polymers' in Polymer Handbook 2nd Edition (Eds. J. Brandrup and E.H. Immergut), III 211, John Wiley and Sons, New York.

Cohen, J., Siegel, R.A. and Langer, R. (1984). 'Sintering technique for the preparation of polymer matrices for the controlled release of macromolecules', J. Pharm. Sci., 73, 1034–1037.

Davis, B.K. (1972). 'Control of diabetes with polyacrylamide implants containing insulin', Experientia, 28, 348.

Davis, B.K. (1974). 'Diffusion in polymer gel implants', Proc. Nat. Ac. Sci., USA, 71, 3120.

Furr, B.J.A. and Hutchinson, F.G. (1985). 'Biodegradable sustained release formulation of the LH–RH analogue 'Zoladex' for the treatment of hormone–responsive tumours' in EORTC Genitourinary Group Monograph 2, Part A: Therapeutic Principles in Metastatic Prostatic Cancer, (Eds. F.H. Schroder and B. Richards), pp. 143–153, Alan R. Liss, Inc.

Hsieh, D.S.T., Chiang, C.C. and Desai, D.S. (1985). 'Controlled release of macromolecules from silicone elastomer', Pharm. Tech., 9, 39–49.

Hsu, T.T. and Langer, R. (1985). 'Polymer for the controlled release of macromolecules: effect of molecular weight of ethylene—vinyl acetate copolymer', J. Biomed. Mater. Res., 19, 445—460.

Hutchinson, F.G. (1982). Continuous—release pharmaceutical compositions, European Patent 58481B.

Kruisbrink, J. and Boer, G.J. (1984). 'Controlled long—term release of small peptide hormones using a new microporous polypropylene polymer: its application to vasopressin in the Brattleboro rat and perinatal use', J. Pharm. Sci., 73, 1713—1718.

Langer, R. (1981). 'Polymer for the sustained release of macromolecules: their use in a single step method of immunisation', Meth. Enzymol., 73, 57—75.

Langer, R. (1984). 'Controlled drug release systems', Biophys. J., 45, A26.

Langer, R. and Folkman, J. (1976). 'Polymers for the sustained release of proteins and other macromolecules', Nature, 263, 797.

Lotz, W. and Syllwasschy, B. (1979). 'Release of oligopeptides from silicone rubber implants in rats over periods exceeding 10 days', J. Pharm. Pharmacol., 13, 649—650.

Mitra, S., Van Dress, M., Anderson, J.M., Peterson, R.V., Gregonis, D. and Feijen, J. (1979). 'Pro—drug controlled release from polyglutamic acid', Polym. Prep. A.C.S., Div. Polym. Chem., 20(2), 32—35.

Okada, H., Yamazaki, I., Ogava, Y., Hirai, S., Yashiki, T. and Mima, H. (1982). 'Vaginal absorption of a potent luteinizing hormone—releasing hormone analogue (Leuprolide) in rats I: absorption by various routes and absorption enhancement', J. Pharm. Sci., 71, 1367—71.

Okada, H., Yamazaki, I., Yashiki, T. and Mima, H. (1983). 'Vaginal absorption of a potent luteinizing hormone—releasing hormone analogue (Leuprolide) in rats II: mechanism of absorption enhancement with organic acids', J. Pharm. Sci., 72, 75—78.

Okada, H., Yashiki, T. and Mima, H. (1983). 'Vaginal absorption of a potent luteinizing hormone—releasing hormone (Leuprolide) in rats III: effect of estrous cycle on vaginal absorption of hydrophilic model compounds', J. Pharm. Sci., 72, 173—76.

Okada, H., Yamazaki, I., Yashiki, T., Shamamoto, T. and Mima, H.
(1984). 'Vaginal absorption of a potent luteinizing hormone–
releasing hormone analogue (Leuprolide) in rats IV: evaluation of
the vaginal absorption and gonadotrophin responses by
radioimmunoassay', J. Pharm. Sci., 73, 298–302.

Petri, W., Seidel, R. and Sandow, J. (1984). 'Pharmaceutical
approach to long–term therapy with peptides', Int. Cong. Ser.–
Excerpta Medica, 656, 63–76.

Pitt, C.G. and Schindler, A. (1980). 'The design of controlled
drug delivery systems based on biodegradable polymers', in
Progress in Contraceptive Delivery Systems (Eds. E.S.E. Hafez and
W.A.A. van Os), Vol. 1, pp. 17–46, MTP Press, Lancaster, UK.

Robinson, M.R.G., Denis, L., Mahler, C., Walker, K., Stitch, R. and
Lunglmayr, G. (1985). 'An LH–RH analogue ('Zoladex') in the
management of carcinoma of the prostate: a preliminary report
comparing daily subcutaneous injections with monthly depot
injections', Eur. J. Surg. Oncol., 11, 159–165.

Uda, Y. and Yamada, M. (1984). 'Percutaneous pharmaceutical
compositions for external use', European Patent Application
127426.

Walker, K.J., Turkes, A.O., Zwink, R., Beacock, C., Buch, A.C.,
Peeling, W.B. and Griffiths, K. (1984). 'Treatment of patients
with advanced cancer of the prostate using a slow–release (depot)
formulation of the LHRH agonist ICI 118630', J. Endocrinol., 103,
R1–R4.

Williams, M.R., Walker, K.J., Turkes, A., Blamey, R.W. and
Nicholson, R.I. (1986). 'The use of an LH–RH agonist (ICI
118630; 'Zoladex') in advanced premenopausal breast cancer',
Brit. J. Cancer, 53, 629–636.

Wise, D.L., Fellman, T.D., Sanderson, J.E. and Wentworth, R.L.
(1979). 'Lactic/Glycolic Acid Polymers' in Drug Carriers in
Biology and Medicine, (Ed. G. Gregoriadis), pp. 237–270, Academic
Press, London.

Drug Carrier Systems
Edited by F.H.D. Roerdink and A.M. Kroon
© 1989 John Wiley & Sons Ltd.

MICROSPHERES AS DRUG CARRIERS

Stanley S. Davis and Lisbeth Illum

The University of Nottingham,
University Park,
Nottingham NG7 2RD, United Kingdom

The growing importance of the use of microspheres as carriers for drugs in recent years has resulted in the appearance of several books and review articles dealing with the various pharmaceutical, immunological and medical aspects of micro-spheric delivery systems in drug therapy (Davis et al, 1984; Poste and Kirsh, 1983, Sezaki and Hashida, 1985; Juliano 1985, Gregoriadis et al., 1982, 1986; Buri and Gumma, 1985; Guiot and Couvreur, 1986).

SOME DEFINITIONS

The term microsphere can be used to describe small particles intended as carriers for drugs or other therapeutic agents. Their size can range from tens of nanometres up to one hundred microns or more. The smaller particles (i.e. those below 500nm), are sometimes termed nanospheres or nanoparticles. Strictly speaking, a microsphere should be a monolithic device and perhaps also solid in nature. Microcapsules are similar in many respects to microspheres but comprise small spheres that have an outer layer or membrane enclosing a core material that could be the drug itself (Tomlinson, 1983). Because of the way in which microcapsules are prepared, they are normally large in comparison to the usual type of microsphere and consequently, can only be administered by a restricted number of routes into the body (Ishizaka et al., 1985, Thies, 1983, Chen et al., 1986).

Recently, workers in Japan have described studies on lipid microspheres intended for drug targeting. These particles are far better defined as emulsions and will not be considered in detail here. Davis and others (1987) have recently reviewed aspects of drug delivery using emulsions of different physical characteristics. While the emulsion is different to a solid microsphere in being the dispersion of one liquid in another liquid, stabilised by a third component (the emulsifying agent) there are some common features governing the behaviour of all colloidal carriers, especially with regard to their interaction with the biological milieu. This is also true for liposomes, that could be thought of as a special category of microspheres or microcapsules. The main advantages of microspheres over liposomes are a larger drug loading capacity, greater stability of the system in vitro and in vivo, (particularly with respect to drug leakage and control of release over longer time periods) and an easier sterilization procedure (Chen et al., 1986).

SOME MICROSPHERE SYSTEMS

Microspheres intended for drug delivery can be prepared from a variety of different materials and are of different physical characteristics depending on the use to which systems are put (Tomlinson, 1983; Chen et al., 1986; Juni and Nakano, 1987; Douglas et al., 1987). Some recent examples are given in table 1. The choice of the material will be dictated by factors such as the drug, the intended destination, disease condition to be treated, duration of action, etc. For example, in some cases, the microsphere will need to have been chosen as a carrier vehicle to maximize the uptake of an agent at a designated site in the body, while in other cases, the carrier aspects will be of lesser importance and the role of the delivery systems will be one associated with slow or sustained release of the drug. Thus, for example, microspheres intended to deliver drugs to the bone marrow following intravenous administration, need to be small (probably below 100nm) and have appropriate surface characteristics that will exploit biological recognition processes. They should also degrade rapidly at the desired site to release the entrapped drug (Davis and Illum, 1986). In contrast, microspheres for the controlled release of hormones will be, for example, injected (implanted) into muscle tissue to release the entrapped drug slowly (and hopefully) uniformly over a period of weeks (Juni and Nakano, 1987).

The choice of the material for the microsphere will also be influenced by toxicity considerations that include immune responses. Clearly, a microsphere intended for a non-parenteral application (e.g. intranasal, gastrointestinal) will not need to satisfy such rigorous constraints as one intended for parenteral use. Questions of biocompatibility, biodegradation and bioresorption (the uptake of metabolites into the endogenous pool) have to be addressed (Vert, 1986). More often than not, early consideration should be given to the drug to be delivered since the physiochemical characteristics of the therapeutic agent, dose and required release pattern can largely dictate feasibility for a given system. As is often the case in selective

TABLE 1
Examples of microsphere systems described in the recent literature.

Material	Drug	Drug loading (%)	Size range (μm)	Route	Reference
Albumin	Mitomycin	5	45	IA	Fujimoto et al. 1985
Fibrinogen	5-fluorouracil	6-16	1-4	IM	Miyazaki et al. 1986
Poly (lactic acid)	Triamcinolone acetonide	3-8	< 1	IV	Krause et al. 1985
Poly (butyl cyanoacrylate)	Progesterone	0.03	0.25	Ocular	Li et al. 1986
Polyacryl Starch	Various proteins	Up to 40	0.5	IV	Artursson et al. 1984
Collagen	Adriamycin	3	0.23	IV	El-Samaligy and Rohdewald 1983

Material	Drug	Drug loading (%)	Size (range) (μm)	Route	Reference
Monoglycerides waxes	Floxuridine	Up to 50	50-350	IA	Bechtel et al. (1986)
Ethylene -vinyl acetate copolymer	Proteins	10-50	1110-1370	IM	Sefton et al. (1984)
Poly (hydroxy butyric acid)	Sulphamethizole	50	425-600	Oral	Regina Brophy and Deasy (1986)
Starch	5-fluorouracil	–	40	IA Co-admin- istration of cytostatics	Lindell et al. (1978)
Gelatin	5-fluorouracil	8	< 1	SC	Oppenheim et al. (1984)
Ethyl- cellulose	Cisplatin	60	.396	IA	Okamoto et al. (1986)

drug delivery, the important questions are, where?, how much? and how often?

PREPARATION OF MICROSPHERES

The methods available for the preparation of microspheres are various. Once again, these are dictated to lesser or greater extents by the nature of the microsphere material, the drug to be delivered and the required size or loading and release specifications. In general terms, the two main approaches are either to "grow" particles of the required size, through processes such as the polymerization of appropriate monomers, and precipitation using solvent replacement, or to disperse materials through processes of size reduction that often involve emulsification/homogenization steps. As illustration three different approaches are given in Figure 1 and an example of resulting polyalkylcyanoacrylate nanoparticles is shown in Figure 2. Greater detail on the preparation of microsphere systems for therapeutic application can be found in the volume edited by Widder and Green (1985) on Drug and Enzyme Targeting in the Methods in Enzymology series, or in the books edited by Davis et al., (1984) and Illum and Davis (1987).

DRUG LOADING AND DRUG RELEASE

It goes without saying that a microsphere system, that will be successful therapeutically, must be able to carry a sufficient load of the required drug to its site of action and then release it according to a designated time profile. The major factors that constrain the nature of the final system will be the drug and its destination. Some (stable) drugs can be usefully incorporated into the microsphere during the manufacturing process. This normally leads to high loading and a subsequent slow release (Juni and Nakano, 1987). An alternative procedure is to sorb the drug into a preformed system (Illum et al., 1986b; Harmia et al., 1986), but the disadvantages here are lower levels of loading and more rapid release. Indeed, many authors have commented on the problem of the so called "burst effect", where a loaded microsphere can release a significant proportion of its drug content during the initial stages of administration, (Tomlinson, 1983). Strategies to reduce or even prevent the "burst effect" include cross-linking of the microsphere matrix, coating microspheres with polymer layers or covalent attachment of the drug to the matrix material (Tokes et al., 1984; Desoize et al., 1986). In the assessment of such drug loading and release characteristics, it is important to have in vitro models that are good predictors of the in vivo situation. Thus it is important to know the probable release mechanism in vivo, be it one of diffusion, desorption,

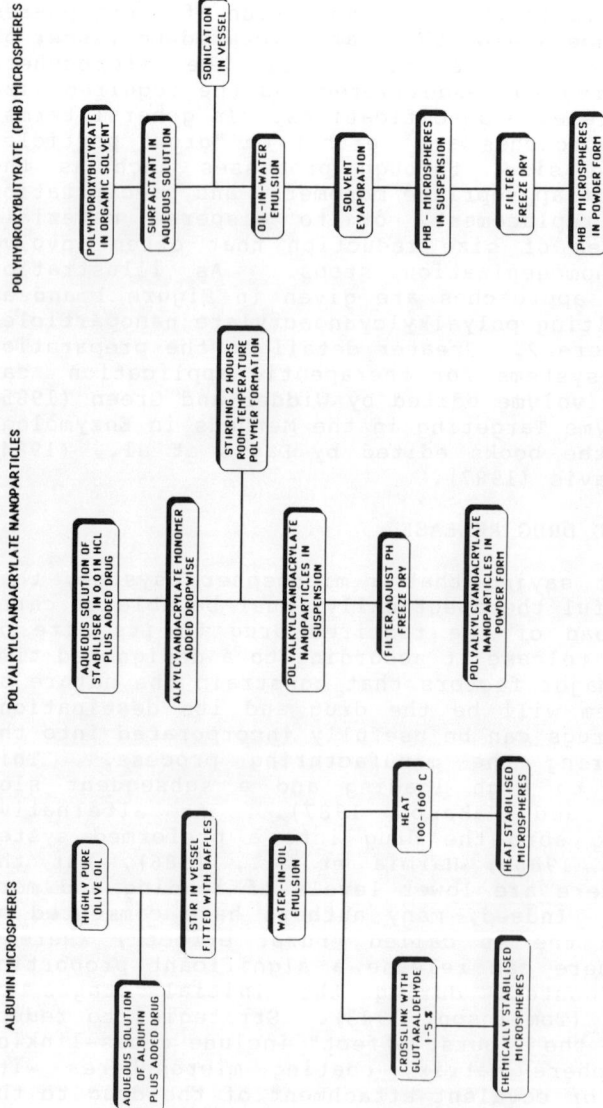

Figure 1. Preparation of albumin microspheres, polyalkylcyanoacrylate nanoparticles and polyhydroxybutyrate (PHB) microspheres.

erosion or biodegradation or as is more likely, a combination of these. For instance, dramatic differences in release can occur in vitro when one changes from a simple release medium, such as a buffer, to one containing plasma proteins (Douglas et al., 1986). The attachment of drugs to microspheres can be achieved through conventional chemical means as well as options involving prodrugs and macromolecular conjugates (Douglas et al., 1986, Davis et al., 1987).

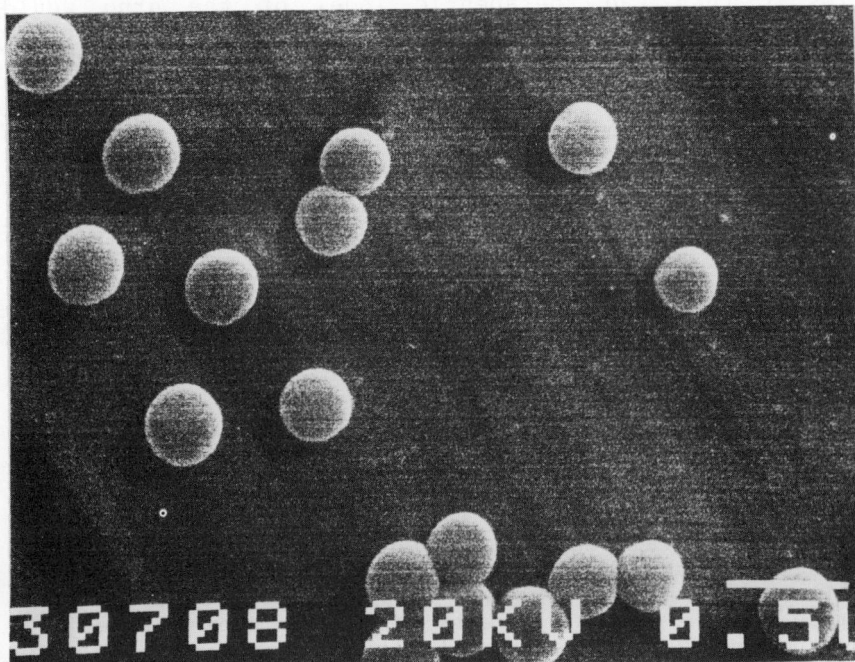

Figure 2.   Scanning electron micrograph of poly-
(butyl 2-cyanoacrylate) nanoparticles.

The drug-microsphere combination can lead to dramatic differences in the pharmacokinetic properties of the entrapped drug, since it is no longer in a free form or is released as such, only slowly. This is to be expected. Less obvious are associated problems that can involve toxicity and immunogencity. For instance, the incorporation of a drug in a microsphere may lead to an enhanced uptake of a therapeutic agent at an appropriate target site but there may also be an enhancement of

uptake elsewhere at a non-target site (Poste, 1985). As
will be discussed below, colloidal particles can often be
sequestered by elements of the reticuloendothelial system
following parenteral administration. An increased
uptake, say at a site of inflammation or even in a
necrotic (tumor) site, could be associated with a
corresponding increase in liver and spleen uptake. If
the microspheres were to carry a cytostatic or cytotoxic
agent, any benefit gained in increased selective delivery
could be offset by iatrogenic effects elsewhere with the
result that the therapeutic index of the drug would
remained unchanged (Poste, 1985). A less obvious problem
can arise with complex molecules such as peptides and
proteins. The sorption of these macromolecules into
microspheres has the potential of rendering the molecule
itself more immunogenic and even to create an immune
response to a carrier that was hitherto non-immunogenic
(Artursson et al., 1986). Under such circumstances, it
could be more sensible to exploit this effect, for
example, to produce a selective response for vaccine
systems. Ultrastructural alterations in cells can occur
after phagocytosis of certain microspheres. Edman et al.
(1984) demonstrated, using mouse peritoneal macrophages,
that high doses of acrylic microspheres could induce
cellular damage with pronounced alteration of the
lysosomal vacuome.

APPLICATION OF MICROSPHERE SYSTEMS

Microspheres can be used as drug carriers for a wide
spectrum of applications and by administration by almost
all parenteral and non-parenteral routes. Detailed
descriptions can be found elsewhere (Davis et al., 1984;
Illum and Davis, 1987) and this chapter will concentrate
more on the general aspects of such systems, especially
with regard to the interaction of carriers with the
biological environment and the assessment of the
different strategies available for both active and
passive targeting.

Sometimes the rationale for incorporating a drug into a
microsphere system is to provide some degree of
protection, either of the drug from the host organism
(degradation) or vice versa (toxicity). Thus, Edman et
al., (1980) described how biologically labile materials
such as enzymes could be protected from a hostile
environment by entrapment into acrylic microspheres. For
the alternative aspect, Morimoto et al., (1981) and
Verdun et al., (1986) have demonstrated that the
incorporation of the anticancer agent doxorubicin into
albumin microspheres and polyalkylcyanoacrylate
nanoparticles, respectively, reduced cardiotoxicity.
Similar host protection (so called toxicity buffering)
has been demonstrated for liposome and emulsion systems

and the drug Amphotericin B.

Recent reports in the literature describing increased
uptake of (complex) drugs from the gastrointestinal tract
following incorporation into microspheres (Maincent et
al., 1986) can best be rationalised in terms of a similar
protective role for the carrier, rather than direct
uptake of the particles across the epithelial lining of
the intestines. The latter explanation is considered
highly unlikely unless membrane integrity has been
perturbed by means of a surface active agent or similar
'absorption promoters' (O'Hagan et al., 1987). The
specific uptake of colloidal particles by the M-cells
residing in the Peyer's patch regions of the small
intestine is well documented (Lefevre et al., 1985).
Such capture does not lead to an encapsulated drug
reaching the systemic circulation but to its presentation
to the underlying lymphoid tissue and the possible
generation of a secretory immune response. Thus,
advantage could be taken of colloidal carriers for oral
immunization where the microsphere could act to protect
the antigenic material from degradation in the lumen of
the gut. Furthermore, proper attention to the size and
surface characteristics of the particle could establish
some degree of selectivity in the recognition and uptake
process.

CONTROLLED RELEASE

Microspheres and microcapsules intended for controlled
release applications have been fabricated from both
natural and synthetic materials (Wakiyama et al., 1981).
The period required for the controlled release effect can
vary from a matter of hours for nasal and ocular
applications, to weeks and months for implanted systems.
Bioadhesive microspheres administered to mucosal surfaces
can provide a beneficial combination of controlled
release and clearance characteristics. Swelling
microspheres made from starch and other natural materials
have been shown to modify mucociliary clearance processes
in the nose, thereby allowing better contact between the
delivery system and absorptive surfaces (Illum et al.,
1987). In this way (Illum et al., 1987) have improved
dramatically the biological availability of poorly
absorbed molecules such as gentamicin and insulin after
intranasal administration to the sheep and rabbit,
respectively.

Long acting injectable microsphere systems made from
poly(DL-lactide-co-glycolide) containing norethindrone
have been administered as contraceptive agents
intramuscularly to women (Beck et al., 1983). This
treatment supressed ovarian functions and inhibited

ovulation for 3 months. Similar types of microspheres have been developed for analogues of the peptide hormone drug LHRH by Sanders and Co-Workers (1987). Microcapsule preparations designed for the same purpose of supressing testicular secretion have been described by (Roger et al., 1985). The release pattern obtained from such microsphere systems can be somewhat eratic and represents the initial release of surface bound material, followed by slow degradation of the microsphere particles and finally total dispersion of the eroded mass. With lactide-glycolide systems and other poly(hydroxy acids), the breakdown products enter the metabolic pool and consequently the polymers are termed bioresorbable.

Albumin microspheres have often been suggested as potential controlled release systems. These degrade much more rapidly than the poly(hydroxy acid) systems, even when the albumin has been modified by chemical cross-linking or heat treatment (Gupta et al., 1986a,b, Davis et al., 1987). Drug release is due to a combination of diffusion and matrix degradation. The relative contributions of these separate processes depends largely on the size of the drug molecule incorporated into the microsphere and whether it is free to diffuse. In many cases, it is difficult to predict the degradation of a microsphere system in a tissue mass by in vitro experimentation because of the presence of enzymes and local (transient) inflammation. Sometimes, the invasion of macrophages and attendant tissue response can lead to the formation of granulomas and even encapsulation of the delivery system thereby changing the release characteristics.

DRUG TARGETING

An often discussed theme in present day programmes in drug delivery is the selective targeting of drugs to designated sites in the body. These sites can be at three levels; organ (e.g. lung, liver, spleen, heart) or cell population within an organ (e.g. for the liver-hepatocytes, endothelial cells, Kupffer cells), or intracellular structures (e.g. golgi apparatus, lysosome, mitochondria, etc.). In the third, most sophisticated level of drug targeting, the site could also be an invading organism such as a virus or parasite.

The ability of a colloidal carrier in the form of a microsphere, to reach its target, will naturally depend on where the target is located and the barriers that need to be circumvented. These barriers can comprise actual epithelial and endothelial structures or can take the form of the capture of particles by the defense system of the body; the reticulendothelial system. A realistic appreciation of the nature of these barriers and the ability to overcome them is an essential step in designing practical drug targeting systems (Posnansky and Juliano, 1984). For example, it is known to be difficult to transport drug molecules across the blood-brain barrier. This is because the cell junctions are extremely tight and thus, the transport of a microsphere into the brain can be considered as highly unlikely, unless there are (as yet undiscovered) selective recognition and uptake processes probably involving endocytosis. Similarly, the use of microspheres for targeting drugs to solid tumours is fraught with difficulties (Poste, 1985). The permeability of many tumours is extremely poor and in addition, some tumours can be very heterogeneous in nature. Microspheres could serve a better and more sensible role in delivering agents to more accessible targets. These include the various cell subsets present in the vascular and lymphatic compartments, especially micrometastases.

COMPARTMENTAL DELIVERY

Some opportunities in targeting can take advantage of the direct administration of the microsphere system into a body compartment. The compartmental delivery of microspheres into joints for the selective treatment of rheumatoid arthritis has been described by Ratcliffe et al., (1987). Here, the objective has been to design and prepare small biodegradable particles containing steroid drugs, that could be injected into the affected joint on a monthly basis. A number of candidate microsphere systems were prepared and albumin was selected as the most suitable material on toxicological grounds (Ratcliffe et al., 1984). Injected microspheres were taken up by the target tissue (the Synovium) provided they had a size of about 6 micron or less. Unfortunately, progress toward a clinically acceptable product has been limited because of the inability to achieve high loadings of drug and a sufficiently slow release thereof.

LYMPHATIC DELIVERY

Delivery of microspheres into the lymphatics after subcutaneous or intraperitoneal injection can also be considered as a form of compartmental delivery. As yet, the movement of injected colloids from the administration site to regional lymph nodes and transnodal spread is not well understood but particle size and probably surface character could be important. Lymph entering a lymph node flows through cortical and medullary sinuses. These sinuses can either be open channels or are traversed by a reticulum consisting of cells that can trap particles either mechanically or by phagocytosis (Bergqvist et al., 1983, 1984). The lymphatic system is known to play an important role in the dissemination of tumours. Consequently, small microspheres carrying cytostatic agents to lymph nodes could have a place in cancer chemotherapy. However, the small sizes probably required for satisfactory spread from the injection site and optimal nodal uptake (i.e. particles less than 100 nm) limits present research on drug carriers. The advent of new processing methods (e.g. microfluidization) as well as the exploitation of microemulsification and phase separation techniques for particle preparation could permit the eventual use of conventional monolithic microsphere systems.

TARGETING TO THE LUNG

The delivery of microspheres to the lung can be achieved in either of two ways; by aerosolization into the pulmonary system or by intravenous injection. Topical administration of particulates into the lungs is governed by the size of the administered particle and its growth (or shrinkage) characteristics under in vivo conditions. Although inert labelled microspheres have been employed as diagnostic agents for the study of lung clearance function, (Newman et al., 1982) drug-containing microspheres have not often been described. Some limited studies using liposomally entraped drugs as sustained release systems for aerosol delivery have been reported (Mihalko et al., 1987).

Microspheres can be easily targeted to the lungs following intravenous administration, simply by a matter of controlling particle size. Capillary diameters are in the range 3-11 microns and therefore, particles greater than this will be removed very effectively by the lungs during their first passage through the tissue (Illum et al. 1982). This removal process is already exploited in diagnostic imaging where albumin microspheres are labelled with the gamma emitter technetium - 99m. Extension of this approach to drug delivery has been described by Yoshioka et al. (1981), Illum and Davis (1982), Burger et al., (1985) and Wilmott et al., (1984). As to be expected, the uptake of the particle was rapid and efficient. Major problems, yet to be solved, include those relevant to drug release (usually too rapid) or to the rationale of treating lung diseases such as emphysema and cancer in this way.

As discussed above, after intravenous administration of microspheres the capillary beds of the lungs are the first mechanical filtration barrier to be reached. However, if the microspheres are injected intra-arterially (a much more difficult procedure), the same type of capillary entrapment process can be obtained for the particular region served by the arterial supply in question (Burger et al., 1985, Fujimoto et al., 1985a,b, Okamoto et al., 1986, Teder et al., 1986, Sigurdson et al., 1986, Pfeifle et al., 1986).

Albumin and ethyl cellulose have been materials of choice for microsphere systems intended for intra-arterial administration and good results, in terms of regression of tumours, have been reported for drug-loaded systems. Biodegradable starch microspheres have been used in a similar fashion but in order to retain a co-administered drug at an organ or tissue site for an extended period of time (Sigurdson et al., 1986, Teder et al., 1986, Pfeifle et al., 1986). The incorporation of drugs into the starch microspheres would appear to be a logical extension of this concept.

TARGETING AND AVOIDING THE RETICULOENDOTHELIAL SYSTEM

Particles that are small enough to escape the capillary beds of the lungs are usually removed rapidly by the phagocytic cells of the reticuloendothelial system (polynuclear macrophage system) (Illum et al., 1982). These cells are found in the liver, spleen and bone marrow. Under normal circumstances, the liver (Kupffer) cells (and to a lesser extent the spleen macrophage cells) will sequester the majority of available particles (Lenaerts et al., 1984, Edman and Sjoholm, 1983, Illum et al., 1982), although by a change of size and surface properties it is possible to retain significant quantities of particles in the circulation or to divert them into the bone marrow (Illum et al., 1986a, 1987, Illum and Davis, 1987). Other strategies for diversion include "blockage" of the RES using doses of placebo microspheres (Illum et al., 1986).

The uptake of particles by macrophages does provide certain therapeutic opportunities to include the treatment of intracellular parasites and enzyme storage disorders (Davis and Illum, 1986, Arturrson et al., 1984). Particles removed by phagocytosis will be quickly directed to the lysosomal region of the cell where constituent enzymes can be exploited to cause degradation of the particle or drug-particle linkage (if it is biodegradable in this milieu). Unwanted degradation of drug can perhaps occur especially if it is a complex macromolecule such as a polypeptide and protein. Incomplete degradation of the carrier in the lysosomal compartment can be expected to lead to adverse effects (Laakso et al., 1986b).

The passive targeting of a drug to the liver using a microsphere may not necessarily result in any therapeutic benefit if the target site is not associated with the phagocytic Kupffer cells. For example, studies in man

using labelled particles for targeting to hepatic tumours
have normally demonstrated that the tumour region is
almost devoid of particle uptake (Poste, 1985).

The rapid and efficient uptake of microspheres by
elements of the reticuloendothelial system is usually
mediated through the coating of the administered
particles by blood components, (e.g. fibrinogen,
fibronectin, complement, immunoglobulin) that render the
particle recognizable as foreign (Davis et al., 1986).
Such uptake of proteinaceous material (a process known
as opsonization) can be controlled apparently by altering
the surface characteristics of the administered
microsphere. In this way Illum and others have been able
to keep microspheres (and emulsions) away from the liver
and spleen and to retain them in the general circulation
(Illum et al., 1987), Figure 3. More recently, the same
group have been able to target small microspheres to the
bone marrow in a highly selective manner where liver
uptake was less than 10% (Illum and Davis, 1987). The
mechanism responsible for this selectivity in recognition
and delivery has yet to be elucidated but the enhanced
uptake of microspheres observed in the bone marrow could
open up a number of interesting therapeutic applications
that involve drug delivery and diagnostic imaging.

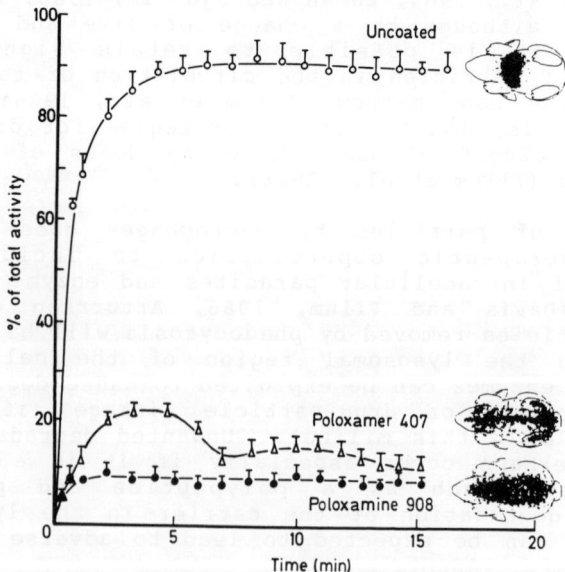

Figure 3.   The uptake in the liver/spleen of rabbits of
            uncoated and poloxamer 407 or poloxamine 908
            coated polystyrene microspheres (60 nm in
            diameter) as determined by gamma scintigraphy.

Coated particles that remain unrecognised in the general
circulation might be useful for passive targeting to
inflammatory lesions. Extravasation of colloidal
particles can occur in such regions because of an
increased permeability of the vascular endothelium and
an incomplete basement membrane (Fuchs, 1977, Schoefl,
1964). Possibilities for active targeting also exist
provided that suitable ligands can be attached to the
particles without affecting their incognito status and
that the attached ligands can be "seen" by the relevant
receptors on target cells (Davis and Illum, 1986).
Ligands that would be sensible for this objective include
antibodies, lectins, oligosaccharides and
apolipoproteins. Recently, Laakso et al., (1986a) were
able to eliminate circulating cells by such an approach
using specific monoclonal antibodies attached to acrylic
microspheres. This form of active targeting would seem
to hold more promise than esoteric methods relying on
external factors such as magnetic fields (and magnetic
particles) (Ranney 1986, Ibrahim et al., 1983, Widder and
Senyei, 1983).

CONCLUSIONS

Microsphere systems that can be prepared from various
synthetic and natural polymers have a number of important
applications as drug carriers. The system appropriate
for a given purpose needs to be designed with careful
consideration as to drug incorporation and intended
destination in the body.

REFERENCES

Artursson, P., Edman, P., Laakso, T. and Sjoholm, I.
  (1984). Characterization of Polyacryl Starch
  Microparticles as Carriers for Proteins and Drugs, J.
  Pharm. Sci., 73, 1507-1513.

Artursson, P., Martensson, I.-L. and Sjoholm, I.
  (1986). Biodegradable Microspheres III: Some
  Immunological Properties of Polyacryl Starch
  Microparticles, J. Pharm. Sci., 75, 697-701.

Artursson, P., Edman, P. and Sjoholm, I. (1984).
  Biodegradable Microspheres.I. Duration of Action of
  Dextranase Entrapped in Polyacrylstarch Microparticles
  in vivo, J. Pharmacol. Exp. Ther., 231, 705-712.

Bechtel, W., Wright, K.C., Wallace, S., Mosier, B.,
  Mosier, D. Mir, S. and Kudo, S. (1986). An
  Experimental Evaluation of Microcapsules for Arterial
  Chemoembolization, Radiology 161, 601-604.

Beck, L.R., Flowers, Jr. C.E., Pope, V.Z., Wilborn, W.H. and Tice, T.R. (1983). Clinical evaluation of an improved injectable microcapsule contraceptive system, Am. J. Obstet. Gynecol. 147, 815-820.

Bergqvist, L., Strand, S.E., Persson, B.R.R. (1983). Particle sizing and biokinetics of interstitial lymphoscintigraphic agents, Semin. Nucl. Med. 13, 9-19.

Bergqvist, L., Strand, S.E., Jonsson, P.E. (1984). 'The characterization of radiocolloids used for administration to the lymphatic system' in Microspheres and Drug Therapy. Pharmaceutical, Immunological and Medical Aspects. (Eds. S.S. Davis, L. Illum, J.G. McVie, E. Tomlinson), pp 263-279, Elsevier Science Publishers B.V., Amsterdam.

Burger, J.J., Tomlinson, E., Mulder, E.M.A. and McVie, J.G. (1985). Albumin microspheres for intra-arterial tumour targeting. I. Pharmaceutical aspects, Int. J. Pharm., 23, 333-344.

Buri, P. and Gumma, A., (Eds). Drug Targeting, (1985). Elsevier Science Publishers, Amsterdam.

Chen, T., Lausier, J.M. and Rhodes, C.T. (1986). Possible Strategies for the Formulation of Antineoplastic Drugs, Drug Devel. Ind. Pharm., 12, 1041-1106.

Davis, S.S., Illum, L., McVie, J.G. and Tomlinson, E. (1984), (Eds.): Microspheres and Drug Therapy, Pharmaceutical, Immunological and Medical Aspects, Elsevier Science Publishers B.V. Amsterdam.

Davis, S.S., Mills, S.N. and Tomlinson, E. (1987). Chemically cross-linked albumin microspheres for the controlled release of incorporated rose bengal after intramuscular injection into rabbits. J. Controlled Release, 4, 293-302.

Davis, S.S., Douglas, S.J., Illum, L., Jones, P.D.E., Mak, E. and Muller, R.H. (1986). Targeting of colloidal carriers and the role of surface properties, in Targeting of Drugs with Synthetic Systems, (Eds. G. Gregoriadis, J. Senior and G. Poste), pp 123-146, Plenum Press, New York.

Davis, S.S. and Illum, L. (1986). Colloidal
delivery systems: Opportunities and challenges in Site
Specific Delivery (Cell Biology, Medical and
Pharmaceutical Aspects), (Eds. E. Tomlinson and S.S.
Davis), pp 93-110, John Wiley & Sons, Chichester.

Davis, S.S., Illum, L., Burgess, D., Ratcliffe, J.
and Mills, S.N., Microspheres as controlled release
systems for parenteral and nasal delivery, ASC-
symposium series (in press).

Davis, S.S., Washington, C., West, P., Illum, L.,
Liversidge, G., Sternson, L. and Kirsh, R. (1987).
Emulsions as drug delivery systems. Biological
approaches to the controlled delivery of drugs:
Barriers, technologies and therapies, New York Academy
of Sciences.

Desoize, B., Jardillier, J.C., Kanoun, K., Guerin,
D., Levy, M.C. (1986). In vitro cytotoxic activity of
cross-linked protein microcapsules. J. Pharm.
Pharmacol. 38, 8-13.

Douglas, S.J., Davis, S.S. and Illum, L. (1987).
Poly(alkyl 2-cyanoacrylate) microspheres as drug
carrier systems, in Polymers in controlled drug
delivery, (Eds. L. Illum and S.S. Davis), John Wright,
Bristol, (in press).

Douglas, S.J., Davis, S.S. and Illum, L. (1986).
Biodistribution of poly(butyl 2-cyanoacrylate)
nanoparticles in rabbits. Int. J. Pharm. 34, 145-152.

Edman, P. and Sjoholm, I. (1983). Acrylic
Microspheres In Vivo VIII: Distribution and Elimination
of Polyacryldextran Particles in Mice, J. Pharm. Sci.,
72, 796-799.

Edman, P., Sjoholm, I. and Brunk, U. (1984).
Ultrastructural Alterations in Macrophages after
Phagocytosis of Acrylic Microspheres, J. Pharm. Sci.,
73, 153-156.

Edman, P., Ekman, Bo. and Sjoholm, I. (1980).
Immobilization of Proteins in Microspheres of
Biodegradable Polyacryldextran, J. Pharm. Sci., 69,
838-842.

El-Samaligy, M.S., and Rohdewald, P. (1983). Reconstituted collagen nanoparticles, a novel drug
carrier delivery system, J. Pharm. Pharmacol. 35, 537-
539.

Fuchs, U. (1977). Morphologische Reaktionsmuster
  der terminapen Strohmbahn, in Mikrozirkulation,
  (Ed. H. Meessen), Springer Verlag, Berlin-Heidelberg -
  New York.

Fujimoto, S., Miyazaki, M., Endoh, F., Takahashi,
  O., Okui, K., Sugibayashi, K. and Morimoto, Y.
  (1985a). Biodegradable Mitomycin C Microspheres Given
  Intra-Arterially for Inoperable Hepatic Cancer, Cancer
  56, 2404-2410.

Fujimoto, S., Miyazaki, M., Endoh, F., Takahashi,
  O., Okui, K. and Morimoto, Y. (1985b). Mitomycin C
  Carrying Microspheres as a Novel Method of Drug
  Delivery, Cancer Drug Delivery, 2, 173-181.

Gregoriadis, G., Senior, J. and Trouet, A., (1982)
  (Eds.), Targeting of Drugs, Plenum Press, London.

Gregoriadis, G., Senior, J. and Poste, G., (1986) (Eds.)
  Targeting of Drugs with Synthetic Systems, Plenum
  Press, London.

Guiot, P. and Couvreur, P. (1986) (Eds.). Polymeric
  Nanoparticles and Microspheres, CRC Press, Inc. Boca
  Raton, Florida.

Gupta, P.K., Hung, C.T. and Perrier, D.G. (1986).
  Albumin Microspheres. I. Release characteristics of
  adriamycin, Int. J. Pharm. 33, 137-146.

Gupta, P.K., Hung, C.T. and Perrier, D.G. (1986).
  Albumin Microspheres. II. Effect of stabilization
  temperature on the release of adriamycin, Int. J.
  Pharm. 33, 147-153.

Harmia, T., Speiser, P. and Kreuter, J. (1986).
  Optimization of pilocarpine loading onto nanoparticles
  by sorption procedures, Int. J. Pharm. 33, 45-54.

Ibrahim, A., Couvreur, P., Roland, M. and
  Speiser, P. (1983). New magnetic drug carrier, J.
  Pharm. Pharmacol. 35, 59-61.

Illum, L. and Davis, S.S. (1986) (Eds.). Polymers
  in controlled drug delivery, John Wright, Bristol, (in
  press).

Illum, L. and Davis, S.S. (1982). Specific intravenous
  delivery of drugs to the lungs using ion-exchange
  microspheres. J. Pharm. Pharmacol. 34, Suppl., 89P.

Illum, L., Davis, S.S., Muller, R.W., Mak, E. and
West, P. (1987). The organ distribution and
circulation time of intravenously injected colloidal
carriers sterically stabilised with a block-copolymer -
poloxamine 908, Life Sciences, 40, 367-374.

Illum, L., Davis, S.S., Wilson, C.G., Thomas, N.,
Frier, M. and Hardy, J.G. (1982). Blood clearance and
organ deposition of intravenously administered
colloidal particles: the effects of particle size,
nature and shape. Int. J. Pharm. 12, 1982, 135-147.

Illum, L. and Hunneyball, I.A. and Davis, S.S.,
(1986a). The effect of hydrophilic coatings on the
uptake of colloidal particles by the liver and by
peritoneal macrophages. Int. J. Pharm. 29, 53-65.

Illum, L., Jorgensen, H., Bisgaard, H., Krogsgaard,
O. and Rossing, N., Bioadhesive microspheres as a
potential nasal drug delivery system, Int. J. Pharm.
(in press).

Illum, L., Khan, M.A., Mak, E. and Davis, S.S.
(1986b). Evaluation of the carrier capacity and
release characteristics for poly(butyl 2-cyanoacrylate)
nanoparticles. Int. J. Pharm. 30, 17-28.

Illum, L., Thomas, N. and Davis S.S. (1986c). The
effect of a selected suppression of the
reticuloendothelial system on the distribution of model
carrier particles. J. Pharm. Sci. 75, 16-22.

Ishizaka, T., Ariizumi, T., Nakamura, T. and Koishi, M.
(1985). Preparation of Serum Albumin Microcapsules,
J. Pharm. Sci., 74, 342-344.

Juliano, R.L. (1985). Microparticulate Drug
Carriers, in Directed Drug Delivery, (Eds. R. T.
Borchardt, A.J. Repta and N.J. Stella), pp 147-170,
Humana Press, Clifton.

Juni, K. and Nakano, M. (1987). Poly(hydroxy
Acids), Drug Delivery, in CRC Critical Reviews,
Therapeutic Drug Carrier Systems, CRC Press Inc., Boca
Raton, 3, 209-232.

Krause, H.-J., Schwarz, A. and Rohdewald, P. (1985).
Polylactic acid nanoparticles, a colloidal drug
delivery system for lipophilic drugs, Int. J. Pharm.
27, 145-155.

Laakso, T., Andersson, J., Artursson, P., Edman, P.
and Sjoholm, I. (1986a). Acrylic Microspheres in vivo.
X. Elimination of circulating cells by active targeting
using specific monoclonal antibodies bound to
microparticles, Life Sciences, 38, 183-190.

Laakso, T., Artursson, P. and Sjoholm, I. (1986b).
Biodegradable Microspheres IV: Factors Affecting the
Distribution and Degradation of Polyacryl Starch
Microparticles, J. Pharm. Sci., 75, 962-967.

Lefevre, M.E., Joel, D.D. and Schidlovsky, G.
(1985). Retention of Ingested Latex Particles in
Peyer's Patches of Germfree and Conventional Mice,
Proc. Soc. Exp. Biol. Med., 179, 522-528.

Lenaerts, V., Nagelkerke, J.F., Van Berkel, T.J.C.,
Couvreur, P., Grislain, L., Roland, M. and Speiser, P.
(1984). In vivo uptake of Polyisobutyl Cyanoacrylate
Nanoparticles by Rat Liver Kupffer, Endothelial and
Parenchymal Cells, J. Pharm. Sci., 73, 980-982.

Li, V.H.K., Wood, R.W., Kreuter, J., Harmia, T. and
Robinson, J.R. (1986). Ocular drug delivery of
progesterone using nanoparticles, J. Microencapsulation
1986, 3, 213-218.

Lindell, B., Aronsen, K.-F., Nosslin, B. and
Rothman, U. (1978). Studies in Pharmacokinetics and
Tolerance of Substances Temporarily Retained in the
Liver by Microsphere Embolization, Ann. Surg., 187,
95-99.

Maincent, P., Le Verge, R., Sado, P., Couvreur, P.
and Devissaguet, J.P. (1986). Disposition Kinetics
and Oral Bioavailability of Vincamine-Loaded Polyalkyl
Cyanoacrylate Nanoparticles, J. Pharm. Sci., 75, 955-
958.

Mihalko, P.J., Schreier, H., Radhakrishnan, R. and
Abra, R.M. (1987). Liposomes as sustained-release
drug-carriers in the lung, Pharm. Res., PD-407, S-36.

Miyazaki, S., Hashiguchi, N., Sugiyama, M., Takada
and Morimoto, Y. (1986). Fibrinogen Microspheres as
Novel Drug Delivery Systems for Antitumor Drugs, Chem.
Pharm. Bull., 34, 1370-1375.

Morimoto, Y., Sugibayashi, K. and Kato, Y. (1981).
Drug-carrier Property of Albumin Microspheres in
Chemotherapy. Antitumor Effect of Microsphere-
entrapped Adriamycin on Liver Metastasis of AH 7974
Cells in Rats. Chem. Pharm. Bull., 29, 1433-1438.

Newman, S.P., Pavia, D. and Clarke, S.W. (1982).
Therapeutic Aerosol Deposition in Radionuclide Imaging
in Drug Research (Eds. C.G. Wilson, J.G. Hardy with M.
Frier and S.S. Davis), pp 203-220, Croom Helm, London.

O'Hagan, D., Palin, K.J. and Davis, S.S. Intestinal
absorption of proteins and macromolecules and the
immunological response, in CRC Critical reviews in
Therapeutic Drug Carrier Systems (in press).

Okamoto, Y., Konno, A., Togawa, K., Kato, T.,
Tamakawa, Y. and Amano, Y. (1986). Arterial
chemoembolization with cisplatin micro-capsules, Br. J.
Cancer, 53, 369-375.

Oppenheim, R., Gipps, E.M., Forbes, J.F. and
Whitehead, R.H. (1984). Development and testing of
proteinaceous nanoparticles containing cytotoxics, in
Microspheres and Drug therapy. Pharmaceutical
Immunological and Medical Aspects, (Eds. S.S. Davis,
L. Illum, J.G. McVie and E. Tomlinson), pp 117-128,
Elsevier Science Publishers B.V., Amsterdam.

Pfeifle, C.E., Howell, S.B., Ashburn, W.L., Barone,
R.M. and Bookstein, J.J. (1986). Pharmacologic
Studies of Intra-Hepatic Artery Chemotherapy with
Degradable Starch Microspheres, Cancer Drug Delivery,
3, 1-14.

Poste, G. (1985). Drug Targeting in Cancer Therapy, in
Receptor-Mediated Targeting of Drugs, (Eds. G.
Gregoriadis, G. Poste, J. Senior and A. Trouet), pp
427-474, Plenum Press, London.

Poste, G. and Kirsh, R. (1983). Site-Specific
(Targeted) Drug Delivery in Cancer Therapy,
Biotechnology, 1, 869-878.

Poznansky, M.J. and Juliano, R.L. (1984). Biological
Approaches to the Controlled Delivery of Drugs: A
Critical Review, Pharmacol. Reviews, 36, 277-336.

Ranney, D.F. (1986). Drug targeting to the Lungs,
Biochemical Pharmacol., 35, 1063-1069.

Ratcliffe, J.H., Hunneyball, I.M., Smith, A.,
Wilson, C.G. and Davis, S.S. (1984). Preparation and
evaluation of biodegradable polymeric systems for the
intra-articular delivery of drugs. J. Pharm.
Pharmacol. 36, 431-436.

Ratcliffe, J.H., Hunneyball, I.M., Wilson, C.G., Smith, A. and Davis, S.S. (1987). Albumin microspheres for intra-articular drug delivery: investigation of their retention in normal and arthritic knee joints of rabbits, J. Pharm. Pharmacol. 39, 290-295.

Regina Brophy, M. and Deasy, P.B. (1986). In vitro and in vivo studies on biodegradable polyester microparticles containing sulphamethizole, Int. J. Pharm. 29, 223-231.

Roger, M., Duchier, J., Lahlou, N., Nahoul, K. and Schally, A.V. (1985). Treatment of Prostatic Carcinoma with D-Trp-6-LH-RH: Plasma Hormone Levels After Daily Subcutaneous Injections and Periodic Administration of Delayed-Release Preparations, The Prostate 7, 271-282.

Sanders, L.M., Vitale, K.M., McRae, G.I. and Mishky, P.B. (1987). Controlled Delivery of Nafarelin, an Agonistic Analogue of LHRH, from microspheres of poly (D, L Lactic-co-Glycolic) acid, In: Delivery Systems for Peptide Drugs (Eds. S.S. Davis, L. Illum and E. Tomlinson), pp 125-138, Plenum Press, London.

Schoefl, G.I. (1964). Electron microscopic observations on the regeneration of blood vessels after injury, N.Y. Acad. Sci. 116, 789-802.

Sefton, M.V., Brown, L.R. and Langer, R.S. Ethylene-Vinyl Acetate Copolymer Microspheres for Controlled Release of Macromolecules, J. Pharm. Sci., 53, 1859-1861.

Sezaki, H., Hashida, M. (1985). Macromolecule-Drug Conjugates in Targeted Cancer Chemotherapy, In CRC Critical Reviews in Therapeutic Drug Carrier Systems, 1, 1-38.

Sigurdson, E.R., Ridge, J.A., Daly, J.M. (1986). Intra- arterial Infusion of Doxorubicin with Degradable Starch Microspheres, Arch. Surg. 121, 1277-1281.

Teder, H., Aronsen, K.F., Bjorkman, S., Lindell, B., Ljungberg, J., The influence of degradable starch microspheres on liver uptake of 5-fluorouracil after hepatic artery injections in the rat, J. Pharm. Pharmacol., 38, 939-941.

Thies, C. (1983). Microcapsules as Drug delivery Devices, CRC Critical Reviews. Biomedical Engineering, 8, 335-383.

Tokes, Z.A., Ross, K.L. and Rogers, K.E. (1984).
Use of Microspheres to direct the cytotoxic action of
adriamycin to the cell surface, in Microspheres and
Drug Therapy, Pharmaceutical Immunological and Medical
Aspects, (Eds. S.S. Davis, L. Illum,   J.G. McVie and
E. Tomlinson), pp 139-149, Elsevier Science Publishers
B.V., Amsterdam.

Tomlinson, E. (1983).   Microsphere Delivery Systems
for Drug Targeting and Controlled Release, Int. J.
Pharm. Tech. & Prod. Mfr., 4, 49-57.

Verdun, C., Couvreur, P., Vranckx, H., Lenaerts, V.
and Roland, M. (1986).   Development of a Nanoparticle
Controlled-Release   Formulation for Human Use, J.
Control. Rel., 3, 205-210.

Vert, M. (1986).   Polyvalent Polymeric Drug
Carriers, CRC Critical Reviews Therapeutic Drug Carrier
Systems, 2, 291-327.

Wakiyama, N., Juni, K. and Nakano, M. (1981).
Preparation and Evaluation in vitro of Polylactic Acid
Microspheres Containing Local Anesthetics, Chem. Pharm.
Bull. 29, 3363-3368.

Widder, K.J. and Senyei, A.E. (1985).   Magnetic
Microspheres.   A Vehicle for Selective Targeting of
Drugs, Pharmac. Ther., 20, 377-395.

Widder, K.J. and Green, R. (1985) (Eds).   Drug and
Enzyme Targeting, Methods in Enzymology, 112, Academic
Press, Inc., New York.

Willmott, N., Kamel, H.M.H., Cummings, J., Stuart,
J.F.B. and Florence, A.T.   Adriamycin-loaded albumin
microspheres: Lung entrapment and fate in the rat, in
Microspheres and Drug Therapy, Pharmaceutical
Immunological and Medical Aspects, (Eds. S.S. Davis, L.
Illum, J.G. McVie and E. Tomlinson), pp 205-215,
Elsevier Science Publishers B.V. Amsterdam, 1984.

Yoshioka, T., Hashida, M., Muranishi, S. and Sezaki,
H. (1981).   Specific Delivery of Mitomycin C to the
Liver, Spleen and Lung: Nano- and Microspherical
Carriers of Gelatin, Int. J. Pharm. 81, 131-141.

Drug Carrier Systems
Edited by F.H.D. Roerdink and A.M. Kroon
© 1989 John Wiley & Sons Ltd.

PHARMACOLOGICAL USES OF RESEALED AND MODIFIED RED CELL CARRIERS

GARRET M. IHLER

Department of Medical Biochemistry and Genetics
Texas A&M College of Medicine
College Station, TX 77840 USA

INTRODUCTION

Red cells have interesting biological properties which permit their use as carriers for drugs, enzymes, DNA and other macromolecules, and particles. In this context, the most important of these properties is that red cells can be readily opened and resealed without serious damage necessarily resulting to the cells, so that substances can be entrapped inside. This has lead to several different strategies for the use of red cells in drug carrier technology. The first suggestion was the use of red cells as vehicles to carry entrapped enzymes or drugs to lysosomes of erythrophagocytic cells (Ihler et al, 1973; Ihler and Glew, 1977). Since entrapped drugs or enzymes would be taken up only by erythrophagocytic cells, the result would be delivery targeted to a specific cell types. Delivery of glucocerebrosidase in Gaucher disease to erythrophagocytic cells which contained intracellular deposits of undegradable glucocerebroside derived from ingested red and white cells was the first use suggested for resealed red cells (Ihler et al, 1973). Using this delivery system, glucocerebrosidase would presumably be taken up by the very cells affected by the enzyme deficiency. Clinical trials of this procedure have been reported (Beutler et al, 1977a,b, 1980). The results were not encouraging, but it is probably unreasonable to expect reversal of extensive pre-existing damage to liver and spleen; the possibility that repeated infusion of red cell entrapped enzyme might arrest the disease has not been adequately tested. This and other approaches to Gaucher disease have been reviewed in Desnick et al (1982), but the rapid progress of gene therapy suggests that genetic manipulation may presently present the best prospects (Sorge et al, 1987).

Entrapment of enzymes or drugs is accomplished by reversible hypotonic lysis of red cells in the presence of the substances to be entrapped, followed by resealing of the membranes after restoration of the ionic strength. The cells become spherical as the result of osmotic swelling and then pores form which can persist indefinitely and which are sufficiently large to permit most small to medium-sized molecules to come to equilibrium and so to be present inside the resealed cells at approximately the same concentration as outside. Resealing traps these substances inside the cells. If damage to the red cells is incurred during the loading operation, upon reinfusion most of the red cells are recognized as abnormal and taken up by phagocytosis in the spleen or liver (Thorpe et al, 1975). More careful control of the level of damage to the resealed red cells enables them to be targeted to the spleen, which recognizes minor amounts of damage, or the liver, which removes most of the badly damaged red cells by virtue of its much larger blood flow (DeLoach et al, 1977b; Smedsrod and Aminoff, 1983).

155

RESEALED RED CELLS CAN HAVE A NORMAL LIFE SPAN IN THE CIRCULATION

In the course of the early experiments, it became apparent that red cells could continue to circulate for substantial periods of time after resealing (DeLoach et al, 1977a). Subsequent work by many groups has amply demonstrated that resealed red cells can continue to circulate for their normal life spans, provided certain fairly simple precautions are taken during the loading process (reviewed in Ihler and Tsang, 1985). Several studies demonstrating extended circulation in humans have been reported (Beutler et al, 1977; Eichler, 1986) and recently it was shown that the life span of resealed human red cells can be identical to that of normal red cells (Ropars et al, 1985b).

This finding considerably extends the possible uses of resealed red cells. Since resealed red cells can have an essentially normal life span, drug loaded cells could permit the escape of drugs over extended periods of time and drugs which modify the physiological response of red cells themselves or which are active against hemotrophic parasites could be entrapped. Currently, the most interesting example of the latter is the compound IHP (inositol hexaphosphate) which acts as an allosteric modifier of oxygen release from hemoglobin and increases the total oxygen delivery possible.

Entrapment in red cells of enzymes whose substrates could permeate or be transported into the red cell is another potential application. Since red cells lack lysosomes and are rather free of proteolytic activity (although reticulocytes have an ATP-, ubiquitin-dependent proteolytic system which requires free amino groups (Katznelson and Kulka, 1983) and certain proteins are appreciably degraded in red cells (Netland and Dice, 1985)) , the entrapped enzymes could potentially remove or degrade unwanted compounds for a considerable period of time. As a prototype example, uricase was entrapped in red cells and its ability to degrade extracellular uric acid was determined and compared with the requirements in hyperuricemic states (Ihler et al, 1975).

MOST DRUGS AND ENZYMES CAN BE ENTRAPPED

Potentially, most water-soluble drugs and most proteins can be entrapped in red cells. Many reports have appeared concerning the entrapment of various drugs and proteins in red cells, motivated by the possibility of using red cell carriers in the treatment of one or another disease state. Many of these reports demonstrate that the drug is encapsulated, does not have any serious effects on the morphology of the red cell, and that it escapes with some observed rate constant. Most drugs which are not membrane active or which do not alter important red cell enzymatic reactions are not likely to be deleterious to the red cells. Drugs will escape with some rate constant that is determined either by their lipid solubility or their ability to be transported by facilitated diffusion or active transport. The red cell ghost system in fact provides a useful and better means of predicting the physiological permeability constants than partitioning between aqueous and organic phases (Korten and Miller, 1979). It must be remembered that it is the circulating level of free drug and not the circulating level of red cell entrapped drug that is significant; in some reports only the level of circulating entrapped drug was determined. It would be expected that most enzymes or proteins can be entrapped in red cells and would not escape except by intravascular hemolysis. Therefore it is hard to understand how red cells could be a useful delivery system for polypeptides, unless the polypeptide can pass through membranes or unless its anticipated use is correlated with intravascular hemolysis, as might be the case for some clotting or anti-coagulant proteins. Resealed red cells provide a possible route for immunization and production of antibodies. For example, Yokoi et al (1983, 1984) have entrapped blood groups substances in red cells ghosts and have demonstrated that antiserum can be prepared in this way. It is worthwhile to determine the rate at which the substrate for an entrapped enzyme enters red cells since this rate will determine whether red cells containing the enzyme in question could possibly be of value in reducing plasma levels of its substrate.

PLASMA CONCENTRATIONS OF FREE DRUG

If the rate of escape of a drug is such that plasma levels of the free drug are in the therapeutic range for an extended period of time, carrier red cells could serve as a delayed release delivery system. In its simplest form, this idea is not as useful as it might seem at first glance because it is not simple to ensure that the drug will escape from the cells at a rate appropriate to maintain desired plasma concentrations over the desired time interval. If a drug is sufficiently soluble, the concentration of entrapped drug can be selected so that the initial rate of escape results in levels of free drug in the plasma at the desired level. If a drug escapes too rapidly (or slowly), in most cases it should be possible to modify the structure of the drug in ways that will permit higher or lower rates of transport or escape, although these modifications may well affect the efficacy of the drug as well. Because the rate of release is proportional to the amount of drug remaining, a continuously decreasing rate of drug release should be observed and the plasma levels of free drug would then decrease from its initial level as more and more drug is escapes, until eventually the plasma level drops below the therapeutic range. The upper limit of plasma concentration is determined by the toxicity of the drug and the lower limit by the minimum that is therapeutically effective. Sustained release is useful only for the period of time that plasma levels can be maintained between these limits; if the half-time for escape were 1 week, after two weeks the plasma level would be only one-quarter the initial level. For a delivery mechanism that obeys first order kinetics, it would perhaps be best to employ drugs with wide levels of therapeutic effectiveness.

A CONSTANT RATE OF DRUG RELEASE IS POSSIBLE WITH RESEALED RED CELLS.

Drug release can be made linear in time using resealed red cells, as suggested in Humphreys and Ihler (1982) and Ihler (1983), because a constant fraction of the red cell population is destroyed each day. Red cells have a fixed life span of about 120 days, so that about 1% of the red cells are destroyed each day and 1% are newly made. A random population of red cells loaded with a drug would therefore deliver a constant amount of drug (1% per day) if the drug is released at the time of red cell destruction. (As a practical matter, the oldest red cells are most likely to be irreversibly damaged during loading, so that the loaded red cell population may be slightly enriched for younger cells. Offsetting this however is the fact that some damage may be done to the cells, thereby shortening their lifespan). Conversion of the drug to a form which can escape from the phagocytes could be accomplished by lysosomal enzymes. For example, the drug could be linked to a protein which would be degraded by lysosomal proteases or liberated by esterases so that it is readily transported out of cells, in this case out of macrophages. For example, Eichler et al (1985) have reported that entrapped Vitamin B12 escapes with a t/2 of 5 h, but presumably it could be converted to a form which would not escape and therefore be released at a slow rate over a considerable period of time.

PHAGOCYTOSIS AND TARGETING

Phagocytosis of drug-loaded red cells in some organ, for example liver, results in delivery of substantial levels of the drug to the organ, but this does not necessarily indicate that the drug is preferentially made available to parenchymal, non-phagocytic cells of the organ. The drug is initially delivered to macrophages and must first escape from the macrophages. After lysosomal disruption of the uptaken red cells, the drug may escape from the macrophages, hopefully without being inactivated, and enter either the interstitial fluid or the blood. If the drug is taken up avidly by other cell types in the same organ before it enters the circulation, targeted delivery to the parenchymal cells of the organ may legitimately be said to have occurred. In the blood, the drug may have a transient, locally high concentration also resulting in enhanced uptake, but once the drug enters the circulation and is distributed generally through the body, delivery to parenchymal cells of the organ in which it was originally sequestered would be no more specific than that which would occur after injecting the drug. Specific delivery to parenchymal cells in an organ cannot be demonstrated simply by finding a high concentration of drug in the organ, because the drug could be still entrapped in macrophages, but must be demonstrated by direct measurement in parenchymal cells or by some specific elevated effect on non-erythrophagocytic cells, for example elevated killing of tumor cells. Clearly each drug will have its own properties in this regard and must be evaluated independently.

FUSION OF RED CELLS WITH NUCLEATED CELLS

Resealed red cells can be fused with other cell types in vitro with polyethylene glycol, fusagenic viruses or in other ways; an important application of the encapsulation and fusion system is the introduction of various molecules, for example proteins, enzymes and DNA, into the cytoplasm of recipient cells. Erythrocytes loaded with arginase have been infused and the effect of circulating cells has been studied (Adriaenssens et al, 1984) , but also they have been fused with fibroblasts from arginase-deficient individuals using Sendai virus or its purified proteins (Kruse et al, 1981). Doxsey et al (1985) have developed an especially efficient system in which the recipient cells (e.g. 3T3) express the hemagglutinin of influenza virus on the cell surface. The hemagglutinin has a high affinity for sialic acid residues, which are abundant on red cells. When the pH is dropped to 5.0, the bound red cells fuse with the recipient and the contents of the red cells are then introduced into the cytoplasm of the recipient cell. Targeting of red cells to specific cell types is possible by coupling antibody against specific cell surfaces to the red cells, followed by fusion (Godfrey et al, 1983). Intracellular functioning of proteins transferred by fusion has been demonstrated, for example by protecting cells from the action of diphtheria toxin by transfer of red cell encapsulated antibody to the toxin. Fusion is conveniently followed by including a fluorescent intercalating dye in the red cells which becomes vividly fluorescent when it binds to the nucleus of the target cell (Drant et al, 1986).

ENTRAPMENT PROCEDURES

Besides direct transport into the red cell and conversion to a non-transported form (Jaffe et al, 1970), there are four procedures for loading substances into red cells that have been successfully employed. These are: 1. reversible rupture of the membrane by osmotic forces generated by dilution or dialysis into hypotonic medium 2. voltage-dependent increase in membrane permeability to small molecules caused by dielectric breakdown of the membrane; 3. fusion with liposomes or other vesicles or cells; and 4. induced endocytosis and vacuolation.

OSMOTIC LYSIS

Hypotonic lysis of red cells is the oldest and simplest of the procedures, requiring only that the ionic strength of the external medium be reduced to the point where the cells cannot withstand the osmotic stress resulting from the more rapid entry of water into the cell than exit of sodium and other ions. Hemolysis by dilution into hypotonic medium has long been a standard clinical assay and red cells are known to be considerably heterogeneous with respect to the concentration at which they lyse. Lysis begins at about 0.55% NaCl and essentially 100% lysis occurs at NaCl concentrations below 0.33%. The cells increase in volume and become spherical; at the point where they are mechanically unable to swell more, the cells lyse. Red cells which are already somewhat spherical of course lyse more readily. Young red cells are more resistant than old red cells, which provides a potential selection procedure for younger red cells. Cells from venous blood are more readily lysed than those from arterial blood. Resistance to hemolysis is substantially increased in sickle cell disease, hypochromic anemias, and in thalassemia major, where the red cells may not lyse in an NaCl concentration as low as 0.03%. The intracellular concentration of the macromolecules is important, at least for creation of pores large enough for hemoglobin to pass through. After resealing, the red cells can be re-lysed by dilution and osmotic fragility is said to be a simple and reliable method to evaluate carrier erythrocytes in vitro.

If red cells are diluted greatly into hypotonic medium and washed extensively before resealing, nearly all the intracellular contents are lost, but the cells can still be resealed. This is the traditional procedure for preparing red cell ghosts (white ghosts), for example for the study of membrane transport. Depending on the procedure, the cytoskeleton may undergo variable amounts of damage. These cells could perhaps best be regarded as sophisticated proteoliposomes.

DIALYSIS

There are several variants on the osmotic lysis procedure. The method that seems to produce cells with the best viability, after resealing, is the dialysis procedure. In this procedure, the cells are suspended within a dialysis bag at a high hematocrit and dialyzed against hypotonic medium until the membrane is made permeable. Then the cells are resealed. This procedure minimizes the amount of unincorporated drug and at the same time ensures that the intracellular concentration of non-dialyzable substances such as enzymes or hemoglobin is relatively normal. Instrumentation for automating this process has been described (DeLoach et al, 1980; DeLoach, 1985; Ropars et al, 1985a).

PRE-SWELL DILUTION

A second procedure, technically easier and more rapid than dialysis, is the preswell dilutional procedure. In this procedure, the cells are first suspended in a hypotonic buffer, equivalent to about 0.6% NaCl, where the cells swell, but do not hemolyze. The cells are then loosely packed, and the minimum volume of hypotonic buffer needed to lyse the cells is added. This procedure results in red cells containing a substantial fraction (generally about 45%) of their original intracellular contents, including the small molecules. These cells do contain lower amounts of macromolecules and the life span of these cells seems to be lower than those prepared by the dialysis procedures.

ISO-IONIC PROCEDURES

In this procedure, glycol (Billah et al, 1976), DMSO (Franco et al, 1984) or some other penetrant, is allowed to diffuse into the red cells. Upon addition of iso-ionic medium lacking the penetrating substance, the cells rapidly lyse . This procedure has the advantage of avoiding any possibility of disrupting the cytoskeleton, which is known to be disrupted by conditions of low ionic strength.

ELECTRIC PULSE

Potentially an excellent procedure, at least for small molecules, is the application of an electric pulse since the pore size can be easily controlled (Kinosita and Tsong, 1977; 1978; Tsong and Kinosita, 1985). In fact, permeability changes can be regulated to fit the molecular dimensions of the drug to be loaded, so that escape (or entry) of larger substances is minimized. In particular this means that gross hemolysis with escape of hemoglobin and enzymes need not occur. Schwister and Deuticke (1985) induced pores (1-10 per cell) with high electric fields (3-20 kV/cm) and estimated pores sizes of 0.6-0.8 nm from movement of polyols and 0.8-1.9 nm from osmotic protection studies with polyethylene glycols. In another study (Sowers, 1986), the efflux of fluorescent markers was used to demonstrate the existence of pores with a diameter at least 17nm which formed primarily in the part of the cell facing the cathode. When the dielectric procedure was employed using conditions that did not produce hemolysis, the morphology of the cells and their survival characteristics upon subsequent infusion were excellent. More vigorous application of the dielectric field procedure however produces abnormal appearing cells (echinocytes, acanthocytes, stomatocytes) which were removed from the circulation rather quickly by spleen and liver.

FUSION WITH LIPOSOMES

Fusion with unilamellar liposomes containing entrapped substances has been used as a procedure for introducing drugs, for example IHP, into red cells. There is perhaps relatively little reason now to use this procedure for introducing soluble small molecules into the red cell cytoplasm. However the procedure has considerable potential importance as a mechanism for introducing proteins into the red cell membrane. Most membrane proteins assume their active configuration in liposomal membranes if they are incorporated into the membrane at the time of formation of the liposome, whereas they do not spontaneously insert into either red cell or liposome membranes. If the liposome fuses with the red cell membrane, the contents of the liposome are introduced into the red cell cytoplasm. If mixing of liposomal lipids and proteins occurs, the red cell membrane would acquire new lipids and membrane proteins. Arvinte et al (1986) have reported that lysozyme covalent abound to the outer surface of liposomes induced fusion at acidic pH.

ENDOCYTOSIS

Endocytosis (and exocytosis) can be induced in red cells by certain membrane active drugs, or by incorporating ATP, Mg^{+2}, and Ca^{+2} into hemolyzed cells. Endocytosis results in the formation of relatively large vacuoles in the cytoplasm containing fluid and whatever else was initially outside the red cell. In this way drugs, enzymes, DNA and particulate structures can be incorporated into the red cell, but in a vacuolar compartment. The endocytotic vacuole itself can be lysed by osmotic shock if it is hyperosmotic with respect to the cell cytoplasm and the contents thus introduced into the cytoplasm. Choe et al (1985) have shown that the membrane lipid distribution is symmetrical for both right-side-out vesicles prepared by incorporation of a calcium ionophore and 1mM Ca^{+2} and inside-out vesicles prepared by endocytosis induced by low ionic strength at high pH followed by release from the cytoplasm by lysis of the plasma membrane. This suggests that endocytotic vacuoles in general, possibly including those which result from invasion by hemotrophic parasites, have symmetrical distributions of lipids in their membranes.

RESEALING

When red cells are diluted into hypotonic medium, the turbidity disappears and the solution becomes clear. When salt is added back, the solution again becomes immediately turbid, showing that resealing has occurred. However the red cells are not necessarily impermeable to small molecules and especially may not have recovered their normal ionic transport properties. For this reason the cells are incubated, usually at 37°C for 30-60 min or longer. Pores formed by electric pulses also can be resealed by incubation at 37°C; for example, pores that admitted Rb^+ were resealed in 40 min but pores that admitted sucrose took 20 h under similar conditions (Serpersu, Kinosita and Tsong, 1985). The properties of resealed ghosts produced by different lysis procedures have been compared (Jausel-Huysken and Dleuticke. 1981). Formation of stable pores may require the presence of membrane proteins (Minetti and Ceccarini, 1982). Optimum conditions both for osmotic lysis and for resealing vary slightly from species to species and especially differences in the stability of the resealed red cells are observed. Papers continue to be published on the conditions for preparing resealed erythrocytes in various species, most recently in sheep (DeLoach et al, 1981a; 1985b; DeLoach and Droleskey, 1986).

PROPERTIES OF RESEALED ERYTHROCYTES

Erythrocytes carefully loaded by either the electrical procedure or the hypo-osmotic dialysis procedure have been shown to retain their normal life span when reintroduced into the circulation. This seems adequate evidence that the properties of the cells are essentially unchanged by the loading process.

Many forms of damage to resealed red cells can result in their premature uptake by erythrophagocytic cells. This damage can fall into several categories, for example: 1. loss of normal concentrations of cytoplasmic components such as enzymes or ATP; 2. attachment of antibodies to the red cell; 3. chemical damage to the membrane, for example by crosslinking agents; and 4. alterations in membrane lipids.

The normal asymmetric distribution of phospholipids in the red cell membrane (outer predominantly phosphatidylcholine and sphingomyelin; inner predominantly phosphatidylethanolamine and phosphatidylserine) is lost if 1mM Ca^{+2} is present during lysis and resealing or if the lysing volume is greater than four times the red cell volume. The significance of the Ca^{+2} result is hard to assess, in view of the multiplicity of calcium-related effects and the lysing volume effect may be related to the requirement of ATP (Seigneuret and Devaux, 1984) in phospholipid translocation which presumably indicates active, enzyme mediated translocation. The observation that asymmetry is maintained if lysed with a 1:1 ratio and lost if lysed with a 4:1 ratio provides an easy way to prepare damaged cells which are more readily phagocytosed (McEvoy et al, 1986). The dye MC540 provides an especially simple procedure to monitor asymmetry. In saline MC540 binds uniformly to red cells without penetrating the membrane, but in the presence of serum, binding is substantially reduced. Thus it appears that most of the MC540 is not bound tightly. In contrast however, in the presence of serum, about three to four times as much binds tightly to cells and vesicles (Choe et al, 1985) which have lost membrane asymmetry.

The signal(s) for phagocytosis of red cells are not well understood at the molecular level. McEvoy et al (1986) suggest that the loss of membrane asymmetry alters surface hydrophobicity which in turn may be related phagocytosis. Kay (1985) on the other hand has presented evidence that a new antigen, referred to as senescent cell antigen, appears on the surface of senescent and damaged cells and is recognized by an IgG autoantibody which initiates removal by macrophages. Resealed red cells have been shown to be effectively removed from circulation by attachment of antibodies (Beutler et al, 1977a; Eichler et al, 1986).

Erythrophagocytosis permits effective delivery of substances entrapped in red cells to macrophages. Carrier red cells therefore could have important uses, for example, in the treatment of infectious diseases in which the infecting organisms are first uptaken into a macrophages and then escape being killed, either because of their resistance to killing in lysosomes, because they prevent fusion with lysosomes or because they escape from the phagocytic vacuole before fusion with lysosomes. In addition, certain other diseases such as Gaucher disease, caused by the accumulation of undigested glucocerebroside derived from circulating red and white cells in lysosomes of phagocytes, could potentially be treated by delivery of drugs or enzymes to phagocytes.

In general, red cells are confined to the circulation and so have limited potential for delivery to parenchymal cells of organs. Red cells may gain entry to lymphatics and they may bind and escape from the circulation at sites of inflammation and so possibly might deliver drugs to interstitial tissue at these sites. Samokhin et al (1984) have suggested that red cells bearing antibody to collagen could be targeted to injured sites which have exposed collagen and have shown (Smirnov et al, 1986) that conjugates of liposomes or red cells are bound by endothelium-free zones of arterial segments and therefore might have some potential in carrier-directed targeting to thrombosis-prone sites.

RED CELLS AS DRUG CARRIERS

DRUGS DELIVERED TO PHAGOCYTES

Intracellular parasites gain entry and grow within cells and are thereby protected from antibodies and other defense mechanisms which are external to cells. In some cases (e.g. M. tuberculosis), the life cycle of the parasite is restricted to phagocytes. In other cases they invade phagocytes such as neutrophils, monocytes, and macrophages and also some other cell types. For example, the hemoflagellate Trypanosoma cruzi, the agent of Chagas disease, invades heart muscle cells, endothelial cells, and macrophages. The motile tropomastigotes invade non-phagocytic cells whereas epimastogotes and amastigotes appear to be infective primarily for macrophages. No known drug will cure T. cruzi infections. The related parasite, Leishmania, infects reticuloendothelial cells and in some cases also other cell types.

The use of liposomes to deliver therapeutic drugs to phagocytic cells has been proposed and extensively investigated, especially for the treatment of leishmaniasis. Erythrocytes also have substantial potential for drug delivery to phagocytic cells, but have not been studied as extensively. It is possible that the combined use of liposomes and erythrocytes could be more successful that the use of either alone. For example, liposomes could be used to provide a bolus of drugs taken up very quickly and erythrocytes could provide a more prolonged delivery of drugs, potentially extending to 120 days. In addition, the target population of phagocytes for liposomes and red cells might not be identical, so that drugs might be delivered to a wider population of phagocytes with combined chemotherapy. The amount of drug that can be encapsulated in erythrocytes is much larger than can be incorporated in the smaller liposomes and so an erythrocyte is capable of delivering a massive drug dose to a target phagocyte.

Formycin B, an inosine analog, is an antileishmanial agent, which is phosphorylated to formycin B monophosphate, converted to formycin A monophosphate, an AMP analog, and then to the triphosphate which is incorporated into RNA by leishmania. Formycin A is efficiently taken up, phosphorlyated to the triphosphate and concentrated by red cells, presumably by the same transport system that transports adenosine and inosine. While phosphorylated, the drug does not escape from the red cells; the observed leakage of 1-5% of the drug per hour when the red cells were incubated may reflect hydrolysis to the nucleoside which could readily escape. In an in vitro system using IgG-coated red cells and leishmania-infected macrophages, about 80% of the macrophages took up red cells and about 80% of the amastigotes were eliminated, suggesting that the drug was completely effective in all macrophage which took up the red cells (Berman and Gallalee, 1985). By comparison, the free drug present at 15-fold greater concentrations than the dose at which it is 50% effective, permitted about 10% survival. Thus even if there is a fraction of amastigotes resistant for reasons other than failure of the macrophage to take up red cells, the encapsulated drug would appear to be no worse than the free drug in this regard. Moreover, if there is some life cycle-dependent resistance, the ability of red cells to maintain continuous delivery over some period of time could be expected to be more advantageous than a single dose of free drug or even multiple infusions of free drug. The encapsulated drug appeared to be effective at concentrations more than 40-fold lower than the free drug. The importance of this lies in the fact that side-effects of the drug, at least those side effects due to the action of the drug in non-phagocytic cells, can be avoided by direct delivery of the drug to phagocytes. Even if all the drug were to escape eventually from the phagocytes and be transported to other tissues, the dose would be much lower. In addition, it is possible that the drug would be broken down in the phagocytes to compounds which were non-toxic and inactive.

This system has the advantage of not requiring loading by any procedure other than normal transport mechanisms. Since transport is complete within two hours, it would be perfectly feasible to merely add the drug to recently drawn autologous red cells which could then be damaged by some relatively easy procedure (heat, sulfhydryl-reactive agents, antibodies, membrane-damaging chemicals, etc) and then reinfused. This potential simplicity would facilitate its use in moderately unsophisticated medical systems.

DRUGS IN CIRCULATING RED CELLS

A principle attraction of red cells for drug delivery is the potential for prolonged delivery in disease states requiring long term drug delivery. For example, Pitt et al (1983a) have used the adjutant-induced arthritis in rats as a model for the use of red cell encapsulated corticosteroids. Cortisol-21-phosphate and prednisolone-21-phosphate were used because they are water-soluble. It was suggested that the drugs are retained in the red cells due to their electrical charge until hydrolysed by an acid phosphatase in the red cell membrane (Pitt et al, 1983b) and then they escape through the membrane. The rats received red cells on day 0 after adjuvant injection and again on day 10 and were compared with rats receiving daily injection of free cortisol-21-phosphate. For both corticosteroid esters, the red cell encapsulated form proved superior direct injection of the free drugs. Since the half life of intravenous cortisol is about 2 h in man, it is likely that the intravenous free drug was present in therapeutic amounts for only part of the day whereas the red cell encapsulated drug presumably was continuously released.

Among the drugs which have been incorporated into red cells are: actinomycin D and bleomycin (Lynch et al, 1980), cytosine arabinoside (Ara C) and Ara CTP (DeLoach, 1982a; Mishra et al, 1981; DeLoach and Barton, 1982b), daunomycin (Kitao et al, 1978; Kitao and Hattori, 1980), diflubenzuron (DeLoach et al, 1980), homidium bromide imidocarb (DeLoach et al, 1981b), methotrexate (Tyrrell and Ryman, 1976; Zimmerman et al, 1978), and tetracycline.

DESFERRIOXAMINE

The potential use of red cell entrapped desferrioxamine as an iron chelator has been extensively investigated (Green et al, 1981). Slow infusion of free desferrioxamine has several disadvantages including the requirement for a more-or-less continual intravenous infusion, the difficulty of removing iron by competing with iron transport proteins, the relatively small amount of body iron that is circulating at any given time, and the relative inability of desferrioxamine to directly mobilize iron from cells. The first suggestion was to entrap it in ghosts (which would need to be relatively free of hemoglobin, so that the resealed red cells would not add to the iron stores), which would be taken up by the reticuloendothelial system. Much of the body stores of iron are present in macrophages and so it was hoped that these deposits at least could be directly mobilized by uptaken desferrioxmine.

More recently the focus of research on entrapped desferrioxamine has been to entrap it within cells that are intended to circulate for substantial periods of time. That is, if red cells given to alleviate a state of chronic anaemia contain enough entrapped desferrioxamine to ultimately chelate as much iron as is in the cells given, the transfusion should not result in a net addition to the iron stores. Thus there might be no net removal of iron from the body, but on the other hand it should be possible to transfuse anaemic patients indefinitely without creating an iron overload state. This principle problem to be overcome is to ensure that the drug-loaded red cells have an adequate life-span.

MODULATION OF OXYGEN DELIVERY

One of the most promising and potentially useful applications of red cell entrapment relates to the modulation of oxygen delivery, the primary task of the red cell. It was observed long ago that purified hemoglobin has a much higher affinity for oxygen than hemoglobin within red cells and it was later shown that small phosphorylated compounds, primarily 2,3 DPG in humans and most mammals, allosterically reduce the affinity of hemoglobin for oxygen to values which are consistent with the observed oxygen tensions in tissues and microcapillaries. Were 2,3 DPG not present, adequate oxygenation of the tissues would not be possible. Fetal hemoglobin has a reduced affinity for DPG and hence a higher oxygen affinity, enabling fetal hemoglobin to capture oxygen from adult hemoglobin.

Coates (1975) has reviewed the evolution of hemoglobin and of ATP, IP5 and DPG as allosteric modifiers of oxygen affinity. Summarizing briefly, hemoglobin first evolved from a monomeric molecule with no cooperative effects (Hagfish) to a monomeric molecule when oxygenated which formed dimers or tetramers on deoxygenation (Lamprey) and later to tetramers in equilibrium with dimers in which oxygenation shifted the equilibrium toward dimers (e.g. Squalus acanthias). ATP was probably the first regulator of oxygen affinity; in the eel Anguillaa anguilla the concentration of ATP decreases in response to hypoxia and the affinity for oxygen increases, which presumably increases the efficiency of extraction of oxygen from water. Reptiles also appear to use ATP. Birds, which evolved from reptiles and which have resting metabolic rates 5- to 10-fold higher, use inositol pentaphosphate (IP_5) which is much more effective than either ATP or DPG in lowering the oxygen affinity. Some mammals have substitutions in the beta chain which decrease the affinity for organic phosphates and so their hemoglobins are not regulated by ATP or DPG.

Ruiz-Ruano et al (1984) have shown that chick embryos red cells use DPG and mature red cells use IP5. DPG levels drop dramatically and IHP levels rise at the time of hatching due to replacement of embryonic cells with mature cells. DPG levels correlate closely with levels of DPG synthase. The mature cells contain high levels of phytase.

2,3 DPG is synthesized from 1,3 DPG in a side pathway from glycolysis and is broken down by a phosphatase to 3 PG (Koler, 1980). In erythrocytes, the major phosphorylated compound is 2,3 DPG. There is at least some regulation of the DPG concentration which increases in individuals at high altitudes or in some individuals with chronic lung disease. The higher concentrations of 2,3 DPG further decrease the affinity of hemoglobin for oxygen which results in better tissue oxygenation under conditions of relative tissue hypoxia. An increase in pH, caused for example by hyperventilation, leads to an increase in the level of DPG.

In stored red cells, the level of 2,3 DPG tends to decrease and so oxygen delivery by stored cells after transfusion is relatively poor. Although the level of DPG is restored within 24 hours, the drop in DPG concentration is undesirable since usually there is a compelling reason for desiring normal oxygen deliver immediately after transfusion. This has lead to a long series of studies to develop a storage media in which the drop in 2,3 DPG concentration is retarded or minimized. Addition of inosine or other purine nucleosides to the storage medium has been shown to be beneficial, presumably because phosphorolysis results in synthesis of ribose 5-phosphate which can be converted to glycolytic intermediates (glyceraldehyde 3-phosphate and fructose

6-phosphate by transaldolase, transketolase, and phosphopentose isomerase) without a requirement for ATP hydrolysis.

The fundamental control of DPG levels presumably is its sequestration by hemoglobin, accounting for the fact that DPG is equimolar with hemoglobin. Thus if there were hemoglobin which was not oxygenated and not saturated wit DPG, any DPG synthesized by mutase would be immediately bound and remove from the pool, thus leading to increased rates of synthesis. If there were an excess of unbound DPG, presumably it would be degraded rapidly by phosphata (A prediction of this argument is that DPG levels ought to be low in cells containing IHP).

In dogs, which lack a high oxygen affinity fetal hemoglobin, levels of 2,3 DPG rise dramatically after birth and the improved oxygen transport actually results in postnatal anemia. This control appears to be the result of a decrease in the activity of pyruvate kinase which leads to an increased concentration of phosphoenolpyruvate and 1,3 DPG (which are in equilibrium) and indirectly to an increase in the concentration of 2,3 DPG (Mueggler and Black, 1982).

It would be desirable to diminish the affinity of hemoglobin for oxygen after acute blood loss, surgery, or in cardiac or pulmonary conditions characterized by inadequate oxygen delivery or excessive demands on cardiac contractility. This could be potentially accomplished by a drug which acted to raise 2,3 DPG levels in red cells, but such a drug is presently not known. Potentially such a drug might exist, for example an unphosphorylated compound which could enter red cells and become phosphorylated, which would prevent it escape, and, by hypothesis, act to inhibit the action of 2,3 DPG phosphatase. Other possibilities are suggested by the observation of Grisolia and Tecson (1967) that a reversible transformation is induced by Hg^{+2} in phosphoglycerate mutase/phosphatases which contain SH groups so that the mutase becomes more active and the phosphatase less active. A less toxic, more specific agent might have the desired properties. Sasaki et al (1971) have reported that a series of compounds--phosphoglycolate, phosphohydroxypyruvate and phosphoenolpuruvate-- stimulated phosphatase activity and inhibited mutase activity--which unfortunately is the opposite of what would be desired. Black et al (1985) have discussed the DPG shunt in terms of a futile cycle, responsible in part for wasting the ATP of stored red cells and have shown that 2-phosphoglycolate, which activates both the mutase and phosphatase, increases the futile cycle activity. Clearly a compound which acted to increase mutase activity only would elevate DPG levels (but at the expense of storage time); however a compound which inhibited phosphatase activity would probably both increase DPG levels and increase the life time of stored red cells. As a less satisfactory alternative to this hypothetical drug, a patient could be transfused with red cells having diminished oxygen affinity.

POTENTIAL USES OF IHP-CONTAINING RED BLOOD CELLS

Red cells having diminished oxygen affinity have been prepared by entrappin inositol hexaphosphate in red cells. IHP is a powerful allosteric effector of oxygen affinity, just as is 2,3 DPG and certain other small phosphorylated molecules such as ATP. In birds, IP_5 and not 2,3 DPG is the natural regulator of oxygen affinity and both IP_5 and IHP are active with mammalian hemoglobins as well. Moreover both IP_5 and IHP are not metabolically active so that their action can persist indefinitely within the red cell, in distinction to 2,3 DPG which would be metabolically degraded relatively quickly.

Incorporation of IHP was initially accomplished by Nicolau and Gersonde

(1979) who fused red cells with small unilamellar liposomes containing IHP and subsequently by Teisseire et al (1984) who used hypotonic hemolysis. Fusion was essentially complete within two hours, as measured either by incorporation of radioactive cholesterol into the red cell membranes from the liposomes or by following the $pO_2(1/2)$ of the red cells. NMR spectroscopy demonstrated that the IHP was quantitatively bound to hemoglobin and was not degraded during 15 days of storage at $4°$, whereas 2,3 DPG reached very low levels within 48 hr. Gersonde and Nicolau (1979) showed that the p_{50} values of red cells could be shifted to values as high as 98 mmHg which is the value observed for free hemoglobin saturated with IHP. This value is probably higher than is desirable since the effective p_{O2} of arterial blood is about 100 mmHg. Thus stored blood containing IHP could be prepared to have the normal p_{50}, or it could be prepared to have an elevated p_{50}.

Cells containing IHP had an altered Bohr effect. As the pH drops in acidosis, the affinity of hemoglobin for oxygen normally diminishes, thereby enhancing oxygen release and potentially alleviating acidosis caused by inadequate oxygen delivery. Hemoglobin partially saturated with IHP (52% and 77%) had a much larger increase in p_{50} as the pH dropped from 7.7 to 7.3 than either fresh normal red cells or stored cells, as is observed for free hemoglobin. The effect of IHP on avian hemoglobin has also been studied by Vandecasserie et al, 1973.

IHP-containing red cells would clearly be superior to 2,3 DPG-depleted stored red cells for transfusion. Under what other physiological conditions would IHP-containing red cells be useful? One would not expect oxygen consumption to increase substantially even if the tissue oxygen tension were to increase since normally under resting conditions only about 25% of the oxygen present is extracted during passage through the microcirculation. Thus one might have predicted for example that a rise in tissue oxygen tension unassociated with any significant changes in cardiovascular physiology; or that cardiac output would decrease to maintain tissue oxygen tensions at their normal values. In piglets, the arterial P_{O2} increased from 82.5 to 97.5 mmHg, while the venous P_{O2} decreased very slightly from 36.7 to 35.1. A major decrease in cardiac stroke volume without any significant change in heart rate was found for the animals exchange transfused with IHP-containing red cells. Presumably physiological mechanisms for autoregulation of oxygen delivery and blood flow lead to a reduction in blood flow under conditions where oxygen delivery was facilitated. Arterial pressure was relatively unaltered, indicating that the mechanism of blood flow reduction was an increase in peripheral vascular resistance. It seems very likely that IHP-containing red cells may have substantial beneficial effects in a variety of medical conditions.

USE OF RESEALED RED CELLS IN TESTING DRUGS ACTIVE AGAINST HEMOTROPHIC PARASITES: REDUCED INVASION OF IHP-CONTAINING CELLS BY PLASMODIUM FALCIPARUM AND BABESIA MICROTI

Plasmodium spp are capable of entry and normal development in resealed red cells (Dluzewski et al, 1981, 1983; Olson, 1981; Olson and Kilejian, 1982). Bartonella bacilliformis can also enter resealed red cells by a process related to endocytosis (Benson et al, 1986) and in fact E. coli can be entrapped in vesicles within red cells after endocytosis induced by the entrapment of ATP, Ca^{+2} and Mg^{+2} in hypotonically lysed red cells (Tsang et al, 1982). Babesia and Bartonella both are eventually found in the red cell cytoplasm and so drugs entrapped in resealed red cells could act directly on these parasites. In addition, even for parasites still within vesicles, it is likely that cytoplasmic drugs might act on the parasites with greater efficiency than the same drugs given external to the red cells so that they must cross both the red cell plasma membrane and the membrane of the endocytotic vacuole.

The resealed red cell system therefore provides an experimental system for studying the effects on parasite development of drugs which might cross the red cell membrane poorly and also an experimental system for determining whether there might be any therapeutic advantage to entrapping drugs in red cells for the treatment of parasitic diseases. Initially we decided to test whether a Ca^{+2}-requiring mechanism is involved in the entry of red cell parasites. Neither EGTA nor calcein entrapped in red cells had any effect on entry or growth of Bartonella bacilliformis, Babesia microti or Plasmodium. Dluzewski et al, 1983c) also reported finding no effect of EGTA on entry of Plasmodium into resealed red cells whereas antibodies to cytoskeletal components did effectively inhibit invasion (Dluzewski et al, 1983a) and antibodies to transmembrane proteins, band 3 and glycophorin A, were shown to inhibit entry whether present externally or internally, when directed against cytoplasmic domains (Dluzewski et al, 1986). At the suggestion of Dr. C. Nicolau who observed that IHP binds Ca^{+2} tightly and does not have deleterious effects on red cells, we also determined the effect of entrapped IHP. Although IHP had no effect on the entry of B. bacilliformis, IHP produced a dramatic lowering of parasitemias for Babesia in vivo and both Babesia and Plasmodium in vitro.

·   Several assays were used to demonstrate this effect. In the first, mice were infected by intraperitoneal injection of 1 x 10^6 Babesia microti. Six days after infection, when the mice had parasitemias of about 60%, they were infused with fluorescent FITC-BSA-containing red cells containing or not containing IHP. At various times after infusion, the percent of red cells, fluorescent or non-fluorescent, containing parasites was determined. The infused cells survived well in the circulation and were readily infected by the parasites in the absence of intracellular IHP. For example, within 18-20 h after infusion, about 76% of the non-fluorescent red cells contained parasites and about 60% of the fluorescent non-IHP-containing cells also contained parasites. This indicates that there is no significant barrier to infection of the infused red cells. In the animals receiving IHP-containing red cells, about 67% of the non-fluorescent red cells contained parasites, but only 10% of the fluorescent cells contained parasites.

This effect could be demonstrated in vitro as well, with both Babesia and Plasmodium. In experiments arranged for us by Dr. Gerald McLaughlin and performed by Dr. Phillipe Deloren at the Malaria Branch, Centers for Disease Control, Atlanta, GA, three strains of Plasmodium falciparum was able to infect fluorescent red cells sent by express mail from our laboratory, but not fluorescent red cells containing IHP. In the former case, 2.35% of the non-fluorescent red cells and 0.77% of the fluorescent red cells contained parasites. In the latter, 1.35% of the non-fluorescent red cells contained parasites and 0.00% of the fluorescent, IHP-containing red cells. In an effort to find parasites in fluorescent cells, the slides were extensively scanned (5-10 times as many cells), but no parasites in fluorescent cells containing IHP were ever seen in any of the experiments.

The in vitro cultivation of Babesia microti is less satisfactory than that for Plasmodium, but the results were qualitatively similar. The mouse red cells seem to lyse readily under the incubation conditions and although human red cells are stable and can be infected by the Babesia, the infection seems to terminate before extensive multiplication of the parasites. This may explain why human Babesiosis is relatively rare. In vitro the IHP-containing red cells appeared about three-fold less susceptible to Babesia, as compared with five-fold less susceptible in vivo.

The mechanism by which this effect is exerted is not clear at present. Simple exposure of red cell to IHP without entrapment is without effect. It is possible that IHP exerts a direct toxic effect on the parasites, or alternately that it affects the red cell in some way. In the latter case, it could for example act by altering the oxygen tension in the cell by decreasing the affinity of the hemoglobin for oxygen; or its effect could be related to the role of phosphatidylinositol phospholipids and inositol tris-phosphate as second messengers.

It remains to be determined whether IHP-containing red blood cells could be life-saving in B. microti infection. Although babesiosis is rare in humans, there is no effective drug therapy. Malaria is imperfectly controlled in humans and IHP-loaded red cells may provide a useful adjunct to current therapeutic regimens. The use of IHP-loaded human red blood cells in surgery or in certain cardiac or pulmonary conditions already seems imminent, so that blood banks prepared to load IHP for increased oxygen delivery could also provide IHP-loaded cells for use in infections by Babesia or Plasmodium.

RED CELLS CONTAINING EXOGENOUS PROTEINS AND ENZYMES

ASPARAGINASE

Experiments with entrappped asparaginase provide an instructive example of the potential use of red cells as carriers for enzymes and proteins. Asparaginase has anti-tumor activity because many types of malignant cells depend on exogenous asparagine. Intravenous injection of the free enzyme is effective, but has many complications. Updike et al (1976, 1983) entrapped E. coli asparaginase and demonstrated that plasma asparagine levels were greatly depressed for at least 20 days. It seemed likely that a low rate of intravascular hemolysis resulted in the slow release of asparaginase and consequent direct action of the enzyme of plasma asparagine.

More recently, the use of asparaginase-loaded red cells has been re-examined by Alpar and Lewis (1985) in mice and rats. Erythrocytes containing a large excess of enzyme (40 units = 40 umole substrate per min) were incubated with radioactive asparagine, the cells centrifuged, lysed, and the amount of asparagine and aspartic acid determined. As no radioactive aspartic acid was found in the supernatant after the red cells were removed, it was concluded that asparagine first entered the red cells and was then deaminated. Since very little asparagine was found in the red cells, it was concluded that the asparagine was rapidly converted to aspartate by the large excess of entrapped enzyme. Using those assumptions, the rate of entry of asparagine in this experiment was 1.2 umole/hr/0.2ml packed red cells or 100 umole/min/lcells. Unfortunately there was no direct determination of the rate of entry of asparagine into these cells. Redetermination of the rate of entry in human cells might be very desirable.

In rats, half the red cells survived 9-10 days and in mice, 14 days (C3H) or 23-24 days (Balb C) and the decrease in asparaginase activity was similar to the survival of the red cells, indicating that the enzyme fully retains its enzymatic activity for these periods of time. Plasma levels of asparagine dropped to undetectable levels very quickly and remained there for at least 14 days for entrapped enzyme, but returned to normal within 48-72 hr for injections of free enzyme.

The therapeutic action of asparaginase was assayed against the 6C3HED lymphoma tumor injected ip into mice, which killed untreated mice in 18-19 days. In one protocol, the mice were injected iv with asparaginase on the same day as the tumor cells and in the other, 8 days later. For each protocol, free enzyme was superior to no treatment and encapsulated enzyme was superior to free enzyme. Since free enzyme depleted plasma asparagine for only a two day period, the better results obtained by delayed injection apparently indicates that depletion of plasma asparagine on days 1 and 2 is less useful than on days 8 and 9. Possibly the fact that the tumor cells are given by ip injection is related to this. However it would appear that three of the mice injected on day a had about the same survival as the average for those injected on day 0 and two survived for at least twice as long ("cured").

Injection of the encapsulated enzyme on day 0 should deplete plasma levels of asparagine at least through day 14 and so, as might be expected, it was observed that encapsulated enzyme was superior to free enzyme injected at either of the two times reported. Unfortunately the day on which plasma asparagine levels began to become normal was not reported for the tumor-bearing animals and so it is not possible to know whether there was a correlation between life span and the length of time the animal was asparaginase-free. Since all of the animals given the encapsulated enzyme at 8 days survived for more than 60 days ("cured"), the presumption is that a second injection of the animals treated on day 0 might have produced equally good results. If, as seems likely, the most effective protocol would be one which maintained plasma asparagine levels at undetectable levels for the entire time, it would seem important to show that the animals which died when injected on day 0 could have been saved by a second injection of encapsulated enzyme. It would also be useful to know whether the tumor was or was not still asparagine-requiring in the animals which died.

ENZYME CYTOCHEMISTRY

Raap and Van Duijn (1981) used alkaline phosphatase loaded red cells to compare the activity of the entrapped enzyme by biochemical and cytochemical methods. They demonstrated that the activity of entrapped alkaline phosphatase could be quantitatively determined within fixed red cell ghosts. They suggest that enzyme-loaded red cells could be a useful model system for development of quantitative cytochemical applications. Previously, red cells loaded with alkaline phosphatase had been used as a cytochemical marker for red cells taken up phagocytically by bone marrow macrophages (DeLoach et al, 1979).

PROSPECTS FOR THE MODIFICATION OF THE RED CELL MEMBRANE

Learning how to modify the composition and properties of cell membranes, especially the plasma membrane, by incorporating new lipids and proteins, seems to me to be an important problem. For example a serious limitation on enzyme-loaded red cells is the requirement that the substrate first be transported into the red cell. In many cases this does not occur at all; in other cases the rate is too slow to be useful. Free enzymes injected into the circulation are removed rather quickly and even normal plasma proteins turn over within a few days. However red cell membrane proteins circulate with the same life span as the entire red cell and are not turned over because the red cell has no biodegrative or biosynthetic capability for proteins. If an enzyme, preferably a naturally occurring human enzyme, could be incorporated into the red cell membrane with its active site facing out, it probably would retain its full activity and hopefully would not decrease the life span of the red cell or stimulate auto-antibody formation.

It is possible to modify the lipid content of red cells. Cholesterol is readily added to or removed from red cell membranes. Swaney (1985) has shown that recombinant lipoproteins containing various phospholipids can remove substantial amounts of red cell cholesterol. Other lipids exchange as well (Brunner et al, 1983). For example fluorescent lipids have been incorporated into red cell membranes and these cells can circulate for extended periods of time without loss of fluorescence (Ihler and Tosi, 1987). It has been known for some time that certain blood group antigens, which are linked to lipids, can be directly transferred from one red cell to another. This means that it is possible to modify both the carbohydrate and lipid composition of red cell membranes.

Insertion of non-erythrocyte membrane proteins into red cells is more difficult, but has been accomplished by transfer from liposomes, fusion of red cells with liposomes and other ways. Red cell non-glycosylated membrane proteins, especially acetylcholinesterase, can be reversibly exchanged by phospholipid vesicles specifically from the outer monolayer (Bouma et al, 1977), along with cholesterol, sphingomyelin and phosphatidylcholine. Acetylcholinesterase, a non-spanning intrinsic protein, is readily detached from red cell membranes by sonication, or incubation with melittin (Maulet et al, 1984), a protein which binds to hydrophobic surfaces and which becomes transbilayer when a voltage is applied across a lipid bilayer (Kempf et al, 1982). Incubation of red cells with small phospholipid vesicles resulted in the transfer of at least four red cell outer surface proteins to the vesicles, including acetylcholinesterase (Bouma et al, 1977), indicating that at least some proteins, presumably those not having highly charged cytoplasmic domains or those not engaged in cytoplasmic interactions, can be removed from the external face of the membrane. In the red cell membrane, acetylcholinesterase is inhibited by butanol or ketamine, but the soluble enzyme is not inhibited. Addition of liposomes resulted in the enzyme becoming inhibitable, suggesting that it has spontaneously been inserted into the red cell membrane. (Zolese et al, 1979). Gamma-glutamyl transpeptidase has also been transferred in a functional state (Sikka et al, 1982). Cytochrome oxidase has been incorporated into liposomes and fused with red cells by incubation in the presence of 20mM $CaCl_2$ (Gad et al, 1979). Only vesicles containing more than 50mol% PE became associated with red cells.

I believe that it may be possible to insert new proteins linked to a universal membrane insertion polypeptide by constructing hybrid proteins, either biochemically by cross-linking reactions or genetically by gene fusion, which have the property of spontaneously inserting into membranes. At present I envision this hybrid protein to contain an insertion cassette of amino acids derived from a protein like cytochrome b_5 which is known under certain circumstances to insert spontaneously into membranes. The insertion cassette would then be chemically or genetically linked to an enzyme, antibody, transport protein, or other interesting polypeptide. A signal sequence apparently would not fulfill these requirements, although these sequences can form amphiphillic helices and interact with membranes (Roise et al, 1986; von Heijne, G, 1986). If the insertion cassette enables the hybrid protein to insert into the red cell membrane in a stable form, modification of the external aspect and perhaps the properties of the membrane of the red cell would result. For example, perhaps the ability of red cells to transport specific compounds could be substantially increased, permitting these compounds to be degraded or sequestered within red cells.

As another example, red cells carrying antibodies on their membranes might function effectively as part of the immune system. For example, many viruses and bacteria bind to red cells, as reflected for example by hemagglutination reactions employed in clinical chemistry, and this binding may facilitate removal of the bound organisms. Siegel et al (1981) have suggested that red cells ordinarily have an important immune function which is not widely appreciated. Re cells adhere to antigen-antibody-complement complexes through C3b or C4b surfac receptors and calculations suggest that over 95% of the C3b receptors in the vascular system are on red cells. Circulating immune complexes are more than 500 fold more likely to encounter red cells than white cells. Red cells also adhere to T cells, suggesting that red cells may bring antigens to T cells. Immune complexes bound to red cells trigger the release of interleukin-1 from human monocytes (Chou et al, 1985). Red cells with antibodies added to their membrane might function in this fashion and, potentially, the antibody would have a life span equal to that of the red cell. Hematogenous spread of microorganisms could be restricted by high concentrations of antibody carried on the surface of red cells. Antibodies to human immunodeficiency virus might be a potential application.

SPONTANEOUS INSERTION OF PROTEINS INTO MEMBRANES

The mechanism of protein localization involves transport of proteins through membranes and also insertion of proteins into specific membranes. Integral membrane proteins in some cases are known to lack signal sequences and to insert spontaneously into their target membranes in vivo; it is clear that this process is in general quite distinct from the process by which proteins are translocated for secretion or are exported to the periplasmic space or the outer membrane of gram negative bacteria. In vitro, most integral membrane proteins can be incorporated into artificial membranes in a functional configuration; for example they can be incorporated into liposomes made by mixing the protein and phospholipid in cholate and then removing the detergent by dialysis. Presumably the protein and phospholipid associate in such a way that when the lipid bilayer forms, the proteins are part of the bilayer and are folded in the correct configuration for them to be functional. (Although they may be oriented in either configuration, often most of the protein molecules have the correct external/internal orientation as well.) Most integral membrane proteins however do not spontaneously insert in vitro either into liposomes or into plasma membranes (although they could in principle be inserted into plasma membranes by fusion of protein-bearing liposomes).

Cytochrome b_5 is an integral membrane protein which spans the membrane and is the best understood example of a protein which spontaneously inserts into membranes. In vivo it is synthesized on cytoplasmic ribosomes and does not require a signal recognition particle; as far as is known it inserts spontaneously into microsomal membranes. It probably inserts into the membrane post-translationally (e.g. it is synthesized on free polysomes and SRP (signal recognition particle) is not required for insertion (Anderson et al, 1983). Cytochrome b_5 has an amino terminal hydrophilic region and a carboxyl terminal hydrophobic region; the hydrophobic region is entirely responsible for anchoring the peptide in the membrane. Like most membrane proteins, cytochrome b_5 can be incorporated into liposomes during formation of the vesicles using the detergent dialysis or other procedures. When so incorporated, cytochrome b_5 does not readily dissociate or exchange between liposomes (Poensgen et al, 1980), indicating that it is anchored in the membrane and is referred to as the non-transferable or "tight" binding form.

The significant property of cytochrome b_5 is that it inserts spontaneously into liposomes. In some cases, it inserts in a form that permits transfer to other liposomes. Probably in this case it is inserted into the outer leaflet of the bilayer only. In other cases cytochrome b_5 inserts in a non-transferable form. In this case, it probably crosses the membrane with the N-terminus outside and the C-terminus inside.

When cytochrome b_5 inserts spontaneously into dimyristyl phosphatidylcholine liposomes or into microsomes, it is non-transferable. However, when it inserts into phosphatidylcholine (PC) liposomes, it is transferrable, indicating that in this case it is not anchored firmly in the membrane (referred to as "loose" binding) (Enoch et al, 1979). Carboxypeptidase Y releases cytochrome b_5 from PC liposomes, but not from microsomes or dimyristyl PC liposomes, indicating that the carboxyl terminus is outside in the PC liposomes and probably inside in dimysteryl PC liposomes. A simple model for the non-transferable configuration is that the protein is anchored in the membrane by the presence of charged residues or other hydrophilic residues on both sides of the membrane, whereas in the transferable configuration the hydrophobic segment of the protein does not span the bilayer.

Dailey and Strittmatter however had previously presented an alternative structure which has the N- and C-termini on the external face in both the transferable and non-transferable configurations. In their model for the non-transferable configuration, the membrane is spanned by an antiparallel beta-sheet in which residues 98-102 are bonded to residues 126-130. Residues 103-112 are present in a beta-turn and residues 115-125 are present in a largely non-polar alpha helix. Dailey and Strittmatter (1981) argued that the C-terminus was outside from the rate of ionization of C-proximal tyrosyl residues and from their reactivity with diazotized sulfanilic acid.

Takagaki et al. (1983a; b) have used photoaffinity labeling with membrane phospholipids to demonstrate cross-linking extending from ser104 to met130 for the transferable form. In their model, the 11 amino acid region from ser104 to ile114 forms an alpha helix confined to the outer leaflet of the bilayer. Pro115 breaks the helix and a second helix from ala116 to tyr126, also confined to the outer leaflet then exits from the bilayer so that both the N- and C-termini are on the external face in the transferable form. Studies on the non-transferable configuration using phospholipids located preferentially in the outer leaflet labeled with one radiolabel and phospholipids in the inner leaflet labeled with an alternate radiolabel, indicated that the COO^- terminus was located internally. They argued that the protein crosses the bilayer in a continuous helix that would be more stable than the loop-back configuration because a greater number of hydrogen bonds can form. Supporting this model is the recent determination (Everett, 1986) that the fluorescent trp is located 0.7 nm below the outer surface of sonicated vesicles (i.e. that it is in the outer leaflet of the membrane).

When cytochrome b_5 inserts spontaneously into small PC liposomes, most of the protein exchanges readily; however a certain fraction (about 10-20%) inserts into large liposomes in the non-transferable configuration. Thus spontaneous insertion into both tight and loose configurations is possible. (Greenhut et al, 1986) show that when cytochrome b_5 equilibrates, it prefers small liposomes (SUV) over large liposomes (LUV) by a factor of 200. This apparent dependence on curvature clearly must be taken in account in models for cytochrome b_5 insertion. Since prolonged incubation of the LUVs does not convert more of the cytochrome b_5 to the non-transferable configuration, the molecular basis for this phenomenon is unclear. It has been speculated that the cytochrome b_5 which inserts in the tight configuration may be in a special (unknown) configuration.

In this connection, it is noteworthy that cytochrome b_5 exists in solution as a oligomer having an apparent molecular weight of 122,000 (which depolymerizes in cholate) while the hydrophilic peptide prepared with trypsin does not aggregate. Only in the monomeric state does cytochrome b_5 insert in the tight configuration (Christiansen and Carlsen, 1985). Cytochrome b_5 associates with NADH-cytochrome b_5 reductase and the hydrophobic segment is also important in this interaction. Complexes of the mixture are larger in size than either separately (Okuda et al, 1972). Triton can be used to permit these two proteins to form a functionally competent complex. Cytochrome b_5 must be added at dilute concentrations in reconstitution of cytochrome P-450 systems, presumably because the aggregated b_5 forms mixed aggregates only poorly with the other components (Gorsky and Coon, 1986). Cytochrome b_5 spontaneously inserts into microsomes in the tight configuration (Bendzko, 1982) and the amount can be increased more than 20-fold above the normal (biological) level, although under similar conditions cytochrome b_5 inserts into lipid vesicles made of microsomal lipid in the loose form. Clearly there is some difference between the microsomes and liposomes that is not understood, possibly the presence of a binding protein such as NADPH-cytochrome b_5 reductase or cytochrome P-450 (Tamburini and Schenkman, 1986) present in the microsomes that facilitates spontaneous insertion (although to rule out this possibility the microsomes were preincubated with trypsin).

The relationship between tight binding (i.e. binding of proteins in a transmembrane configuration) and protein translocation is not clear. In some well studied cases, translocation is linked both to the presence of a transmembrane potential gradient and to ATP (Geller et al, 1986) and in others to ATP only (Waters and Blobel, 1986). This complication seems to be absent with cytochrome b_5.

I believe that a short segment of the protein (the hydrophobic peptide) located at the C-terminus might function as an insertion cassette if linked to other proteins, genetically or biochemically. DNA corresponding to the hydrophobic peptide can be readily synthesized, cloned, and probably expressed, enabling the hydrophobic peptide to be purified. Fusion proteins could be synthesized biochemically by chemical crosslinking, for example using MBS (N-(m-maleimidobenzoyloxy)succinimide) as a crosslinking agent (Kitagawa et al., 1983). This crosslinker acylates amino groups under mild aqueous conditions. There are two amino groups in the hydrophobic peptide, the N-terminus (leu in rabbit and ile in equine, porcine, and bovine) and lys_3 (in all hydrophobic peptides) and linkage to either would be acceptable (more interior locations would of course probably disrupt insertion into membrane). The hydrophobic peptide-MBS conjugate may then be reacted with a free thiol group of the protein to be conjugated, either a naturally existing free thiol or one made by borohydride reduction. Because the hydrophobic peptide does not contain cys, this procedure would result in 1:1 linkage of the hydrophobic peptide and the protein with the hydrophobic peptide located at the carboxyl terminus (which would be more desirable initially).

It has been reported that cytochrome b_5 does not insert into the plasma membrane of mature red cells (Enomoto and Sato, 1977). This seems to be generally true in our hands as well, although we do find binding of fluorescent cytochrome b_5 to intact red cells in a curious punctate pattern. However cytochrome b_5 does insert into the inner aspect of red cell membranes and also into the outer aspect of resealed red cells prepared in certain ways or into the outer aspect of intact red cells which have been subjected to certain shape changes. Thus there seems to be some basis for the supposition that modification of the membranes of red cell membranes and the membranes of other cell types may be possible.

REFERENCES

Adriaenssens, K., Karcher, D., Marescau, B., Van Broeckhoven, C., Lowenthal, A., and Terheggen, H.C. (1984). 'Hyperargininemia: The Rat as a Model for the Human Disease and the Comparative Response to Enzyme Replacement Therapy with Free Arginase and Arginase-loaded Erythrocytes in vivo', Int. J. Biochem., 16: 779-786.

Alpar, H.O. and Lewis, D.A. (1985). 'Therapeutic Efficacy of Asparaginase Encapsulated in Intact Erythrocytes', Biochem. Pharm., 34: 257-261.

Anderson, D.J., Mostov, K.E. and Blobel, G. (1983). 'Mechanisms of Integration of de-novo-synthesized Polypeptides into Membranes: Signal Recognition Particle is Required for Integration into Microsomal Membranes of Calcium ATPase and of Lens MP26 but not of Cytochrome b5', Proc. Natl. Acad. Sci. USA, 80: 7250-7253.

Arvinte, T, Hildenbrad, K., Wahl, P. and Nicolau, C. (1986). 'Lysozyme-induced Fusion of Liposomes with Erythrocyte Ghosts at Acidic pH', Proc. Natl. Acad. Sci. USA, 83: 962-966.

Bendzko, P., Prehn, S., Pfeil, W. and Rapoport, T.A. (1982). 'Different Modes of Membrane Interactions of the Signal Sequence of Carp Preproinsulin and of the Insertion Sequence of Rabbit Cytochrome b5', Eur. J. Biochem., 123: 121-126.

Benson, L.A., Kar, S., McLaughlin, G. and Ihler, G.M. (1986). 'Entry of Bartonella bacilliformis into Erythrocytes', Infection and Immunity, 54: 347-353.

Berman, J.D. and Gallalee, J.V. (1985). 'Antileishmanial Activity of Human Red Blood Cells Containing Formycin A', J. Infect. Dis., 151: 698-703.

Beutler, E., Dale, G.L. and Kuhl, W. (1977a). 'Enzyme Replacement with Red Cell', New Eng. J. Med., 296: 942-943.

Beutler, E., Dale, G.L. and Kuhl, W. (1977b). 'Enzyme Replacement Therapy in Gaucher's Disease: Preliminary Clinical Trial of a New Enzyme Preparation', Proc. Natl. Acad. Sci. USA, 74: 4620-4623.

Beutler, E., Dale, G.L. and Kuhl, W. (1980). 'Replacement Therapy in Gaucher's Disease', Birth Defects: Original Article Series, 16: 369-381.

Billah, M.M., Finean, J.B., Coleman, R., and Michell, R.H. (1976). 'Preparation of Erythrocyte Ghosts by a Glycol-Induced Osmotic Lysis Under Isoionic Conditions', Biochem. Biophys. Acta., 433: 54-62.

Black, J.A., Acott, K.M. and Burton, L. (1985). 'A Futile Cycle in Erythrocyte Glycolysis', Biochem. Biophys. Acta., 810: 246-251.

Bouma, Drisland and Huestis. (1977). 'Selective Extraction of Membrane-Bound Proteins by Phospholipid Vesicles', J. Biol. Chem., 252: 6759-6763.

Brunner, J., Spiess, M., Aggeler, R., Huber, P., and Semenza, G. (1983). 'Hydrophobic Labeling of a Single Leaflet of the Human Erythrocyte Membrane', Biochemistry, 22: 3812-3820.

Choe, H.R., Williamson, P., Rubin, E., and Schlegel, R.A. (1985). 'Disruption of Phospholipid Asymmetry in Erythrocyte Vesicles Deficient in Spectrin', Cell Biol. Intnatl Rpts, 9: 597-605.

Chou, Y.K., Sherwood, T. and Virella, G. (1985). 'Erythrocyte-Bound Immune Complexes Trigger the Release of Interleukin-1 from Human Monocytes', Cell. Immunol., 91: 308-314.

Christiansen, K. and Carlsen, J. (1985). 'Reconstitution of Cytochrome b5 into Lipid Vesicles in a Form Which is Nonsusceptible to Attack by Carboxypeptidase Y', Biochem. Biophys. Acta., 815: 215-222.

Coates, M.L. (1975). 'Hemoglobin Function in the Vertebrates: An Evolutionary Model', J. Mol. Evol., 6: 285-307.

Dailey, H.A. and Strittmatter, P. (1981). 'Orientation of the Carboxyl and NH2 Termini of the Membrane-Binding Segment of Cytochrome b5 on the Sam Side of Phospholipid Bilayers', J. Biol. Chem., 256: 3951-3955.

DeLoach, J. and Ihler, G.M. (1977a). 'A Dialysis Procedure for Loading Erythrocytes With Enzymes and Lipids', Biochem. Biophys. Acta., 496: 136-145.

DeLoach, J., Peters, S.P., Pinkard, O., Glew, R. and Ihler, G. (1977b). 'Effect of Glutaraldehyde Treatment on Enzyme-Loaded Erythrocytes', Biochem. Biophys. Acta., 496: 507-515.

DeLoach, J., Widnell, C. and Ihler, G.M. (1979). 'Phagocytosis of Enzyme-Containing Carrier Erythrocytes by Macrophages', J. App. Biochem., 1: 95-103.

DeLoach, J., Harris, R. and Ihler, G.M. (1980). 'An Erythrocyte Encapsulator Dialyzer Used in Preparing Large Quantities of Erythrocyte Ghosts and Encapsulation of a Pesticide in Erythrocyte Ghosts', Anal. Biochem., 102: 220-227.

DeLoach, J.R., Barton, C., and Culler, K. (1981a). 'Preparation of Resealed Carrier Erythrocytes and in vivo Survival in Dogs', Am. J. Vet. Res., 42: 667-669.

DeLoach, J.R., Wagner, G.G., and Craig, T. (1981b). 'Imidocarb Dipropionate Encapsulation and Binding to Resealed Carrier Bovine Erythrocyte for Potential Babesiasis Chemotherapy', J. Appl. Biochem., 3: 254.

DeLoach, J.R. (1982a). 'Comparative Encapsulation of Cytosine Arabinoside Monophosphate in Human Canine Erythrocytes with in vivo Drug Efflux', J. Appl. Biochem., 4: 533.

DeLoach, J.R. and Barton, C. (1982b). 'Circulating Carrier Erythrocytes: Slow-Release Vehicle for an Antileukemic Drug, Cytosine Arabinoside', Am. J. Vet. Res., 43: 2210.

DeLoach, J.R. (1985a). 'Continuous-Flow Hollow-Fiber Dialysis System for Preparation of Bovine Carrier Erythrocytes', Am. J. Vet. Res., 46: 1089-1091.

DeLoach, J.R. (1985b). 'Effects of Storage Conditions on Bovine Carrier Erythrocytes', Am. J. Vet. Res., 46: 1092-1094.

DeLoach, J.R. and Droleskey, R. (1986). 'Preparation of Ovine Carrier Erythrocytes: Their Action and Survival', Comp. Biochem. Physiol., 84A: 441-445.

Desnick, R.J. (1982). 'Gaucher Disease: A Century of Delineation and Understanding', Prog. Clin. Biol. Res., 95: 1-30.

Dluzewski, A.R., Rangachari, K., Wilson, R.J.M., and Gratzer, W.B. (1981). 'Entry of Malaria Parasites into Resealed Ghosts of Human and Simian Erythrocytes', Brit. J. Hematology, 49: 97-101.

Dluzewski, A.R., Rangachari, K., Gratzer, W.B., and Wilson, R.J.M. (1983a). 'Inhibition of Malarial Invasion of Red Cells by Chemical and Immunochemical Linking of Spectrin Molecules', Brit. J. Haematol., 55: 629-637.

Dluzewski, A.R., Rangachari, K., Wilson, R.J.M., Gratzer, W.B. (1983b). 'Properties of Red Cell Ghost Preparations Susceptible to Invasion by Malaria Parasites', Parasitol., 87: 429-438.

Dluzewski, A.R., Rangachari, K., Wilson, R.J., Gratzer, W.B. (1983c). 'A Cytoplasmic Requirement of Red Cells for Invasion by Malarial Parasites', Mol. Biochem. Parasitol., 9: 145-160.

Dluzewski, A.R., Rangachari, K., Tanner, M.J.A., Anstee, D.J., Wilson, R.J.M., and Gratzer, W.B. (1986). 'Inhibition of Malarial Invasion by Intracellular Antibodies Against Intrinsic Membrane Proteins in the Red Cell', Parasitol., 93: 427-431.

Doxsey, S.J., Sambrook, J., Helenius, A., and White, J. (1985). 'An Efficient Method for Introducing Macromolecules into Living Cells', J. Cell Biol., 101: 19-27.

Drant, S., Montestruque, S., Bradley, G., Spira, A., and Bramhall, J. (1986). 'The Use of DNA-Intercalating Dye to Monitor Cell Fusion and Microinjection', J. Biochem. Biophys. Methods., 12: 253-264.

Eichler, H.G., Raffesber, W., Gasic, S., Korn, A. and Bauer, K. (1985). 'Release of Vitamin B_{12} from Carrier Erythrocytes in vitro', Res. Exp. Med., 185: 341-344.

Eichler. H.G., Rameis. H., Bauer, K., Korn, A., Bacher, S., and Gasic, S. (1986). 'Survival of Gentamicin-Loaded Carrier Erythrocytes in Healthy Human Volunteers', Eur. J. Clin. Invest., 16: 39-42.

Enoch, H.G., Fleming, P.J. and Strittmatter, P. (1979). 'The Binding of Cytochrome b5 to Phospholipid Vesicles and Biological Membranes. Effect of Orientation on Intermembrane Transfer and Digestion by Carboxypeptidase Y', J. Biol. Chem., 254: 6483-6487.

Enomoto, K. and Sato, R. (1977). 'Asymmetric Binding of Cytochrome b5 to the Membrane of Human Erythrocyte Ghosts', Biochem. Biophys. Acta., 466: 136-147.

Everett, J., Zlotnick, A., Tennyson, J. and Holloway, P.W. (1986). 'Fluorescence Quenching of Cytochrome b5 in Vesicles With an Asymmetric Transbilayer Distribution of Brominated Phosphatidylcholine', J. Biol. Chem., 261: 6725-6729.

Franco, R.S., Wagner, K., Weiner, M., and Martelo, O.J. (1984). 'Preparation of Low-Affinity Red Cells with Dimethyl Sulfoxide-Mediated Inositol Hexaphosphate Incorporation: Hemoglobin and ATP Recovery Using a Continuous-Flow Method', Am. J. Hematol., 17: 393-400.

Gad, A.E., Broza, R. and Eytan, G.D. (1979). 'Fusion of Cells and Proteoliposomes: Incorporation of Beef Heart Cytochrome Oxidase into Rabbit Erythrocytes', FEBS Lett., 102: 230-234.

Geller, B.L., Movva, N.R., and Wickner, W. (1986). 'Both ATP and the Electrochemical Potential are Required for Optimal Assembly of Pro-OmpA into E. coli Inner Membrane Vesicles', Proc. Natl. Acad. Sci. USA, 83: 4219-4222.

Gersonde, K. and Nicolau, C. (1979). 'Incorporation of Inositol into Intact Red Blood Cells. II Enhancement of Gas Transport in Inositol Hexaphosphate-Loaded Red Blood Cells', Naturwissenschaften., 66: 567-570.

Gisolia, S. and Tecson, J. (1967). 'Mercury-Induced Reversible Increase in 2,3-Diphosphoglycerate Phosphatase and Concomitant Decrease in Mutase Activity of animal Phosphoglycerate', Biochem. Biophys. Acta., 132: 56-67.

Godfrey, W., Doe, B. and Wofsy, L. (1983). 'Immunospecific Vesicle Targeting Facilitates Microinjection into Lymphocytes', Proc. Natl. Acad. Sci. USA, 80: 2267-2271.

Gorsky, L.D. and Coon, M.J. (1986). 'Effects of Conditions for Reconstitution With Cytochrome b5 on the Formation of Products in Cytochrom P-450-Catalyzed Reactions', Drug Metab. Dispos., 14: 89-96.

Green, R., Miller, J. and Crosby, W. (1981). 'Enhancement of Iron Chelation by Desferrioxamine Entrapped in Red Blood Cell Ghosts', Blood, 57: 866-872.

Greenhut, S.F., Bourgeois, V.R. and Roseman, M.A. (1986). 'Distribution of Cytochrome b5 Between Small and Large Unilamellar Phospholipid Vesicles,' J. Biol. Chem., 261: 3670-3675.

Humphreys, J. and Ihler, G.M. (1982). 'Encapsulation of Drugs, Enzymes and DNA within Human and Mouse Erythrocytes', in Alfred Benzon Symposium 1 Optimization of Drug Delivery (Bundgaard, H. et al., eds.), Munksgaard, Copenhagen.

Ihler, G.M., Glew, R.H. and Schnure, F.W. (1973). 'Enzyme Loading of Erythrocytes', Proc. Natl. Acad. Sci. USA, 70: 2663-2666.

Ihler, G.M., Lantzy, A., Purpura, J., and Glew, R. (1975). 'Enzymatic Degradation of Uric Acid by Uricase-Loaded Human and Blood Cells', J. Clin. Inv., 56: 595-602.

Ihler, G.M. and Glew, R. (1977). 'Enzyme-Loaded Erythrocytes', in Biomedical Applications of Immobilized Enzymes and Proteins,.(T.M.S. Chang, ed.) Plenum Publishing Corp.

Ihler, G.M. (1983). 'Erythrocyte Carriers', Pharmac. Ther., 20: 151-169.

Ihler, G.M. and Tsang H.C. (1985). 'Erythrocyte Carriers', CRC Critical Reviews in Therapeutic Drug Carrier Systems, 1: 155-187.

Ihler, G.M. and Tosi, P.F. (1987). 'IHP Dramatically Reduces Babesia Microti and Plasmodium Falciparum Parasitemias: Observations of Fluorescent Red Cells and Fluorescent Vacuoles', in the 2nd Int'l. Meeting on Red Blood Cells as Carriers for Drugs Potential Therapeutic Applic. Symposium, in press.

Jaffe, J.J., Meymarian, E., Doremus, H.M. (1970). 'Antischistosomal Action of Tubercidin Administered after Absorption into Red Cells', Nature, 230: 408-409.

Jausel-Husken, S. and Deuticke, B. (1981). 'General and Transport Properties of Hypotonic and Isotonic Preparations of Resealed Erythrocyte Ghosts', J. Membr. Biol., 63: 61-70.

Katznelson, R. and Kulka, R.G. (1983). 'Degradation of Microinjected Methylated and Unmethylated Proteins in Hepatoma Tissue Culture Cells', J. Biol. Chem., 258: 9597-9600.

Kay, M.M.B. (1985). 'Aging of Cell Membrane Molecules Leads to Appearance of an Aging Antigen and Removal of Senescent Cells', Gerontology, 31: 215-235.

Kempf, C., Klausner, R.D., Weinstein, I.N., van Renswoude, J., Pincus, M. and Blumenthal, R. (1982). 'Voltage-Dependent Trans-Bilayer Orientation of Melittin', J. Biol. Chem., 257: 2469-2476.

Kinosita, K. and Tsong, T.Y. (1977). 'Formation and Resealing of Pores of Controlled Sizes in Human Erythrocyte Membrane', Nature, 268: 438-440.

Kinosita, K. and Tsong, T.Y. (1978). 'Survival of Sucrose-Loaded Erythrocytes in the Circulation', Nature, 272: 258-260.

Kitagawa, T., Kawasaki, T. and Munechika, H. (1982). 'Enzyme Immunoassay of Blasticidin S with High Sensitivity: A New and Convenient Method for Preparation of Immunogenic (Hapten-Protein) Conjugates', J. Biochem, 92: 585-590.

Kitao, T., Hattori, K., and Takeshita, M. (1978). 'Agglutination of Leukemic Cells and Daunomycin Entrapped Erythrocytes with Lectin in vitro and in vivo', Experientia., 34: 94.

Kitao, T. and Hattori, K. (1980). 'Erythrocyte Entrapment of Daunomycin by Amphotericin B Without Hemolysis', Cancer Res., 40: 1351.

Koler, R.D., McClung, M.R., Peterson, L.L., and Jones, M.B. (1980). 'Physiologic and Genetic Alterations in Human Red Cell DPGM', Hemoglobin., 4: 593-600.

Korten, K. and Miller, K.W. (1979). 'Erythrocyte Ghost-Buffer Partition Coefficients of Phenobarbital, Pentobarbial, and Thiopental Support the pH-Partition Hypothesis', Can. J. Physiol. Pharmacol., 57: 325-328.

Kruse, C.A., Spector, E.B., Cederbau, S.D., Wisnieski, B.J. and Popjak, G. (1981). Microinjection of Arginase into Enzyme-Deficient Cells with the Isolated Glycoproteins of Sendai Virus as Fusogen', Biochem. Biophys. Acta., 645: 339-345.

Lynch, W.E., Sartiano, G.P., and Ghaffer, A. (1980). 'Erythrocytes as Carriers of Chemotherapeutic Agents for Targeting the Reticuloendothelial System', Am. J. Hematol., 9: 249-259.

Maulet, Y., Brodbeck, U., and Fulpius, B. (1984). 'Selective Solubilization by Melittin of Glycophorin A and Acetylcholinesterase from Human Erythrocyte Ghosts', Biochem. Biophys. Acta., 778: 594-601.

McEvoy, L., Williamson, P., and Schlegel, R.A. (1986). 'Membrane Phospholipid Asymmetry as a Determinant of Erythrocyte Recognition by Macrophages', Proc. Natl. Acad. Sci. USA, 83: 3311-3315.

Minetti, M. and Ceccarini, M. (1982). 'Protein-Dependent Lipid Lateral Phase Separation as a Mechanism of Human Erythrocyte Ghost Resealing', J. Cell. Biochem., 19: 59-75.

Mishra, K.P., Binh, L. D., and Singh, B.B. (1981). 'Effect of Dielectric Discharge on Drug Treated Mammalian Cells', Indian J. Exp. Biol., 19: 520-523.

Mueggler, P.A., and Black, J.A. (1982). 'Postnatal Regulation of Canine Oxygen Delivery: Control of Erythrocyte 2,3-DPG Levels', Am. J. Physiol., 242: H500-506.

Netland, P.A. and Dice, J.F. (1985). 'Red Blood Cell-Mediated Microinjection: Methodological Considerations', Anal. Biochem., 150: 214-220.

Nicolau, C. and Gersonde, K. (1979). 'Incorporation of Inositol Hexaphosphate Into Intact Red Blood Cells', Naturwissenschaften., 66: 563-566.

Okuda, T., Mihara, K. and Sato, R. (1972). 'Interactions Between NADH-Cytochrome b5 Reductase and Cytochrome b5 Purified from Liver Microsomes', J. Biochem., 72: 987-992.

Olson, J.A. (1981). 'In vitro Studies of Malarial Parasites Using Resealed Ghosts of Human Erythrocytes', in The Red Cell: Fifth Ann Arbor Conference, Alan R. Liss, N.Y., 55: 537-548.

Olson, J.A. and Kilejian, A. (1982). 'Involvement of Spectrin and ATP in Infection of Resealed Erythrocyte Ghosts by the Human Malarial Parasite, Plasmodium falciparum', J. Cell Biol., 95: 757-762.

Pitt, E., Lewis, D.A., and Offord, R.E. (1983a). 'The Use of Corticosteroids Encapsulated in Erythrocytes in the Treatment of Adjuvant Induced Arthritis in the Rat', Biochem. Pharm., 32: 3355-3358.

Pitt, E., Johnson, C.M. and Lewis, D.A. (1983b). 'Encapsulation of Drugs in Intact Erythrocytes: An Intravenous Delivery System', Biochem. Pharm., 32: 3359-3368.

Poensgen, J. and Ullrich, V. (1980). 'Transfer of Cytochrome b5 and NADPH Cytochrome c Reductase Between Membranes', Biochem. Biophys. Acta., 596: 248-263.

Raap, A.K., Van-Duijn, P. (1981). 'Enzyme-Incorporated Erythrocyte Ghosts: A New Model System for Quantitative Enzyme Cytochemistry', J. Histochem. Cytochem., 29: 1418-1424.

Roise, D., Horvath, S.J., Tomich, J.M., Richards, J.H. and Schatz, G. (1986). 'Chemically Synthesized Pre-Sequence of an Imported Mitochondrial Protein can Form an Amphiphilic Helix and Perturb Natural and Artificial Phospholipid Bilayers', EMBO J. 5: 1327-1334.

Ropars, C., Chassaigne, M., Villereal, M. Avenard, M.C. and Nicolau, C. (1985a). 'Resealed Red Blood Cells as a New Blood Transfusion Product', Bibl. Hematol., 51: 92-107.

Ropars, C., Teisseire, B., Avenard, G., Chassaigne, M., Hurel, C., Girot, R. and Nicolau, C. (1985b). 'Improved Oxygen Delivery to Tissues and Iron Chelator Transport Through the Use of Lysed and Resealed Red Blood Cells: A New Perspective on Cooley's Anemia Therapy'. Annuals of the NY Acad. Sci., 445: 304-315.

Ruiz-Ruano, A., Martin, M., Luque, J. (1984). 'Synthesis and Levels of Organic Phosphates in Erythrocytes During Avian Development: Specific Formation of BPG and IPS in Two Distinct Populations From Young Chicks', Cell. Biochem. Funct., 2: 257-262.

Samokhin, G.P., Smirnov, M.D., Muzykantov, V.R., Domogatsky, S.P. and Smirnov, V.N. (1984). 'Effect of Flow Rate and Blood Cellular Elements on the Efficiency of red Blood Cell Targeting to Collagen-Coated Surfaces', J. Appl. Biochem. 6: 70-75.

Sasaki, R., Hirose, M., Sugimoto, E., and Chiba, H. (1971). 'Studies on a Role of the 2,3-Diphosphoglycerate Phosphatase Activity in the Yeast Phosphoglycerate Mutase Reaction', Biochem. Biophys. Acta., 227: 595-607.

Schwister, K. and Deuticke, B. (1985). 'Formation and Properties of Aqueous Leaks Induced in Human Erythrocytes by Electrical Breakdown', Biochem. Biophys. Acta. 816: 332-348.

Seigneuret, M. and Devaux, P.F. (1984). 'ATP-Dependent Asymmetric Distribution of Spin-Labeled Phospholipids in the Erythrocyte Membrane: Relation to Shape Changes', Proc. Natl. Acad. Sci. USA, 81: 3751-3755.

Serpersu, E.H., Kinosita, K., Jr., and Tsong, T.-Y. (1985). 'Reversible and Irreversible Modification of Erythrocyte Membrane Permeability by Electric Field', Biochem. Biophys. Acta., 812: 779-785.

Siegel, I., Liu, T.L. and Gleicher, N. (1981). 'The Red Cell Immune System', The Lancet, 257: 556-559.

Sikka, S.C., Green, G.A., Chauhan, V.P.S. and Kalra, V.K. (1982). 'Proteoliposome Interaction with Human Erythrocyte Membranes. Functional Implantation of Gamma-glutamyl Transpeptidase', Biochemistry, 21: 2356-2366.

Smedsrod, B. and Aminoff, D. (1983). 'Studies on the Sequestration of Chemically and Enzymatically Modified Erythrocytes', Amer. J. of Hemat., 15: 123-133.

Smirnov, V.N., Domogatsky, S.P., Dolgov, V.V., Hvatov, V.B., Klibanov, A.L., Koteliansky, V.E., Muzykantov, V.R., Repin, V.S., Samokhin, G.P., and Shekhonin,-B-V. (1986). 'Carrier-Directed Targeting of Liposomes and Erythrocytes to Denuded areas of Vessel Wall', Proc. Natl. Acad. Sci. USA., 83: 6603-6607.

Sorge, J., Kuhl, W., West, C., and Beutler, E. (1987). 'Complete Correction of the Enzymatic Defect of Type I Gaucher Disease Fibroblasts by Retroviral-Mediated Gene Transfer', Proc. Natl. Acad. Sci. USA, 84: 906-909.

Sowers, A.E. (1986). 'A Long-Lived Fusogenic State is Induced in Erythrocyte Ghosts by Electric Pulses', J. Cell Biol., 102: 1358-1362.

Swaney, J.B. (1985). 'Membrane Cholesterol Uptake by Recombinant Lipoproteins', Chemistry and Physics of Lipids, 37: 317-327.

Takagaki, Y., Ramachandran, R., Wirtz, K.W. and Khorana, H.G. (1983a). 'Th Membrane-Embedded Segment of Cytochrome b5 as Studied by Cross-Linking with Photoactivatable Phospholipids. I. The Transferable Form', J. Biol. Chem., 258: 9128-9135.

Takagaki, Y., Ramachandran, R., Wirtz, K.W. and Khorana, H.G. (1983b). 'II. The Nontransferable Form', J. Biol. Chem., 258: 9136-9142.

Tamburini, P.P. and Schenkman, J.B. (1986a). 'Mechanism of Interaction Between Cytochromes P-450 RLM5 and b5: Evidence for an Electrostatic

Mechanism Involving Cytochrome b5 Heme Propionate Groups', Arch. Biochem. Biophys., 245: 512-522.

Tamburini, P.P., MacFarquhar, S. and Schenkman, J.B. (1986b). 'Evidence of Binary Complex Formations Between Cytochrome P-450, Cytochrome b5, and NADPH-Cytochrome P-450 Reductase of Hepatic Microsomes', Biochem. Biophys. Res. Commun., 134.

Teisseire, B.P., Ropars, C., Nicolau, C., Vallez, M.O., Chassaigne, M. (1984). 'Enhancement of P50 by Inositol Hexaphosphate Entrapped in Resealed Erythrocytes in Piglets', Adv. Exp. Med. Biol., 180: 673-677.

Thorpe, S.R., Fiddler, M.B. and Desnick, R.J. (1975). 'Enzyme Therapy, V. in vivo Fate of Erythrocyte-Entrapped ß-Glucuronidase in ß-Glucuronidase-Deficient Mice', Pediatr. Res., 9: 918-923.

Tsang, H., Mollenhauer, H., and Ihler, G.M. (1982). 'Entrapment of Proteins, Viruses, Bacteria, and DNA in Erythrocytes during Endocytosis', J. App. Biochem., 4: 418-435.

Tsong, T.-Y. and Kinosita, K., Jr. (1985). 'Use of Voltage Pulses for the Pore Opening and Drug Loading, and the Subsequent Resealing of Red Blood Cells', Bibl. Haematol., 51: 108-114.

Tyrrell, D.A. and Ryman, B.E. (1976). 'The Entrapment of Therapeutic Agents in Resealed Erythrocyte Ghosts and Their Fate in vivo', Biochem. Soc. Trans,. 4: 677.

Updike, S.J., Wakamiya, R.T. and Lightfoot, E.N., Jr. (1976). 'Asparaginase Entrapped in Red Blood Cells Action and Survival', Science, 193: 681.

Updike, S.J. and Wakamiya, R.T. (1983). 'Infusion of Red Blood Cell-Loaded Asparaginase in Monkey', J. Lab. Clin. Med., 101: 679-691.

Vandecasserie, C., Paul, C., Schnek, A.G., and Leonis, J. (1973). Oxygen affinity of avian hemoglobins. Comp. Biochem. Physiol. 44: 711-718.

von Heijne, G. (1986). Mitochondrial targeting sequences may form amphiphilic helices. EMBO J. 5: 1335-1342.

Waters, M.G. and Blobel, G. (1986). 'Secretory Protein Translocation in a Yeast Cell-Free System Can Occur Posttranslationally and Requires ATP Hydrolysis', Cell Biol., 102: 1543-1550.

Yokoi, T., Iwasa, M. and Sagisaka, K. (1983) A new method to prepare rabbit immune anti-M and -N sera using blood group substance trapped in autologous red cell ghost as an immunogen. Tokoku J. Exp. Med. 140:, 289-296.

Yokoi, T., Sagisaka, K, and Iwasa, M. (1984). Preparation of Anti-Le[a] and Anti-Rh$_o$(D) Sera by immunization with blood group substance trapped in autologous red cell ghost. Tohoku. J. Exp. Med. 144: 321-325.

Zimmerman, U., Pilwat, G., and Esser, B. (1978). 'The Effect of Encapsulation in Red Blood Cells on the Distribution of Methotrexate in Mice', J. Clin. Chem. Clin. Biochem., 16: 135.

Zolese, G., Curatola, G., Mazzanti, L., Leporoni, B., and Lenza, G. (1979). 'Molecular Mechanism of General Anesthesia: III. Kinetic Studies on Erythrocyte Ghost Acetylcholinesterase', Boll. Soc. Ital. Biol. Sper., 55: 511-516.

Drug Carrier Systems
Edited by F.H.D. Roerdink and A.M. Kroon
© 1989 John Wiley & Sons Ltd.

LIPOSOMES AS A DRUG DELIVERY SYSTEM IN CANCER CHEMOTHERAPY

Alberto Gabizon

Cancer Research Institute, University of California,
San Francisco, and Liposome Technology, Inc.,
Menlo Park, California, USA

Contents

Abbreviations: ACD = Actinomycin D; Ara-C = Cytosine Arabinoside;
cDDP = cis-Dichloro-Diammine-Platinum II; CH = Cholesterol;
CL = Cardiolipin; DCP = Dicetylphosphate; DNR = Daunorubicin;
DPPA = Dipalmitolyl Phosphatidyc Acid; DPPG = Dipalmitoyl
Phosphatidylglycerol; DSPC = Distearoyl Phosphatidylcholine;
DXR = Doxorubicin; 5-FO = 5-Fluoorotate; ID = Intradermal;
IP = Intraperitoneal; IV = Intravenous; MTX = Methotrexate;
MTX-Asp = MTX-gamma-Aspartate; PA = Phosphatidic Acid;
PC = Phosphatidylcholine; PG = Phosphatidylglycerol; PS =
Phosphatidylserine; RES = Reticuloendothelial System; SA =
Stearylamine; SC = Subcutaneous; SULF = Sulfatides

Correspondence: A. Gabizon, Cancer Research Institute, University
of California, San Francisco, CA 94143-0128 (USA).

INTRODUCTION

Most of the cytotoxic drugs used in cancer treatment have a very narrow therapeutic index and result in serious effects that reduce compliance to therapy and impair quality of life. Although the availability of new drugs and the use of aggressive regimen protocols have improved the rates of response and survival in various types of cancer, morbidity remains high and the course of many of the most common types of cancer is virtually unaffected. Out of 550,000 patients who present each year in the USA with metastatic disease, only between 20,000 to 30,000 are cured by chemotherapy (Frei, 1985). Except for a fraction of patients with breast cancer, chemotherapy has not significantly affected the overall survival in patients with the other most frequent types of cancer (lung, colorectal, prostate) (Bailar and Smith, 1986). Despite a huge effort in drug development, cancer chemotherapy remains largely nonspecific and most drugs are toxic for tumor cells as well as for normal cells. Thus, improving the selectivity of cytotoxic drugs with the use of carriers is an attractive option with potentially great applications in cancer.

Cancer chemotherapy presents serious challenges for the design of drug delivery systems. Treatment is generally directed at a systemic disease since metastatic dissemination is usually already present when chemotherapy is started. To reach efficiently all tumor sites, the intravenous route has to be used and the interaction of the carrier system with plasma proteins, blood cells and the reticuloendothelial system (RES) has to be dealt with. Tumor heterogeneity, with regard to patterns of metastatic dissemination, microvascularization, phagocytic capacity, and mechanisms of drug resistance, adds an additional difficulty in predicting the therapeutic outcome.

Liposomes have raised considerable interest as drug carriers in cancer chemotherapy. As a non-covalently bound, biocompatible and biodegradable carrier, liposomes are attractive candidates for drug delivery systems (Gregoriadis, 1977). Nonetheless, liposomes remain a controversial subject regarding their potential usefulness as carriers of cancer chemotherapeutic agents (Poste, 1983; Weinstein, 1984). In contrast to the enthusiastic notes in the initial literature on this subject, a more realistic view of liposome application is now emerging. The predominant uptake of liposomes by the RES, the difficulty in predicting patterns of liposome extravasation, and the problems involved in achieving long-term physicochemical stability constitute the major obstacles to the liposome approach. Notwithstanding these obstacles, there is wide recognition that drug delivery systems, and among them, liposomes, may be useful in anticancer therapy by exploiting controlled drug release and changes in tissue distribution leading to reduced toxicity and increased efficacy.

In this discussion, we have not attempted to produce an exhaustive review of literature data on this subject but to restrict ourselves to a few examples which are helpful in gaining a perspective of this field. Various comprehensive reviews and books on liposomes have been published and the reader is addressed to them for basic aspects. For two recent review books on liposomes, see Ostro (1987), and Gregoriadis (1988). For article reviews devoted to medical applications of liposomes, see Yatvin and Lelkes (1982); and Weinstein and Leserman (1984).

We will first review some aspects of liposome biodistribution relevant to cancer treatment. This will be followed by a section which attempts to identify those areas in chemotherapy with a strong rationale for the use of liposomes. Finally, we will discuss the present status of liposome work with a few selected drugs or groups of drugs, which may be considered representative of the research in this field.

BIODISTRIBUTION OF LIPOSOMES -- IMPLICATIONS ON DRUG DELIVERY TO TUMOR CELLS

Any serious approach to cancer chemotherapy using liposomes as drug carriers should take into consideration the pharmacokinetics and tissue distribution of liposomes after parenteral injection. Essentially, most applications of liposomes in cancer chemotherapy will aim at altering tissue distribution and various pharmacokinetic parameters of the drug in question in such a way that toxicity can be reduced and/or efficacy increased (Mayhew and Papahadjopoulos, 1983). A further and more sophisticated level of application, not discussed here, is to enhance delivery of drugs to the target cells using liposomes with fusogenic ability and/or specific ligands, such as antibodies (Heath, 1982).

Intravenous injection is the most practical and widely-used route of administration for cancer chemotherapy. Using various liposome-encapsulated markers, it has been well established that IV-injected liposomes are taken up predominantly by liver and spleen. The first use of liposomes in humans dates to 1974 when Gregoriadis et al. (1974) injected sonicated liposomes containing 131-I radiolabeled albumin in three cancer patients to study their tissue distribution. The majority of the label, up to 81% of the injected dose in one patient, was found in the liver. A report on two additional patients (Segal et al., 1976) receiving liposome-encapsulated 111-In labeled bleomycin pointed again at the large hepatic uptake. Within the liver, radioactivity was concentrated predominantly in normal rather than neoplastic tissue. Another interesting observation was the considerable liposome localization in the bone marrow in a patient with hepatocellular carcinoma and cirrhotic liver. More extensive studies in humans were reported on 1979 by Richardson et al. (1979). Fifteen patients were examined using a lipid dose of 20 to 30 mg and a label of 99mTc. Fluid-phase, sonicated liposomes consisting of egg phosphatidylcholine (PC), phosphatidic acid (PA) and cholesterol (CH) were used. In all cases, excluding one patient with Polycytemia Vera, the major site of uptake was in the liver and spleen. In the patient with Polycytemia Vera, the majority of the radioactivity was apparently localized in the bone marrow. There was no evidence of significant liposome localization in tumors.

The pharmacokinetics and organ distribution of 99mTc-labeled large
multilamellar liposomes have been studied by Lopez-Berestein et
al. (1984) in seven cancer patients. Besides the high uptake in
liver and spleen (~70%), a significant fraction was accumulated in
the lungs (14.5%), which is a specific feature of these large
particles (up to 5 um diameter). Clearly these observations
diminished the enthusiasm for liposomes as a tumor-localizing
carrier. These observations in humans were in agreement with
animal data indicating that the homing of liposomes after IV
injection is very much restricted to cells of the RES with liver
and spleen being responsible for an early clearance from
circulation (Segal et al., 1974; Roerdink et al., 1981; Poste et
al., 1982).

Several investigators have attempted to increase the blood
half-life of liposomes and reduce the RES uptake. One approach is
the blockade of the RES with large doses of plain liposomes,
subsequently followed by injection of the liposomes containing the
relevant label (Abra et al., 1980; Poste, 1983; Allen et al.,
1984). RES blockade is in general of short duration and requires
injection of large amounts of lipids. The use of a cholesterol
derivative (6-aminomannose-cholesterol) as a liposome component
appears to result in a more efficient blockade (Mauk et al.,
1980). The potential toxicity and pharmacokinetic complexity of
this approach are however serious drawbacks. The most direct
approach at RES avoidance is obviously the design of liposome
formulations that would not be readily removed from circulation.
Extensive work by Gregoriadis et al. (reviewed in Gregoriadis and
Senior, 1986) indicates that sonicated neutral liposomes
consisting of distearoyl-PC (DSPC) and CH remain long in
circulation with an apparent half-life of 10 hours, using 111- In
labeled bleomycin as a marker. These findings together with those
from other laboratories (Hwang et al., 1980; Allen and Everest,
1983) suggest that high-phase transition lipids confering rigidity
to the bilayer at 37°C synergize with CH and small size to prolong
the vesicle half-life in the blood stream. Using small
unilamellar DSPC:CH liposomes, Profitt et al. (1982, 1983) showed
enhanced liposome accumulation in a transplantable mouse tumor. A
recent report from our laboratory points also at a significant
accumulation of IV-injected liposomes in SC-implanted tumors using
glycolipid-containing formulations which could be further boosted
by conjugation of specific antibodies (Papahadjopoulos and
Gabizon, 1987). Moreover, in further studies, a significant
correlation between liposome half-life and localization in tumors
was found (Gabizon and Papahadjopoulos, submitted for
publication). Given the broad versatility of liposomes in
composition, charge and size, the impact of liposome formulation
on the pharmacokinetics and pharmacodynamics of liposome-
encapsulated drugs is still an open area for manipulations.

An additional obstacle to liposome targeting is the limited
ability of liposomes and other nanoparticles to reach
extravascular tissues. As pointed out by Poste (1983) and
Weinstein (1984), extravasation of liposomes is limited by the

permeability of the microvascularization of each specific tissue. Extravasation of liposomes can be certainly envisaged in tissues with sinusoidal capillaries, such as liver, spleen, and bone marrow. It seems, however, an unlikely event in other normal body tissues. Evidence that small-sized liposomes (<100 nm) can cross through liver sinusoids and reach the parenchymal cells has been obtained by several groups of investigators (Poste et al., 1982; Rahman et al., 1982; Scherphof et al., 1983). This finding is compatible with electron microscopy observations in the liver pointing at the presence of up to 0.1 um-gaps between sinusoidal endothelial cells (Wisse, 1970). Thus, it appears that, while large liposomes (>100 nm) have no significant access to cells other than circulating or sinusoidal macrophages, smaller vesicles can penetrate to a certain extent into the liver parenchyma. The critical issue remains the extravasation of liposomes in tumors. If prolonged liposome half-life in blood is achieved by inhibiting RES uptake, then, liposome extravasation in the tumor area becomes the rate-limiting factor for targeting. The broad heterogeneity of tumors with regard to microvascular architecture (Shubik, 1982) makes difficult any predictions on accessibility and uptake of liposomes. In general, the permeability of tumor vascularization is increased as compared to normal tissues (Peterson, 1979; Jain and Gerlowski, 1986). Yet, it is unclear whether these differences can be exploited to achieve a preferential accumulation of liposomes in tumors as opposed to normal tissues. Obviously, physical hindrance to extravasation should diminish by reducing vesicle size. In this context, as with RES avoidance, small-sized liposomes would be the more advantageous ones for drug delivery to tumors. There are also disadvantages in the use of small vesicles in the range of 30 to 70 nm diameter. Their low drug-carrying capacity (Gabizon and Barenholz, 1988), the potential instability resulting from elastically-stressed bilayers (Wetterau and Jonas, 1982), and the difficulty in large-scale production are the most serious problems.

Using the IP route of injection, Parker et al. (1981, 1982a, 1982b) have obtained evidence indicating that liposomes are absorbed from the peritoneal cavity by lymphatics and concentrate selectively in the draining lymph nodes. Using the intrathecal route, Kimelberg et al. (1978), found a significant alteration of the pharmacokinetics of methotrexate (MTX) encapsulated in liposomes, suggesting that sustained drug release in the cerebrospinal fluid may take place. This line of work has been pursued by other investigators with cytosine-arabinoside (Ara-C) (Kim et al., 1987), and bleomycin (McKeran et al., 1985). The latter drug was tested in three patients with malignant gliomas injected through Ommaya reservoirs.

Attempts at modifying liposome biodistribution using temperature-sensitive liposomes (Weinstein et al., 1979) or pH-sensitive liposomes (Connor and Huang, 1986) are of interest, although much more work is required to assess their feasibility. One practical application could be the use of the intravesical route in combination with temperature-sensitive liposomes. Bassett et al. (1986) have shown that hyperthermia of bladder tumors resulted in release of liposome contents and enhanced tumor uptake.

RATIONALE FOR THE USE OF LIPOSOMES IN CANCER CHEMOTHERAPY

Currently, the main reasons for failure of response to chemotherapy appear to be the following:
a.   Primary or acquired drug resistance of neoplasms.
b.   Threshold of toxicity to normal tissues lower than threshold of tumoricidal effect.
c.   Inefficient drug delivery to tumor cells.
Let us examine what contribution may be expected from liposomes in each of these areas.

Despite some in vitro observations on increased sensitivity of resistant cells to Ara-C (Richardson et al., 1982), MTX (Todd et al., 1982) and actinomycin D (ACD) (Poste and Papahadjopoulos, 1976) encapsulated in liposomes, there is no in vivo evidence that liposomes can overcome drug resistance. Rustum et al. (1981), and Richarson and Ryman (1982) found that the in vivo sensitivity of L1210 cells resistant to Ara-C could not be modified by liposome encapsulation of Ara-C or Ara-CTP, its active metabolite. A similar observation for an ACD-resistant cell line was reported by Kaye et al. (1981).

Reduction of toxicity by liposome-mediated delivery is well founded by preclinical experimentation with regard to doxorubicin (DXR), daunorubicin (DNR), and ACD (Mayhew and Papahadjopoulos, 1983). At least two factors may contribute to this reduction in toxicity: site avoidance of drug-sensitive tissues, such as the heart muscle in the case of anthracyclines; and slow release of drug avoiding the high plasma concentrations of bolus injection of free drug. In contrast, the toxicity of strictly phase-specific drugs, such as Ara-C and MTX, tends to increase when administered in liposome-encapsulated form (Rustum et al., 1979; Kaye et al., 1981).

As discussed in the preceding section, systemic targeting of liposomes to increase drug delivery to tumor cells faces two serious obstacles: fast and predominant uptake by the RES, and little or no ability of particulate carriers to extravasate other than through sinusoidal capillaries. Although there are some encouraging notes suggesting that these problems may be overcome, it would seem premature at this point to claim a strong rationale for this approach. Passive targeting of liposome-encapsulated drugs to liver and spleen represents a distinct and attractive possibility to increase drug concentration in tumor cell colonies residing in these organs. This approach is supported by preclinical evidence in various mouse metastatic tumor models using DXR-containing liposomes (Gabizon et al., 1983; Mayhew et al., 1983). Its potential relevance is stressed by the fact that the liver is a common site of metastatic spread with a most detrimental effect in the prognosis of cancer patients.

In some instances, cancer therapy has to be directed to a specific anatomic compartment or body cavity This may be either for purely symptomatic reasons or to treat with curative intent anatomically confined cancer. By retarding the clearance of liposome-entrapped drugs from the cavity of inoculation and slowly releasing the drug to free form, liposomes may provide with significant pharmacological advantages, such as increased area under the curve in the site of action and reduced systemic exposure to the drug. In the previous section, we have discussed the works of Parker et al. (1981, 1982a, 1982b) using the intraperitoneal route of injection with Ara-C and DXR, and Kimelberg et al. (1978) using the intrathecal route for MTX. In both cases, the potential benefits of liposomes for drug delivery in intracavitary therapy are foreseeable. Another important advantage of this approach is that, by avoiding the RES, and the need for extravasation, sophisticated methods of delivery such as fusogenic liposomes and antibody-conjugated liposomes (Connor and Huang, 1986) can be envisaged . Besides intracavitary therapy to the peritoneal cavity, pleural space, and subarachnoidal space, other especial routes of administration are of interest for liposome-mediated delivery: intralymphatic (Kaledin et al., 1981), intravesical (Tacker and Andersen, 1982), and bronchoalveolar (Juliano and McCullough, 1980).

Table I presents, in synopsis, potential directions for the role of liposomes in chemotherapy.

TABLE I   POTENTIAL APPLICATIONS OF LIPOSOMES TO CYTOTOXIC THERAPY

o   Reduction of tissue-specific toxicity by site-avoidance delivery.

   Examples: DXR cardiotoxicity (dogs, rodents).   Other forseeable applications - Bleomycin (pulmonary fibrosis); Vincristine (peripheral neuropathy); cDDP (nephrotoxicity)

o   Passive targeting to some liver tumors, possibly depending on patterns of tumor infiltration and vascularization.

o   Therapy of Systemic cancer - New liposome formulations with decreased uptake by the RES and improved tumor-homing properties are required.

o   Increase of locoregional Cxt values in intracavitary therapy by retarding drug diffusion into the systemic circulation.

o   Parenteral administration of water-insoluble compounds in detergent-free and alcohol-free form.
   Examples: Nitrosoureas, Etoposide

INVESTIGATIONS WITH LIPOSOME-ENTRAPPED CYTOTOXIC DRUGS

The examples we will discuss are the following: DXR, an amphipathic, membrane bilayer-intercalating drug; Ara-C, a water-soluble S phase-specific drug; two strictly lipophilic drugs, a quinazolone derivative (NSC 251635) and a lipophilic derivative of cis-diammine-dichloro-platinum (cDDP); and, two examples of "liposome-dependent" drugs, MTX-gamma aspartate (MTX-Asp) and 5-fluorotate (5-FO). We will deal in more detail with DXR, a research subject in which we have been directly involved.

A.   Doxorubicin (DXR)

One of the most encouraging and prolific areas in the liposome-anticancer drug field is the work with anthracyclines. DXR is an anthracycline antibiotic with a broad spectrum of antitumor activity including a variety of human and animal solid tumors and leukemias (reviewed by Young et al., 1981). Its administration in humans and animals is known to cause an irreversible cardiotoxic effect which is a major clinical handicap limiting its cumulative dosage. The severity of this effect is dose related in humans and experimental animals. Therefore, the development of a carrier system which will decrease uptake by the heart muscle without diminishing its antitumor activity may reduce cardiotoxicity and significantly improve the therapeutic index of this drug. Liposomes could adequately fulfill this task given their relative inability to cross continuous capillaries and the lack of phagocytic reticuloendothelial cells in the myocardial tissue. Since 1979, when the first report on the pharmacology of liposome-associated DXR (L-DXR) was published (Forssen and Tokes, 1979), more than 30 articles on this subject have appeared in the scientific literature. For a reference list, see Table II. Most of these reports point at a decreased toxicity of L-DXR with regard to various parameters: median lethal dose, cardiotoxicity, dermal toxicity, and immunosuppression. In addition significant antitumor activity of L-DXR comparable and, in some cases, superior to free DXR has been found in various murine models. Based on these preclinical data, pilot clinical studies have been recently initiated to test L-DXR in humans (Gabizon et al., 1986c; Treat et al, 1987; Sells et al., 1987).
We have addressed the tailoring of liposomes as DXR carriers following a step-wise approach that can be generalized to the development of other anti-cancer drug carrier systems for parenteral administration (Table III). This approach allows to discard unsuitable liposome types through an initial screening which is quick and inexpensive. To give an example, a formulation that is found to have a good loading capacity for DXR, and similar in vitro IC50 (50% growth inhibition) to free DXR, but shows a very poor retention of the drug in the presence of plasma, should be withdrawn from further testing. Obviously, such a formulation would not be useful for in vivo studies, since one may anticipate

that   most of the drug will leak from the vesicles soon after
injection into the blood system.
   We will attempt to summarize the work done at the Hadassah
Medical Center (Gabizon et al., 1982-in press; Gabizon and
Barenholz, in press; Peretz et al., 1987), and, as we go along,
discuss the analogies and discrepancies with the results obtained
by other groups of investigators. Our approach was designed at
engineering an optimal liposome for DXR delivery by selecting one
or various compositions that would fulfill the following criteria:
high efficiency of drug capture; preservation of the full
biological activity of the drug after liposome encapsulation as
shown by _in vitro_ cytoxicity tests; stability in the presence of
plasma; favorable tissue distribution pattern with reduced uptake
of L-DXR in the heart muscle and increased levels in the liver;
decreased toxicity including cardiotoxicity; increased therapeutic
index in a variety of relevant tumor models.

TABLE II   IN VIVO PRECLINICAL WORK WITH DOXORUBICIN
            IN LIPOSOMES

Reference		Lipids[1]	Animal/Route	Comments[2]
Abra et al.	(1983)	PC,PS	Mouse/IV	Lung Targeting
Cervato et al.	(1986)	PC,SULF	Mouse/IP	Phar, Ther
Fichtner et al.	(1981)	PC,PA,SA	Mouse/IV	Tox, Ther(DNR)
ibid	(1984)	PC,DCP,SA	Mouse/IV	Cardiotoxicity(DNR)
Forssen and Tokes	(1979)	PC,PS	Mouse/IV	Phar
ibid	(1981)	PC,PS	Mouse/IV,IP	Cardiotoxicity
ibid	(1983a)	PC,PS	Mouse/IV,IP	Ther,Immune function
ibid	(1983b)	PC,PS	Mouse/ID	Dermal Toxicity
Gabizon et al.	(1982)	PC,PS,CL	Mouse/IV	Phar
ibid	(1983)	PC,PS	Mouse/IV	Phar
ibid	(1985)	PC,PS,PG	Mouse/IV	Ther
ibid	(1986a)	PC,PG	Mouse/IV	Tox
ibid	(1986b)	PC,PS,PG	Mouse/IV	Ther, Phar
ibid	(in press)	PC,PS,PG	Mouse/IV,IP	Ther, Phar
Ganapathi and Krishan	(1984)	PC,DCP	Mouse/IP	Tox
Herman et al.	(1983)	PC,SA,CL	Dog/IV	Cardiotoxicity
Kojima et al.	(1986)	PC,SULF	Mouse/IV	Phar, Ther
Konno et al.	(1987)	PC,DPPA	Mouse/IV	Antibody Targeting
Litterst et al.	(1982)	PC,SA	Rat/IP	Tox
Mayhew et al.	(1983)	PC,PG	Mouse/IV	Ther
Mayhew and Rustum	(1983)	PC,PG	Mouse/IV	Tox, Ther
ibid	(1985)	PC,PG	Mouse/IV	Tox, Ther
Olson et al.	(1982)	PC,PG	Mouse/IV	Phar, Tox, Ther
Onuma et al.	(1986)	PC,DPPA	Mouse/IV	Antibody Targeting
Parker et al.	(1981)	PC,SA	Rat/IP	Phar
ibid	(1982b)	PC,SA	Rat/IP,IV	Phar
Rahman et al.	(1980)	PC,PS,SA	Mouse/IV	Phar, Ther, Tox
ibid	(1982)	PC,PS,SA	Mouse/IV	Cardiotoxicity
ibid	(1984)	PC,SA,CL	Mouse/IV	Phar, Ther, Tox
ibid	(1985)	PC,SA,CL	Mouse/IP	Phar, Ther, Tox
ibid	(1986a)	PC,SA,CL	Rat/IV	Phar
ibid	(1986b)	PC,SA,CL	Mouse/IV	Ther, Tox
ibid	(1986c)	PC,SA,CL	Mouse/IV	Immune function
Roza and Clementi	(1983)	PC,DCP	Mouse/IV,IP	Phar
Shinozawa et al.	(1981)	PC,SA,DCP	Mouse/IP	Phar, Ther
Storm et al.	(1987)	PC,DSPC,PS,DPPG	Rat/IV	Phar, Ther
ibid	(in press)	PC,PS	Rat/IV	Phar
Van Hoesel et al.	(1984)	PC,PS,SA	Rat/IV	Phar, Ther, Tox

---

[1] - Liposome composition excluding cholesterol and other minor components.

[2] - Indicates main feature of the paper. Tox=toxicity; Ther=therapy;
      Phar=pharmacokinetics and/or drug biodistribution. Some of the papers
      labeled "Tox" may also contain information on cardiotoxicity.

Efficiency of capture of DXR can be significantly increased to more than 50% by negatively-charged phospholipids (Gabizon et al., 1982). This has led most of the investigators in this field to include lipids such as phosphatidylserine (PS) (Van Hoesel et al., 1984), phosphatidylglycerol (PG) (Olson et al., 1982), cardiolipin (CL) (Rahman et al., 1985), or sulfatides (SULF) (Kogima et al., 1986; Cervato et al., 1986). We favor the use of PG, because it can be readily obtained from egg PC by a simple enzymatic conversion, as opposed to brain-derived lipids; and it is less sensitive to lipid peroxidation than lipids highly rich in polyunsaturated lipids such as PS and CL. We have also found that decreasing the size of liposomes to 50 nm or less, results in a significantly diminished drug capture per mol phospholipid (Gabizon and Barenholz, 1988). Most of the liposome-entrapped DXR (>90%) is intercalated in the lipid bilayer by specific interactions resulting from the affinity of DXR for negatively-charged phospholipids (Goormaghtigh and Ruysschaert, 1984). An interesting technique of remote loading of DXR into preformed liposomes has been reported by Mayer et al. (1985). This method results in more than 90% of drug capture, purportedly in the water phase, and does not require the inclusion of negatively charged lipids. No pharmacological and toxicological data comparing this method to the former and conventional method of bilayer intercalation have yet been published.

TABLE III   ENGINEERING OF LIPOSOMES AS CARRIERS OF CYTOTOXIC DRUGS

	1.	Efficiency of drug capture
	2.	Stability of liposome--Drug association in plasma
PRECLINICAL	3.	Preservation of _in vitro_ cytotoxic activity
PHASE	4.	In _vivo_ tissue distribution and pharmacokinetics in normal and tumor-bearing animals
	5.	Anti-tumor activity in representative animal models
	6.	Toxicological evaluation
	7.	Critical analysis of factors involved in pharmaceutical development (e.g., shelf-life, batch reproducibility, liquid vs lyophilized preps)

CLINICAL PHASE with optimally defined preclinical formulation		Phase I study - M.T.D.
		                       - Toxicity
		                       - Pharmacokinetics
		Phase II etc. ...

Liposome entrapment does not interfere with the cytotoxic activity of DXR in in vitro tests with non-phagocytic cells lines (Gabizon, et al., 1986b). It is as yet unclear whether the mechanism of action involves leakage of DXR in the extracellular medium and subsequent entry into cells or direct membrane-to-membrane transfer of the drug.

Regarding plasma stability of liposome-associated DXR we have found a moderate effect of CH increasing drug retention (Gabizon, et al., 1986b). CL-containing liposomes were very unstable releasing a large fraction of DXR, even in the presence of CH. However, Rahman et al. (1983) have reported that stable liposomes are formed when CL and stearylamine are both present. Because of the significant binding of DXR to various plasma proteins, accurate determination of the fraction of released drug requires separation of liposomes from plasma proteins. This may be difficult in the case of small unilamellar liposomes. Most of the results reported in the literature based on gel filtration or dialysis probably underestimate the leakage.

Tissue distribution studies point at increased DXR concentrations in liver and spleen and at decreased DXR concentrations in the heart muscle (Gabizon et al., 1982). These changes in drug distribution have been found by most investigators and have potential clinical implications with regard to reduction of cardiotoxicity and increased therapeutic activity against liver metastases. This is stressed by liver cell fractionation experiments indicating a higher drug concentration in tumor cells infiltrating the liver when DXR was administered in liposome-associated form (Gabizon et al., 1983).

The pharmacokinetics of L-DXR in rodents has been studied by Rahman et al. (1986a). Recently, we have obtained pharmacokinetic data in humans in the frame of a pilot clinical study with a formulation made of PG, PC, and CH (Gabizon et al., submitted for publication). Plasma clearance of L-DXR is significantly diminished, resulting in a greater area under the curve. Liposome delivery also results in a smaller apparent volume of distribution of DXR. Although DXR blood levels after L-DXR injection are generally higher and more sustained than those obtained after an equal dose of free DXR, liposome-medicated delivery probably results in reduced DXR bioavailability, since toxicity is actually decreased. We have proposed the scheme shown in Figure 1, as a pharmacokinetic model for L-DXR, which is probably applicable to other liposome-entrapped drugs. This model points at the additional complexity introduced by liposomes in the pharmacokinetic analysis of DXR. Two important aspects have to be considered:

1. Drug levels measured in plasma represent the pooled data of three fractions; liposome-bound (L-DRUG), protein-bound (P-DRUG), and free drug (F-DRUG). Since toxicity is best correlated with the free diffusible fraction of drug it is obvious that understanding the pharmacodynamics of L-DXR will require discriminating between the various drug plasma compartments.

2.   The RES may function as a drug depot. The work of
Martin et al. (1982) suggests that macrophages can store and
subsequently release DXR or active metabolites into the
extracellular fluid and neighboring cells. This mechanism of
action is strongly supported by the work of Storm (1987). Using
an in vivo model, peritoneal macrophages harvested after IP
injection of L-DXR were shown to release DXR into the medium in
active form. A similar finding was obtained when Kupffer cells
were fed with DXR-containing liposomes in vitro.

PHARMACOKINETICS OF IV–ADMINISTERED LIPOSOME–ENTRAPPED DRUGS

PLASMA

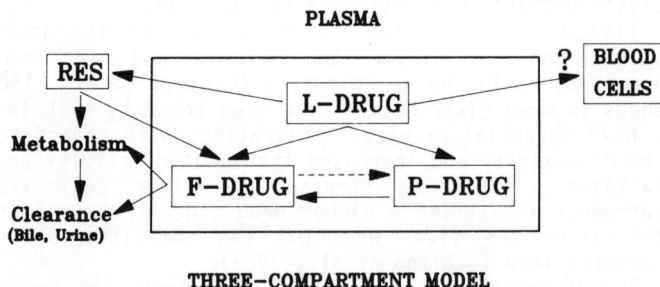

THREE–COMPARTMENT MODEL

FIGURE 1: As a result of leakage and exchange with proteins, the
drug present in plasma may be distributed in the three
fractions shown (L-Drug, P-Drug, and F-Drug).   The
percent of each of these fractions will vary depending
on the nature of the drug and the stability of the drug-
carrier complex. Drug will be removed from circulation
either through liposome uptake by the RES or as free
drug following standard metabolic and clearance
pathways.   The RES may also release intact or
metabolized drug in free form back into circulation.
The possibility of a direct uptake of liposome-
encapsulated drugs by some types of blood cells exists,
but its quantitative significance is questionable. The
proposed model addresses especially liposomes of ≥ 0.2
um to 1 um diameter. With smaller liposomes, fluid-
phase pinocitic uptake by hepatocytes becomes
significant.   With larger liposomes, arrest in the
pulmonary capillary bed may occur.

In addition, quantitative changes in metabolic pathways resulting from a different pattern of drug distribution are likely to occur. With regard to this last point, we have observed a greater area under the curve for doxorubicinol (an active DXR metabolite) than for DXR in a group of five patients receiving L-DXR. Administration of free DXR, on the contrary, results in a greater area under the curve for DXR, the parent compound.

Systemic toxicity, cardiotoxicity and nephrotoxicity are all significantly decreased when DXR is administered in liposome-associated form (Van Hoesel et al., 1984; Gabizon et al., 1986a). The mouse LD50 of DXR is increased approximately two-fold using PG:PC:CH liposomes (Olson et al., 1982; Gabizon et al, 1986a). Decreased cardiotoxicity has also been confirmed in dogs by the Georgetown University group. A possible drawback of L-DXR is that it does not appear to offer any advantages over free DXR in terms of myelosuppression.

We have recently reviewed our preclinical work on the anti-tumor efficacy of L-DXR (Gabizon et al, in press). Briefly, our studies indicate that L-DXR is more active than free DXR on tumors infiltrating the liver and spleen, and equally effective on bone marrow-residing leukemic cells. In contrast, free DXR was more effective than milligram-equivalent doses of L-DXR when tested by the IV route against ascitic and SC-implanted tumors. The anti-tumor effect correlates well with differences in drug levels in the relevant anatomic areas. Our observations are in general agreement with those of Mayhew and Rustum (1985). Studies of Rahman et al. (1986b) have indicated that using the respective maximal tolerable doses of free DXR and L-DXR in various tumor models, including SC-growing tumors, a superior therapeutic index is obtained with L-DXR. Regarding liposome composition, we have observed no difference in antitumor activity using PG or PS as negatively-charged lipid and reducing the CH content from a CH:phospholipid molar ratio of 100:100 to 25:100 (Gabizon et al., 1985). In a recent report by Storm et al. (1987), the dependence of anti-tumor activity on the phase-transition temperature of liposome components was studied. It was found that inhibition of tumor growth was delayed, albeit effective, when "rigid" liposome compositions were compared to "fluid" ones, thus suggesting that, in the former, the in vivo release of drug is slower than in the latter.

Scanty information is available from the literature on stability of L-DXR. Crommelin and Van Bloois (1983) studied the stability of various L-DXR preparations upon storage and found that the most stable ones are 0.2 um-extruded, negatively-charged vesicles. Our observations with a formulation of L-DXR prepared for clinical use indicate stable physicochemical parameters for at least several months (Gabizon et al., unpublished results).

Human studies with L-DXR are ongoing in Israel, USA, and Britain (Gabizon et al., 1986c; Treat et al, 1987; Sells et al., 1987). We have summarized our observations on 20 patients who received escalating doses of L-DXR from 20 to 70 mg/m^2 (Peretz et al., 1987). Treatment was generally well tolerated and acute toxic effects such as nausea and vomiting were mild and infrequent. Although the maximal tolerated dose has not yet been reached, the results point at bone marrow toxicity as the most likely single dose-limiting factor. Our studies also indicate that the variable fraction of free DXR in the L-DXR batches is the most significant technical problem. The issue of removal of free drug from liposome preparations by simple and quick methods has been addressed by Storm et al. (1985), who have reported on a useful method based on the interaction of DXR with cation-exchange resins. A critical prerequisite for clinical trials with liposome-encapsulated drugs will be the use of well-standardized and stable preparations with reproducible physicochemical properties. Until those become available, clinical observations should be cautiously assessed as pilot studies.

B.   Cytosine Arabinoside (Ara-C)

Unlike DXR, Ara-C does not interact with the liposome bilayer and is entrapped in the water phase only. Therefore, the efficiency of capture is relatively low (1 to 20%, depending on the liposome size) when lipid concentrations of approximately 60 umol/ml are used (Rustum et al., 1979). Technological improvements allowing the processing of high concentrations of lipid (300 umol/ml) result in excellent capture efficiency surpassing 70% of the total drug at start (Mayhew et al., 1984).

Several groups of investigators have studied the pharmacology of Ara-C containing liposomes (Kobayashi et al., 1977; Rustum et al., 1979; Ganapathi et al., 1980). Ellens et al. (1982) measured the concentration of Ara-C and its metabolites after injection of liposome-entrapped Ara-C in mice. Sustained levels of Ara-C and Ara-C-triphosphate in liver and spleen were found, suggesting that liposome delivery protects Ara-C against deamination and may act as a local depot for Ara-C.

Liposomal entrapment enhances the antitumor activity of Ara-C against L1210 leukemia (Kobayashi et al., 1977; Rustum et al., 1979), although toxicity is also increased (Rustum et al., 1979). The therapeutic efficacy of the liposome-entrapped drug is similar to that of a five day IV infusion of free drug (Mayhew et al., 1982), supporting the conclusion that Ara-C liposomes act by slowly releasing the drug into the extracellular space. It would seem that the main advantage that liposome-entrapped Ara-C can offer in future clinical trials is a logistical one: a single injection would be much simpler for administration purposes than prolonged infusions of free drug in the treatment of human leukemia.

C.   Liposomes As Vehicles for the Administration of Lipophilic (Water-Insoluble) Drugs

Many cytotoxic drugs are difficult to administer by the intravenous route because of solubility problems. This is the case of some nitrosoureas, Hexamethylmelamine, Etoposide (VP16),

Bisantrene, and other compounds which require the co-administration of ethanol or detergents for solubilization. In some instances, the oral route, which often results in erratic absorption, is the only way of administration. For a review publication dealing with the administration and pharmacokinetics of cytotoxic drugs, see Ames et al., 1983. The activity of a water-insoluble, quinazolone derivative (NSC 251635), in liposomes has been studied at the Jules Bordet Institute (Belgium). Using sonicated vesicles containing PC, CH, and stearylamine, it was found that the drug was active against L1210 leukemia only when encapsulated in liposomes. A suspension of free drug in 0.3% hydroxypropylcellulose was totally inactive (Brassinne et al., 1983). Intravenous infusion of liposomes containing NSC 251635 has been reported in 14 patients (Sculier et al., 1986). Up to 12 g lipid and 456 mg drug per m^2 body surface were given in a single infusion. Side effects were observed during some infusions (sedation, fever, chills, lumber pain, urticarial rash, bronchospasm), but no dose-limiting toxicity was observed. No objective tumor regressions were detected. An interesting observation was an important activation of the complement system in all patients studied. The presence of stearylamine, a relatively toxic lipid confering to the vesicles a net positive charge may account for this phenomenon.

This approach approach to liposome-mediated delivery exemplifies two serious problems in the strategy for clinical testing of liposome-encapsulated drugs:

1.   No baseline reference of toxicity and anti-tumor activity with free drug in humans is available. This makes difficult to assess the real value of liposome delivery. A drug-directed effort in liposome development can be frustrating with the choice of a drug which, a priori, may have no activity against human cancer.

2.   Methods for the production of liposomal drug batches with reasonable stability and reproducibility have not been developed. Clinical data obtained with material prepared separately for each individual patient (Sculier et al., 1986) can hardly be validated.

Perez-Soler et al. (1986) and Lautersztain et al. (1986) have used liposomes as vehicles of lipophilic cDDP analogues. Using multilamellar vesicles consisting of dimirystoyl-PC and dimiristoyl-PG, Platinum levels were increased in lung, liver and spleen and reduced in small bowel following liposome encapsulation. This approach also results in a high entrapment efficiency of the cDDP derivative, reduced toxicity and nephrotoxicity, and similar anti-tumor activity against L1210 leukemia as compared to free cDDP. The IP route was used in toxicity and anti tumor studies (Perez-Soler et al., 1986). Given the poor encapsulation efficiency and low solubility of cDDP, the use of a lipophilic derivative provides a useful alternative, especially if these results are extended to the IV route and to solid tumor models.

D.   Liposome-Dependent Cytotoxic Drugs

The term "liposome-dependent drug" was first coined by Heath et al. (1983), and refers to drugs with poor ability to enter cells and whose toxicity can be significantly enhanced by intracytoplasmic delivery. MXT-Asp (Heath et al., 1985a) and 5-FO (Heath et al., 1985b) have been described as liposome-dependent drugs. MTX-Asp is an effective dihydrofolate reductase inhibitor that enters cells 100 times less effectively and is 160 times less toxic than MTX. When encapsulated in liposomes, it becomes up to 150 times more potent than the free drug, depending on the ability of the cell line to take up liposomes. 5-FO is a derivative of 5-Fluouracil for which no cellular transport systems exist. Both drugs may be metabolized to fluorodeoxyuridine monophosphate, an inhibitor of thymidilate synthetase. The cytotoxic potency of 5-FO is increased up to 35-fold by liposome encapsulation. In contrast to these examples of liposome-dependent drugs, most known drugs are liposome-independent and show an in vitro potency that is equal or less than free drug. Their activity relies on rapid flux through the cell membrane following leakage from liposomes, as shown for Ara-C (Allen et al., 1981).

Liposome-dependent drugs represent an interesting approach to cancer chemotherapy, provided sufficient localization of liposomes in the tumor area can be achieved as in the case of small, long-circulating vesicles (Proffitt et al., 1982; Papahadjopoulos and Gabizon, in press). However, as long as crucial in vivo toxicity studies to assess median lethal doses and damage to Kupffer cells and hepatocytes have not been performed, the advantages of this approach remain speculative.

CONCLUDING REMARKS

Although significant advances in the field of liposomes as drug carriers have been made in the last few years, its future applicability in cancer therapy remains uncertain. It is imperative to adopt an experimental approach and a clinical strategy with a sound rationale. The development of pharmaceutically acceptable liposome preparations (Fildes, 1981) is an absolute requirement for further progress in this area. Although there are encouraging notes in this direction, a huge effort is still needed to overcome the technological barrier.

Despite much controversy around liposomes, no other particulate carriers currently offer a useful alternative for systemic drug delivery. When the clinical perspective is considered, it is likely that a significant impact on the therapy of systemic metastatic disease will require the development of liposome formulations with at least some degree of localization in extra-RES tumor sites. However, a clinical strategy aiming at the therapy of primary and metastatic liver tumors and some hematopoietic malignancies, with concomitant reduction of toxicity to specific tissues, already appears as a feasible and well-founded approach with a reasonable probability of success.

ACKNOWLEDGEMENTS

I would like to thank the Office Support Center at Liposome Technology, Inc. for their excellent work in typing and formatting the manuscript.

REFERENCES

Abra R.M., Bosworth, M.E. and Hunt C.A. (1980). "Liposome Disposition In Vivo: Effects of Pre-Dosing with Liposomes", Res. Commun. Chem. Pathol. Pharmacol., 29, 349-360.

Abra R.M., Hunt C.A., Fu K.K. and Peters J.H. (1983). "Delivery of Therapeutic Doses of Doxorubicin to the Mouse Lung Using Lung-Accumulating Liposomes Proves Unsuccessful", Cancer Chemother. Pharmacol., 11, 98-101.

Allen T.M., McAllister L., Mausolf S. and Gyorffy E. (1981). "Liposome-Cell Interactions: A Study of the Interaction of Liposomes Containing Entrapped Anticancer Drugs with the EMT6, S49 and AE1 (transport-deficient) Cell Lines", Biochim. Biophys. Acta., 643, 346-362.

Allen T.M. and Everest J. (1983). "Effect of Liposome Size and Drug Release Properties on Pharmacokinetics of Encapsulated Drug in Rat", J. Pharmacol. Exp. Ther., 226, 539-544.

Allen T.M., Murray L., MacKeigan S., and Shah M. (1984). "Chronic Liposome Administration in Mice: Effects on Reticuloendothelial Function and Tissue Distribution", J. Pharmacol. Exp. Ther., 229, 267-275.

Ames M.M., Powis G. and Kovach J.S. (1983). "Pharmacokinetics of Anticancer Agents in Humans", Elsevier, Amsterdam.

Bailar J.C. and Smith E.M. (1986). "Progress Against Cancer?", N. Engl. J. Med., 314, 1226-1232.

Bassett J.B., Anderson R.U. and Tacker J.R. (1986). "Use of Temperature-Sensitive Liposomes in the Selective Delivery of Methotrexate and Cis-Platinum Analogues to Murine Bladder Tumor", J. Urol., 135, 612-615.

Brassinne C., Atassi G., Fruhling J., Penasse W., Coune A., Hildebrand J., Ruysschaert J.M. and Laduron C. (1983). "Antitumor Activity of a Water-Insoluble Compound Entrapped in Liposomes on L1210 Leukemia in Mice", J. Natl. Cancer Inst., 70, 1081-1086.

Cervato G., Viani P., Galatulas I., Bossa R., and Cestaro B. (1986). "In Vitro and In Vivo Studies with Anionic Sulfatide-Liposomes Containing Adriamycin", Anticancer Res., 6, 1287-1290.

Connor J. and Huang L. (1986). "PH-Sensitive Immunoliposomes as an Efficient and Target-Specific Carrier for Antitumor Drugs", Cancer Res., 46, 3431-3435.

Crommelin D.J. and Van Bloois L. (1983). "Preparation and Characterization of Doxorubicin Containing Liposomes. II Loading Capacity, Long-Term Stability and Doxorubicin-Bilayer Interaction Mechanism", Int. J. Pharm., 17, 135-144.

Ellens H., Rustum Y., Mayhew E. and Ledesma E. (1982). "Distribution and Metabolism of Liposome-Encapsulated and Free Ara-C in Dog and Mouse Tissues", J. Pharmacol. Exp. Ther., 222, 324-330.

Fichtner I., Reszka R., Elbe B. and Arndt D. (1981). "Therapeutic Evaluation of Liposome-Encapsulated Daunoblastin in Murine Tumor Models", Neoplasma, 28, 141-149.

Fitchner I., Arndt D., Elbe B. and Reszka R. (1984). "Cardiotoxicity of Free and Liposomally Encapsulated Rubomycin (Daunorubicin) in Mice", Oncology, 41, 363-369.

Fildes F.J. (1981). "Liposomes: The Industrial Viewpoint", In Liposomes: From Physical Structure to Therapeutic Applications (Ed., C.G. Knight), pp. 465-485, Elsevier, Amsterdam.

Forssen E.A. and Tokes Z.A. (1979). "In Vitro and In Vivo Studies with Adriamycin Liposomes", Biochem. Biophys. Res. Commun., 91, 1295-1301.

Forssen E.A., and Tokes Z.A. (1981). "Use of Anionic Liposomes for the Reduction of Chronic Doxorubicin-Induced Cardiotoxicity", Proc. Natl. Acad. Sci. U.S.A., 78, 1873-1877.

Forssen E.A. and Tokes Z.A. (1983a). "Improved Therapeutic Benefits of Doxorubicin by Entrapment in Anionic Liposomes", Cancer Res., 43, 546-550.

Forssen E.A. and Tokes Z.A. (1983b). "Attenuation of Dermal Toxicity of Doxorubicin by Liposome Encapsulation", Cancer Treat. Rep., 67, 481-484.

Frei E. (1985). "Curative Cancer Chemotherapy", Cancer Res., 45, 6523-6537.

Gabizon A., Dagan A., Goren D., Barenholz Y. and Fuks Z. (1982). "Liposomes as In Vivo Carriers of Adriamycin: Reduced Cardiac Uptake and Preserved Anti-Tumor Activity in Mice", Cancer Res, 42, 4734-4739.

Gabizon A., Goren D., Fuks Z., Dagan A., Barenholz Y. and Meshorer A. (1983). "Enhancement of Adriamycin Delivery to Liver Metastatic Cells with Increased Tumoricidal Effect Using Liposomes as Drug Carriers", Cancer Res., 43, 4730-4735.

Gabizon A., Goren D., Fuks Z., Meshorer A., and Barenholz Y. (1985). "Superior Therapeutic Activity of Liposome-Associated Adriamycin in a Murine Metastatic Tumor Model", Br. J. Cancer, 51, 681-689.

Gabizon A., Meshorer A., and Barenholz Y. (1986a). "Comparative Long-Term Study of the Toxicities of Free and Liposome-Associated Doxorubicin in Mice After Intravenous Administration", J. Natl. Cancer Inst., 77, 459-469.

Gabizon A., Goren D., Ramu A., and Barenholz Y. (1986b). "Design, Characterization and Anti-Tumor Activity of Adriamycin-Containing Phospholipid Vesicles," in, Targeting of Drugs with Synthetic Systems (Eds., G. Gregoriadis, J. Senior and G. Poste), pp. 229-238, Plenum, New York.

Gabizon A., Peretz T., Ben-Yosef R., Catane R., Biran S. and Barenholz Y. (1986c). "Phase I Study with Liposome-Associated Adriamycin: Preliminary Report", Proc. Am. Soc. Clin. Oncol., 5, 43 (Abstract).

Gabizon A., Goren D. and Barenholz Y. (in press). "Investigations on the Antitumor Efficacy of Liposome-Associated Doxorubicin in Murine Tumor Models", Isr. J. Med. Sci.

Gabizon A. and Barenholz Y. (1988). "Adriamycin-Containing Liposomes in Cancer Chemotherapy", in Liposomes as Drug Carriers: Trends and Progress (Ed., G. Gregoriadis), pp. 365-379, Wiley, London.

Ganapathi R., Krishan A., Wodinsky I., Zubrod C.G. and Lesko L.J. (1980). "Effect of Cholesterol Content on Antitumor Activity and Toxicity of Liposome-Encapsulated 1-Beta-D-Arabinofuranosylcytosine In Vivo", Cancer Res., 40, 630-633.

Ganapathi R. and Krishan A. (1984). "Effect of Cholesterol Content of Liposomes on the Encapsulation, Efflux and Toxicity of Adriamycin", Biochem. Pharmacol., 33, 698-700.

Goormaghtigh E., and Ruysschaert J.M. (1984). "Anthracycline Glycoside Membrane Interactions", Biochim. Biophys. Acta, 779, 271-288.

Gregoriadis G., Wills E.F., Swain C.P., and Tavill A.S. (1974). "Drug Carrier Potential of Liposomes in Cancer Chemotherapy", Lancet, I, 1313-1317.

Gregoriadis G. (1977). "Targeting of Drugs", Nature, 265, 407-411.

Gregoriadis G. and Senior J. (1986). "Liposomes In Vivo: A Relationship between Stability and Clearance", in Targeting of Drugs with Synthetic Systems (Eds., G. Gregoriadis, J. Senior, and G. Poste), pp. 183-192, Plenum, New York.

Gregoriadis G. (1988). "Liposomes as Drug Carriers: Trends and Progress", Wiley, London.

Heath T. (1982). "Antibody-directed Liposomes: Achievements and Potential", Cancer Surveys, 1, 417-427.

Heath T.D., Montgomery J.A., Piper J.R. and Papahadjopoulos D. (1983). "Antibody-Targeted Liposomes: Increase in Specific Toxicity of Methotrexate-gamma-Aspartate", Proc. Natl. Acad. Sci. USA, 80, 1377-1381.

Heath T.D., Lopez N.G., and Papahadjopoulos D. (1985a). "The Effects of Liposome Size and Surface Charge on Liposome-Mediated Delivery of Methotrexate-gamma-Aspartate to Cells In Vitro", Biochim. Biophys. Acta., 820, 74-84.

Heath T.D., Lopez N.G., Stern W.H., and Papahadjopoulos D. (1985b). "5-Fluoorotate: A New Liposome Dependent Cytotoxic Agent", FEBS Lett., 187, 73-73.

Herman E.H., Rahman A., Ferrans V.J., Vick J.A. and Schein P.S. (1983). "Prevention of Chronic Doxorubicin Cardiotoxicity in Beagles by Liposomal Encapsulation", Cancer Res., 43, 5427-5432.

Hwang K.J., Luk K.K. and Beaumier P.L. (1980). "Hepatic Uptake and Degradation of Unilamellar Sphingomyelin/Cholesterol Liposomes: A Kinetic Study", Proc. Natl. Acad. Sci. U.S.A., 77, 4030-4034.

Jain R.K. and Gerlowski L.E. (1986). "Extravascular Transport in Normal and Tumor Tissues", Crit. Rev. Oncol. Hematol., 5, 115-170.

Juliano R.L. and McCullough H.N. (1980). "Controlled Delivery of an Antitumor Drug: Localized Action of Liposome-Encapsulated Cytosine Arabinoside Administered Via the Respiratory System", J. Pharmacol. Exp. Ther., 214, 381-387.

Kaledin V.I., Matienko N.A., Nikolin V.P., Gruntenko Y.V. and Budker V.G. (1981). "Intralymphatic Administration of Liposome-Encapsulated Drugs to Mice: Possibility for Suppression of the Growth of Tumor Metastases in the Lymph Nodes", J. Natl. Cancer Inst., 66, 881-887.

Kaye S.B., Borden J.A. and Ryman B.E. (1981). "The Effect of Liposome Entrapment of Actinomycin D and Methotrexate on the In Vivo Treatment of Sensitive and Resistant Solid Murine Tumors", Europ. J. Cancer, 17, 279-289.

Kim S., Kim D.J., Geyer M.A. and Howell S.B. (1987). "Multivesicular Liposomes Containing 1-Beta-D-Arabinofuranosyl-cytosine for Slow Release Intrathecal Therapy", Cancer Res., 47, 3935-3937.

Kimelberg H.K., Tracy T.F., Watson R.E., Kung D., Reiss F.L. and Bourke R.S. (1978). "Distribution of Free and Liposome-Entrapped 3H-MTX in the Central Nervous System after Intracerebroventricular Injection in a Primate", Cancer Res., 38, 706-712.

Kobayashi T., Kataoka T., Tsukagoshi S. and Sakurai Y. (1977). "Enhancement of Antitumor Activity of 1-Beta-D-Arabinofuranosylcytosine by Encapsulation in Liposomes", Int. J. Cancer, 20, 581-587.

Kojima N., Ueno N., Takano M., Yabushita H., Nogushi M., Ishihara M. and Yagi K. (1986). "Effect of Adriamycin Entrapped by Sulfatide-Containing Liposomes on Ovarian Tumor-Bearing Nude Mice", Biotechnol. Appl. Biochem., 8, 471-478.

Konno H., Suzuki H., Tadakuma T., Kumai K., Yasuda T., et al. (1987). "Antitumor Effect of Adriamycin Entrapped in Liposomes Conjugated with Anti-Human Alpha-Fetoprotein Monoclonal Antibody", Cancer Res., 47, 4471-4477.

Kumai K., Takahashi T., Tsubouchi K., Yoshino K., Ishibiki K., and Abe O. (1985). "Selective Hepatic Arterial Infusion of Liposomes Containing Anti-Tumor Agents", Jpn. J. Cancer Chemother., 12, 1946-1948.

Lautersztain J., Perez-Soler R., Khokhar A.R., Newman R.A. and Lopez-Berestein G. (1986). "Pharmacokinetics and Tissue Distribution of Liposome-Encapsulated Cis-Bis-N-Decyl-Iminodiacetato-1,2-Diaminocyclohexane-Platinum (II)", Cancer Chemother. Pharmacol., 18, 93-97.

Litterst C.L., Sieber S.M., Copley M. and Parker R.J. (1982). "Toxicity of Free and Liposome-Encapsulated Adriamycin Following Large Volume, Short-Term Intraperitoneal Exposure In the Rat", Toxicol. Appl. Pharmacol., 64, 517-528.

Lopez-Berestein G., Kasi L., Rosenblum M.G., Haynie T., Jahns M., et al. (1984). "Clinical Pharmacology of 99mTc-labeled Liposomes in Patients with Cancer", Cancer, Res., 44, 375-378.

Martin F., Caignard A., Olson O., Jeanin J.F., and Leclerc A. (1982). "Tumoricidal Effect of Macrophages Exposed to Adriamycin In Vivo and In Vitro", Cancer Res., 42, 3851-3855.

Mauk M.R., Gamble R.C., and Baldeschwieler J.D. (1980). "Targeting of Lipid Vesicles: Specificity of Carbohydrate Receptor Analogues for Leukocytes in Mice", Proc. Natl. Acad. Sci. USA, 77, 4430-4434.

Mayer L.D., Bally M.B., Hope M.J., and Cullis P.R. (1985). "Uptake of Antineoplastic Agents into Large Unilamellar Vesicles in Response to a Membrane Potential", Biochim. Biophys. Acta, 816, 294-302.

Mayhew E., Rustum Y.E., Szoka F. and Papahadjopoulos D. (1979). "Role of Cholesterol in Enhancing the Antitumor Activity of Cytosine Arabinoside Entrapped in Liposomes", Cancer Treat. Rep., 63, 1923-1928.

Mayhew E., Rustum Y.M. and Szoka F. (1982). "Therapeutic Efficacy of Cytosine Arabinoside Trapped in Liposomes", in Targeting of Drugs (Eds., G. Gregoriadis, J. Senior, A. Trouet), pp. 249-260, Plenum, New York.

Mayhew E. and Rustum Y. (1983). "Effect of Liposome Entrapped Chemotherapeutic Agents on Mouse Primary and Metastatic Tumors", Biol. Cell, 47, 81-86.

Mayhew E., Rustum Y. and Vail W.J. (1983). "Inhibition of Liver Metastases of M5076 Tumor by Liposome-Entrapped Adriamycin", Cancer Drug Deliv., 1, 43-58.

Mayhew E. and Papahadjopoulos D. (1983). "Therapeutic Applications of Liposomes", in Liposomes (Ed. M.J. Ostro), pp. 289-341, Marcel Dekker, New York.

Mayhew E., Lazo R., Vail W.J., King J. and Green A.M. (1984). "Characterization of Liposomes Prepared Using a Microemulsifier", <u>775</u>, 169-174.

Mayhew E. and Rustum Y.E. (1985). "The Use of Liposomes as Carriers of Therapeutic Agents", Prog. Clin. Biol. Res., <u>172B</u>, 301-310.

McKeran R.O., Firth G., Oliver S., Uttley D. and O'Laoire S. (1985). "A Potential Application for the Intracerebral Injection of Drugs Entrapped Within Liposomes in the Treatment of Human Cerebral Gliomas", J. Neurol. Neurosurg. Psychiatry, <u>48</u>, 1213-1219.

Olson F., Mayhew E., Maslow D., Rustum Y., and Szoka F. (1982). "Characterization, Toxicity and Therapeutic Efficacy of Adriamycin Encapsulated in Liposomes", Eur. J. Cancer Clin. Oncol., <u>18</u>, 167-176.

Onuma M., Odawara T., Watarai S., Aida Y, Ochiai K., et al. (1986). "Anti-tumor Effect of Adriamycin Entrapped in Liposomes Conjugated with Monoclonal Antibody Against Tumor-Associated Antigen of Bovine Leukemia Cells", Jpn. J. Cancer Res. (Gann), <u>77</u>, 1161-1167.

Ostro M.J. (1987). "Liposomes-From Biophysics to Therapeutics", Marcel Dekker, New York.

Papahadjopoulos D. and Gabizon A. (1987). "Targeting of Liposomes to Tumor Cells In Vivo", Ann. N.Y. Acad. Sci., <u>504</u>:64-74.

Parker R.J., Hartman K.D. and Sieber S.M. (1981). "Lymphatic Absorption and Tissue Disposition of Liposome-Entrapped [14C]Adriamycin Following Intraperitoneal Administration to Rats", Cancer Res., <u>41</u>, 1311-1317.

Parker R.J., Priester E.R. and Sieber S.M. (1982a). "Comparison of Lymphatic Uptake, Metabolism, Excretion, and Biodistribution of Free and Liposome-Entrapped [14C]Cytosine Beta-D-Arabinofuranoside Following Intraperitoneal Administration to Rats", Drug Metab. Dispos., <u>10</u>, 40-46.

Parker R.J., Priester E.R., and Sieber S.M. (1982b). "Effect of Route of Administration and Liposome Entrapment on the Metabolism and Disposition of Adriamycin in the Rat", Drug Metab. Dispos., <u>10</u>, 499-504.

Peretz T., Gabizon A., Catane R., Ben-Yosef R., Biran S., Drukman S. and Barenholz Y. (1987). "Clinical Studies on Liposome-Associated Doxorubicin: Progress Report", Proc. Am. Assoc. Clin. Oncol., <u>6</u>, 43 (Abstract).

Perez-Soler R., Khokhar A.R., Hacker M.P. and Lopez-Berestein G. (1986). "Toxicity and Antitumor Activity of Cis-Bis-Cyclopentenecarboxylato-1,2-Diaminocyclohexane Platinum (II) Encapsulated in Multilamellar Vesicles," Cancer Res., 46, 6269-6273.

Peterson H.I. (1979). Ed., "Tumor Blood Circulation: Angiogenesis, Vascular Morphology, and Blood Flow of Experimental and Human Tumors", CRC Press, Boca Raton.

Poste G. and Papahadjopoulos D. (1976). "Drug-Containing Lipid Vesicles Render Drug-resistant Tumor Cells Sensitive to Actinomycin D", Nature, 261, 699-701.

Poste G., Bucana C., Raz A., Bugelski P., Kirsh R., and Fidler I.J. (1982). "Analysis of the Fate of Systemically Administered Liposomes and Implications for their Use in Drug Delivery", Cancer Res., 42, 1412-1422.

Poste G. (1983). "Liposome Targeting In Vivo: Problems and Opportunities", Biol. Cell, 47, 19-38.

Proffitt R.T., Williams L.E., Presant C.A., Tin G.W., Uliana J.A., Gamble G.C. and Baldeschwieler J.D. (1983). "Liposomal Blockade of the Reticuloendothelial System: Improved Tumor Imaging with Small Unilamellar Vesicles", Science, 220, 502-505.

Proffitt R.T., Williams L.E., Presant C.A., Tin G.W., Uliana J.A., Gamble G.C. and Baldeschwieler J.D. (1983). "Tumor Imaging Potential of Liposomes Loaded with In-111-NTA: Biodistribution in Mice", J. Nucl. Med., 24, 45-51.

Rahman Y.E., Cerny E.A., Patel K.R., Lau E.H. and Wright B.J. (1982). "Differential Uptake of Liposomes Varying in Size and Lipid Composition by Parenchymal and Kupffer Cells of Mouse Liver", Life Sci., 31, 2061-2071.

Rahman A., Kessler A., More N., Sikic B., Rowden G., Woolley P., and Schein P.S, (1980). "Liposomal Protection of Adriamycin-Induced Cardiotoxicity in Mice", Cancer Res., 40, 1532-1537.

Rahman A., More N. and Schein P.S. (1982). "Doxorubicin-Induced Chronic Cardiotoxicity and Its Protection by Liposomal Administration", Cancer Res., 42, 1817-1825.

Rahman A., Fumagalli A., Goodman A. and Schein P.S. (1984). "Potential of Liposomes to Ameliorate Anthracycline-Induced Cardiotoxicity", Semin. Oncol., 11, 45-55.

Rahman A., White G., More N. and Schein P.S. (1985). "Pharmacological, Toxicological, and Therapeutic Evaluation in Mice of Doxorubicin Entrapped in Cardiolipin Liposomes", Cancer Res., 45, 796-803.

Rahman A., Carmichael D., Harris M. and Roh J.K. (1986a). "Comparative Pharmacokinetics of Free Doxorubicin and Doxorubicin Entrapped in Cardiolipin Liposomes", Cancer Res., 46, 2295-2299.

Rahman A., Fumagally A., Barbieri B., Schein P.S., and Casazza A.M. (1986b). "Anti-Tumor and Toxicity Evaluation of Free Doxorubicin and Doxorubicin Entrapped in Cardiolipin Liposomes", Cancer Chemother. Pharmacol., 16, 22-27.

Rahman A., Ganjei A. and Neefe J.R. (1986c). "Comparative Immunotoxicity of Free Doxorubicin and Doxorubicin Encapsulated in Cardiolipin Liposomes", Cancer Chemother. Pharmacol., 16, 28-34.

Richardson V.J., Ryman B.E., Jewkes R.F., Jeyasingh K., Tattersall M.H., Newlands E.S., and Kaye S.B. (1979). "Tissue Distribution and Tumor Localization of Technetium-99M Labeled Liposomes in Cancer Patients", Br. J. Cancer, 40, 35-43.

Richardson V.J., Curt G.A. and Ryman B.E. (1982). "Liposomally Trapped Ara-CTP to Overcome Ara-C Resistance in a Murine Lymphoma In Vitro", Br. J. Cancer, 45, 559-564.

Richardson V.J. and Ryman B.E. (1982). "Effect of Liposomally Trapped Antitumor Drugs on a Drug-Resistant Mouse Lymphoma In Vivo", Br. J. Cancer, 45, 552-558.

Roerdink F., Dijkstra J., Hartman G., Bolscher B., and Scherphof G. (1981). "The Involvement of Parenchymal, Kupffer and Endothelial Liver Cells in Hepatic Uptake of Intravenously Injected Liposomes. Effects of Lanthanum and Gadolinium Salts", Biochim. Biophys. Acta., 677, 79-89.

Rosa P. and Clementi F. (1983). "Absorption and Tissue Distribution of Doxorubicin Entrapped in Liposomes Following Intravenous or Intraperitoneal Administration", Pharmacology, 26, 221-229.

Rustum Y.M., Dave C., Mayhew E. and Papahadjopoulos D. (1979). "Role of Liposome Type and Route of Administration in the Antitumor Activity of Liposome-Entrapped 1-Beta-D-Arabinofuranosylcytosine Against Mouse L1210 Leukemia", Cancer Res., 39, 1390-1395.

Rustum Y.M., Mayhew E., Szoka F. and Campbell J. (1981). "Inability of Liposome-Encapsulated 1-Beta-D-Arabinofuranosylcytosine Nucleotides to Overcome Drug Resistance in L1210 Cells", Eur. J. Cancer Clin. Oncol., 17, 809-817.

Scherphof G., Roerdink F., Dijkstra J., Ellens H., DeZanger R. and Wisse E. (1983). "Uptake of Liposomes by Rat and Mouse Hepatocytes and Kupffer Cells", Biol. Cell, 47, 47-58.

Sculier J.P., Coune A., Brassinne C., Laduron C., Atassi G., Ruysschaert J.M. and Fruhling J. (1986). "Intravenous Infusion of High Doses of Liposomes Containing NSC 251635, a Water-Insoluble Cytostatic Agent. A Pilot Study with Pharmacokinetic Data", J. Clin. Oncol., 4, 789-797.

Segal A.W., Wills E.J., Richmond J.E., Slavin G., Black C.D., and Gregoriadis G. (1974). "Morphological Observations on the Cellular and Subcellular Destination of Intravenously Administered Liposomes", Br. J. Exp. Pathol., 55, 320-327.

Segal A.W., Gregoriadis G., Lavender J.P., Tarin D., and Peters T.J. (1976). "Tissue and Hepatic Subcellular Distribution of Liposomes Containing Bleomycin after Intravenous Administration to Patients with Neoplasms", Clin. Sci. Mol. Med., 51, 421-425.

Sells R.A., Owen R.R., New R.R. and Gilmore I.T. (1987). "Reduction in Toxicity of Doxorubicin by Liposomal Entrapment", Lancet, II, 8559 (Letter).

Shinozawa S., Araki Y. and Oda T. (1981). "Tissue Distribution and Antitumor Effect of Liposome-Entrapped Doxorubicin (Adriamycin) in Ehrlich Solid Tumor-Bearing Mouse", Acta. Med. Okayama, 35, 395-405.

Shubik P. (1982). "Vascularization of Tumors: A Review", J. Cancer Res. Clin. Oncol., 103, 211-226.

Storm G., Van Bloois L., Brouwer M. and Crommelin D.J. (1985). "The Interaction of Cytostatic Drugs with Adsorbents in Aqueous media. The Potential Implications for Liposome Preparation", Biochim. Biophys. Acta., 818, 343-351.

Storm G. (1987). "Liposomes as Delivery System for Doxorubicin in Cancer Chemotherapy", (Ph.D. Thesis), Optimax Press, The Netherlands.

Storm G., Roerdink F.H., Steerenberg P.A., deJong W.H. and Crommelin D.J. (1987). "Influence of Lipid Composition on the Antitumor Activity Exerted by Doxorubicin-containing Liposomes in a Rat Solid Tumor Model", Cancer Res., 47, 3366-3372.

Storm G., Van Gessel H.J., Steerenberg P.A., Speth P.A., Roerdink F.H., et al. (in press). "Investigation of the Role of Mononuclear Phagocytes in the Transportation of Doxorubicin-Containing Liposomes into a Solid Tumor", Cancer Drug Deliv.

Tacker J.R. and Anderson R.U. (1982). "Delivery of Antitumor Drug to Bladder Cancer by Use of Phase Transition Liposomes and Hyperthermia", J. Urol., 127, 1211-1214.

Todd J.A., Modest E.J., Rossow P.W. and Tokes Z.A. (1982). "Liposome Encapsulation Enhancement of Methotrexate Sensitivity in a Transport Resistant Human Leukemic Cell Line", Biochem. Pharmacol., 31, 541-546.

Treat J., Roh J.K., Woolley P.V., Neefe J., Schein P.S. and Rahman A. (1987). "A Phase I Study: Liposome-Encapsulated Doxorubicin", Proc. Am. Soc. Clin. Oncol., 6, 31 (Abstract).

Van Hoesel Q.G., Steerenberg P.A., Crommelin D.J., Van Dijk A., Van Oort W., Klein S., Douze J.M., de Wildt D.J., and Hillen F.C. (1984). "Reduced Cardiotoxicity and Nephrotoxicity with Preservation of Anti-Tumor Activity of Doxorubicin Entrapped in Stable Liposomes in the Lou/M Wsl Rat", Cancer Res., 44, 3698-3705.

Weinstein J.N., Magin R.L., Yatvin M.B. and Zaharko M.B. (1979). "Liposomes and Local Hyperthermia: Selective Delivery of Methotrexate to Heated Tumors", Science, 204, 188-191.

Weinstein J.N. (1984). "Liposomes as Drug Carriers in Cancer Therapy", Cancer Treat. Rep., 68, 127-135.

Weinstein J.N. and Leserman L.D. (1984). "Liposomes as Drug Carriers in Cancer Chemotherapy", Pharmacol. Ther., 24, 207-233.

Wetterau J.R. and Jonas A. (1982). "Effect of Dipalmitoylphosphatidylcholine Vesicle curvature on the Reaction with Human Apolipoprotein A-I", J. Biol. Chem., 257, 10961-10966.

Wisse E. (1970). "An Electron Microscopic Study of the Fenestrated Endothelium Lining of Rat Liver Sinusoids", J. Ultrastruc. Res., 31, 125-150.

Yatvin M.B. and Lelkes P.I. (1982). "Clinical Prospects for Liposomes", Med. Phys., 9, 149-175.

Young R.C., Ozols R.F. and Myers C.E. (1981). "The Anthracycline Antineoplastic Drugs", N. Engl. J. Med., 305, 139-153.

Drug Carrier Systems
Edited by F.H.D. Roerdink and A.M. Kroon
© 1989 John Wiley & Sons Ltd.

THE USE OF LIPOSOMES AS DRUG CARRIERS IN THE IMMUNOTHERAPY OF CANCER

Isaiah J. Fidler
The University of Texas System Cancer Center,
M.D. Anderson Hospital and Tumor Institute,
Texas Medical Center, 6723 Bertner Avenue
Houston, TX 77030, USA

I.   INTRODUCTION

The lethality of most cancers is due to their propensity to spread from their primary site of growth to distant organs, producing metastases (secondary tumors). Despite major advancements in diagnosis and in surgical excision of primary tumors, most deaths of patients with solid cancers are caused by metastases. Our inability to successfully treat most metastases has several causes. First, by the time of surgical excision of many primary neoplasms, metastasis probably has already occurred. Second, even when metastases are diagnosed, the dose of therapeutic agents that can be delivered to the lesions without being toxic to the host is insufficient because of their location and number. The heterogeneous nature of malignant neoplasms and the rapid emergence of metastases resistant to conventional therapeutic regimens constitute additional major obstacles to treatment of metastasis (Review, Fidler and Hart, 1982; Fidler and Poste, 1985; Fidler and Balch, 1987). By the time of diagnosis, human neoplasms contain multiple populations of tumor cells with differences in such biological properties, as growth rate, antigenic and immunoyenic status, cell-surface receptors and products, response to cytotoxic agents, invasion, and metastatic potential (Fidler and Poste, 1985; Fidler and Balch, 1987; Nicolson, 1987). Recent data indicate that metastases can arise from the nonrandom spread of specialized subpopulations of cells that preexist within the primary tumor (Fidler and Kripke, 1977), that they can be clonal in their origin, and that different ones can originate from different progenitor cells (Talmadge et al., 1982). These data explain the clinical observation that, even within the same patient, different metastases often exhibit different susceptibilities to different therapies. The problem of tumor heterogeneity is further complicated by the finding that some metastatic cells are genetically unstable and can readily become resistant to chemotherapy (Cifone and Fidler, 1981; Fidler and Poste, 1985). Collectively, these findings point out that to be successful, the therapy of disseminated metastases must circumvent the problems of biological diversity and the development of resistance by tumor cells.

There is a growing body of evidence that activated macrophages are able to distinguish between tumorigenic and nontumorigenic cells and to kill tumor cells exhibiting various phenotypes, including resistance to other host defense mechanisms and various anticancer drugs. At least in vitro, tumoricidal macrophages function in an immunologically nonspecific manner; that is, they kill tumor cells of syngeneic, allogeneic, and xenogeneic origin without harming normal cells, even in cocultivation experiments (Fidler, 1978, 1985; Fidler and Kleinerman, 1984). In this chapter, I shall review the evidence suggesting that macrophages can be used to destroy tumor cells and concentrate on the use of liposomes to deliver immunomodulators to macrophages.

The Physiological Role of Macrophages

The most recognized physiologic role of macrophages is the clearance and catabolism of debris, which includes effete red blood cells (erythrophagocytosis). In the latter process, hemoglobin is converted to an iron-free pigment, which is subsequently utilized in erythropoiesis. Macrophages are, therefore, involved in the controlled recycling of iron. In addition, the macrophage contains several esterases involved in the metabolism of lipids and the dissolution of atheromas. When host homeostatic mechanisms are stressed, the mononuclear phagocyte system participates in complex interactions that involve cellular and humoral aspects of the inflammatory and immunologic response. The macrophage is also an important component of host defense against bacterial and fungal infections, parasitic infections, and cancer (Fidler, 1985).

Activation of Macrophages to the Tumoricidal State

Continuous functions of macrophages, such as removal of aged or damaged RBC from the circulation, are constituitive. Infrequent functions, such as defense against infections, parasites, and cancer, require the recruitment and "activation" of macrophages. The term

"activated macrophage" is operational, and its use in the literature
has been extended to describe a large number of macrophage
characteristics that may or may not be related to increased capacity
of macrophages to recognize and destroy microorganisms, parasites, or
cancer cells (Fidler, 1985). I reserve this term to describe only
those macrophages capable of lysing neoplastic tumorigenic and not
normal cells.

There are two major physiological pathways for the activation of
macrophages to cytotoxicity against microorganisms, parasites, or
cancer. Macrophages are readily activated by interaction with
microorganisms or their products (endotoxins, cell wall skeletons,
and components of bacterial cell walls). Attempts to systemically
activate macrophages by administration of microorganisms or their
products has suffered from major drawbacks because it was accompanied
by serious toxic reactions, such as allergic reaction and granuloma
formation (Allison, 1974, 1979). For this reason, little progress
was made until the discovery of muramyl dipeptide (Review, Lederer,
1980), a small component of the bacterial cell wall that is capable
of activating macrophages (Sone and Fidler, 1980, 1981; Fogler and
Fidler, 1984).

Muramyl dipeptide is a water-soluble, low-molecular-weight (M_r
459), synthetic moiety of N-acetylmuramyl-L-alanyl-D-isoglutamine
(MDP), that has potent effects on a variety of host defense cells,
including the activation of macrophages (Chedid et al., 1979; Fidler
et al., 1983; Fidler and Kleinerman, 1984; Sone and Fidler, 1980,

1981). A lipophilic MDP derivative, N-acetyl- muramyl-L-alanyl-D-isoglutamyl-L-alanyl-2-(1',2', dipalmitoyl)-sn glycero-3'-phosphoryl-ethylamide (MTP-PE), has also been synthesized (Fidler et al., 1981, 1982). Although muramyl peptides influence several macrophage functions in vitro, comparable effects have not been observed in vivo because these molecules are rapidly cleared after parenteral administration (Parant et al., 1979; Fogler et al., 1985). Even when injected at very high doses, these molecules do not induce significant macrophage-mediated antitumor activity (Fidler et al., 1981; Fogler et al., 1985).

The other category of agents that activate macrophages in vivo are the lymphokines. Antigen- and mitogen-stimulated T lymphocytes release diffusible mediators that specifically interact with target cells bearing appropriate receptors. In the case of macrophages, activation is produced by a family of lymphokines generally referred to as macrophage activation factors (MAF) (Fidler and Poste, 1982; Fidler, 1984; Kleinerman et al., 1983a,b), which include gamma-interferon (IFN- γ) (Saiki and Fidler, 1985; Saiki et al., 1985; Fidler et al., 1985; Sone et al., 1986).

Regardless of whether macrophages are activated by lymphokines or by bacterial products, they acquire the ability to recognize and destroy neoplastic cells both in vitro and in vivo while leaving nonneoplastic cells unharmed. The mechanisms is unknown but appears to be nonimmunologic in nature and to require intimate cell-to-cell contact (Hibbs, 1974a,b; Bucana et al., 1976, 1981; Fidler and Kleinerman, 1984).

Many attempts to activate macrophages _in vivo_ with lymphokines

or bacterial products to enhance host defense against infections or

cancer have not been successful. The major reason for this failure

is that lymphokines or MDP injected into the circulation have a very

short half-life (2-60 min). These molecules are, therefore,

eliminated before they can begin the pleiotropic reaction that yield

activation in macrophages.

Efforts to activate macrophages _in vivo_ by treatment with agent

that stimulate T lymphocytes to produce lymphokines have been

disappointing. We have shown that the lymphocytes of animals bearin

large progressive tumors are deficient in their ability to release

lymphokines that recruit and activate macrophages (Kripke et al.,

1979). We have recently reported similar findings in patients with

colorectal carcinomas (Fidler et al., 1986). Indeed, if such a

defect is found to be common to all tumor-bearing hosts,

administration of agents to augment production of lymphokines _in sit_

may be of little value.

Liposmes as Carriers for Immunomodulating Agents to Macrophages

Fortunately, many of the problems in activating properties of

macrophages _in vivo_ with diffusible immunomodulators can be overcome

with liposome delivery systems. Liposomes provide a unique carrier

vehicle for the delivery of biologically active materials to

phagocytic cells _in vivo_. Like any other circulating particle,

liposomes are rapidly cleared by fixed and free phagocytic cells

(Fidler et al., 1980; Fidler and Poste, 1982; Poste et al., 1982;
Schroit et al., 1983a,b; Fidler, 1985, 1986a). We have taken
advantage of this natural physiological fate of liposomes to deliver,
albeit passively, a variety of agents, including immunomodulators, to
phagocytic cells that belong to the reticuloendothelial system
(RES). This specific delivery results in the activation of the
phagocytes to a bactericidal, fungicidal, viricidal, and tumoricidal
state and has also been associated with enhanced host destruction of
established lung, lymph node, and liver metastases (Review, Fidler
and Poste, 1982; Fidler, 1985, 1986a). The exploitation of liposome
targeting to cells of the RES may enhance therapeutic efficacy
against a variety of parasitic, fungal, and bacterial macrophage-
associated diseases, (Schroit et al., 1983a), viral diseases (Koff
and Fidler, 1985), and cancer metastasis (Fidler, 1985, 1986).

Following intravenous administration, the majority (80-90%) of
liposomes are taken up by phagocytic cells that belong to the RES in
the liver, spleen, lymph nodes, and bone marrow and by circulating
monocytes. By exploiting this localization pattern, liposome-
encapsulated materials can be passively "targeted" to macrophages _in_
vivo. In order to serve as a useful carrier of biologically active
agents to mononuclear phagocytes _in vivo_, liposomes must: (a) readily
bind to and be phagocytosed by reticuloendothelial cells; (b) retain
the entrapped drug for reasonable periods of time; (c) demonstrate
localization to sites other than those rich in RES activity following
intravenous administration; and (d) have a relatively long shelf

life. To identify liposomes with these optimal characteristics we have evaluated the influence of lipid composition of liposomes on macrophage-liposome interactions (Poste et al., 1979; Fidler et al., 1980; Schroit and Fidler, 1982).

Liposomes, composed of natural phospholipids, are biodegradable and nonimmunogenic and have only limited intrinsic toxicity (Gregoriadis and Allison, 1980; Knight, 1981; Alving, 1983; Poste, 1983; Poste et al., 1984). Indeed, liposomes prepared from phosphatidylcholine (PC), phosphatidylserine (PS), and lysolecithin are well tolerated by mice and dogs after single or multiple intravenous injections (Hart et al., 1981). In these studies, no evidence of gross or microscopic toxicity was detected in any major organ, even after administration of high doses of PC/PS multilamellar vesicles (MLV). Furthermore, liposomes consisting of PC and phosphatidylglycerol (PG) safely distribute to organs rich with reticuloendothelial activity in cancer patients (Lopez-Berestein et al., 1984).

The diversity of phospholipids allows endless manipulation of their biochemical and biophysical properties, which facilitates the design of liposomes with specialized characteristics. For example, MLVs have been found to be superior to small liposomes in inducing macrophage phagocytosis (Raz et al., 1981; Schroit and Fidler, 1982). Leakage rates of entrapped contents can also be manipulated by phospholipid acyl chain composition, surface charge, and cholesterol content (Knight, 1981). Temperature- and pH-sensitive

liposomes have also been designed to trigger release of entrapped
contents under specific conditions (Straubinger et al., 1985;
Sullivan and Huang, 1985; Nayar and Schroit, 1985).

Binding and Phagocytosis of Liposomes by Macrophages

Through the systematic evaluation of MLV phospholipid
composition, we and others have identified certain classes of
phospholipids that are preferentially recognized by macrophages. For
example, the inclusion of negatively charged phospholipids, such as
PS and PG, in PC MLVs greatly enhances their binding to and
phagocytosis by macrophages (Poste et al., 1979; Fidler et al., 1980,
1981; Schroit and Fidler, 1982; Metha et al., 1982). In contrast,
neutral MLVs composed exclusively of PC are not efficiently bound by
macrophages. Indeed, it appears that the inclusion of PS within PC
MLV leads to recognition of these liposomes by every cell of the
RES. Thus, enhancement of phagocytosis has been shown to occur in
mouse peritoneal macrophages (Raz et al., 1981), mouse Kupffer cells
(Xu and Fidler, 1984), rodent alveolar macrophages (Sone and Fidler,
1980, 1981), human peripheral blood monocytes (Metha et al., 1982;
Lopez-Berestein, 1983; Kleinerman et al., 1983a,b), and human
alveolar macrophages (Sone and Tsubura, 1982) (Figure 1).

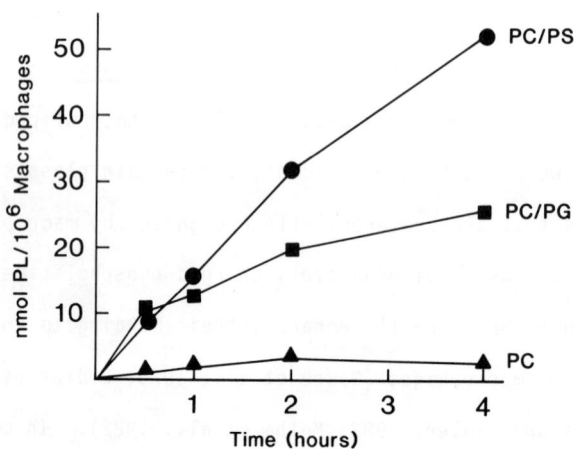

Figure 1. Phagocytosis of MLV by murine pertoneal macorphages. Phagocytosis of MLV liposomes was detemined b incubating 100 nmol of phospholipid with 10^5 macrophages at 37°C. Trace amounts of ^{125}I-labeled phenylpropionyl-PE was incorporated into the MLV preparations as a lipsomal marker.

Regardless of the source of macrophages, development of the activated phenotype requires the phagocytic uptake of liposomes followed by a lag period of several hours before expression of tumoricidal activity (Poste et al., 1982; Raz et al., 1981). Studies on the mechanism of activation by liposome-encapsulated activators indicate that participation of macrophage cell surface receptors is not required and that activation results from the interaction of immunomodulating molecules with intracellular targets (Fidler, 1985; Fidler et al., 1985; Kleinerman et al., 1985; Fogler and Fidler, 1986) (Figure 2).

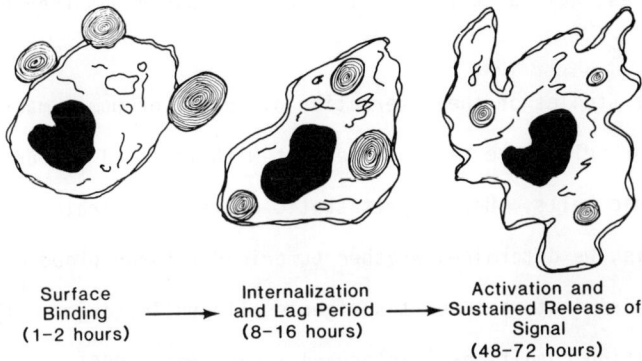

| Surface Binding (1-2 hours) | → | Internalization and Lag Period (8-16 hours) | → | Activation and Sustained Release of Signal (48-72 hours) |

Macrophage – Liposome Interaction

Lysis of Tumor Cells by Macrophages Activated by Liposomes Containing Immunomodulators.

Macrophages activated by liposome-entrapped immunomodulators such as MDP or lymphokines (MAF, IFN-γ) acquire the ability to discriminate between tumorigenic and normal cells. At least in vitro, tumoricidal macrophages destroy neoplastic cells by a process independent of transplantation antigens, species-specific antigens,

tumor-specific antigens, cell cycle, or various phenotypes associated
with transformation (Fidler 1978, 1985). Moreover, data obtained in
various murine systems suggest that the susceptibility of tumor cell
to destruction by tumoricidal macrophages is also independent of the
in vivo biologic behavior of the tumor cells, such as invasiveness,
metastatic potential, growth rate, and resistance to lysis by
lymphocytes, natural killer cells, or cytotoxic drugs (Review Fidler
1985).

Most studies on the interaction of rodent or human macrophages
with tumor cells have been carried out with isolated cultures of
tumorigenic cells. Because metastatic cells proliferate among normal
host cells, we determined whether tumoricidal human blood monocytes
could discriminate between tumorigenic allogeneic target cells under
cocultivation conditions (Fidler and Kleinerman, 1984). Highly
purified preparations of peripheral blood monocytes isolated from
normal human donors were activated in vitro subsequent to incubation
with liposomes containing immunomodulators. The cytotoxic properties
of these monocytes against various combinations of three different
tumorigenic and three different nontumorigenic target-cell
populations, labeled with either [^{3}H]thymidine or [^{14}C]thymidine,
were assessed. In all combinations used, activated monocytes
selectively lysed only neoplastic cells and left nontumorigenic cells
unharmed. The selective lysis of tumorigenic cells was not due to an
inherent resistance of normal cells to lysis mediated by host immune
cells, but rather was associated with activated monocytes, since both

tumorigenic and nontumorigenic cells were equally susceptible to in vitro lysis mediated by mitogen-stimulated peripheral blood lymphocytes (Fidler and Kleinerman, 1984).

The interaction of control and tumoricidal human blood monocytes with human melanoma cells was analyzed by means of light microscopy and scanning and transmission electron microscopy. Activated monocytes that phagocytosed liposomes containing MTP-PE clustered around the melanoma cells at a higher density than did control monocytes. This monocyte clustering was followed by the establishment of numerous focal points of tight binding and areas of discontinuous membranes, a finding confirmed by stereophotography. After 24 hr of cocultivation, many of the target cells exhibited zones of vacuolation in the immediate vicinity of the tumoricidal monocytes. This suggested to us that the target cell was damaged, and time-course cytotoxicity studies were confirmatory (Bucana et al., 1983).

Macrophage lysis of tumor cells appears to be nonselective. This conclusion is based on studies in which we attempted to select in vitro tumor cell variant lines that exhibit a phenotypic resistance to macrophage-mediated lysis. We used techniques similar to those used previously to successfully select a B16 melanoma tumor cell line resistant to lysis by cytotoxic T-lymphocytes or UV-2237 fibrosarcoma resistant to natural killer (NK) cell-mediated lysis. Despite repeated in vitro selections, we failed to isolate tumor cells with increased resistance to macrophage-mediated lysis (Fogler and Fidler, 1985).

The data generated in rodent and human systems indicate that, at least in vitro, tumoricidal macrophages can discriminate between neoplastic and nonneoplastic cells by a process that is independent of transplantation antigens, species-specific antigens, tumor-specific antigens, cell cycle time, or various phenotypes associated with transformation (Fidler, 1978, 1985). Taking into consideration that the major limitations of many cancer therapies are the lack of selectivity and their toxic effect on the patient, the ability of tumoricidal macrophages to distinguish tumnorigenic from normal cells presents an attractive possibility for treatment of disseminated cancer.

In Vivo Organ Localization of Liposomes

As discussed above, the majority of liposomes introduced into the circulation are cleared by cells of the RES (liver, spleen, lymph nodes). Since cancer metastases frequently grow in the lung, an organ poor in RES activity, we searched for liposomes with increased efficiency for localization to the lung microvasculature. The retention of radiolabeled liposomes of differing structure and composition within the lungs of mice after intravenous injection is shown in Table 1. Large multilamellar vesicles (MLV) or reverse evaporation vesicles (REV) were retained in the lungs more efficiently than small unilamellar liposomes (SUV) of identical lipid

composition. The major factor, however, that determined the in vivo
distribution of liposomes is their lipid composition. After i.v.
injection, liposomes of the same structural class are more
efficiently arrested in the lung vasculature when they contain
negatively charged phospholipid (PS) than when they contain only a
neutral phospholipid (PC) (Fidler et al., 1980; Poste et al., 1982;
Fidler and Poste, 1982; Schroit and Fidler, 1982) (Table 1). MLV
consisting of PC and PS (7:3 mole ratio) that contained a trace of
fluorescent lipid analogue were injected intravenously into mice, and
24 hr later, a time sufficient for blood monocytes to extravasate and
differentiate into alveolar macrophages, the latter contained
fluorescent liposomes (Key et al., 1982). About 12% of the total
alveolar macrophage population (as identified by the formation of
rosettes between opsonized erythrocytes and macrophages) contained
MLV, suggesting that the PC/PS liposomes were effective vehicles for
the delivery of encapsulated compounds to blood monocytes in the
lung, which migrated out of the circulation and differentiated into
alveolar macrophages (Poste et al., 1982; Key et al., 1982).

TABLE 1

Lung retention of intravenously administered liposomes of

different phospholipid composition and size[a]

Liposome structure and size	Lipid composition (mol % ratio)	Percent of injected liposomes retained in lungs
MLV	PC	2.2
SUV	PC	0.5
REV	PC/PS (70/30)	7.2
SUV	PC/PS (70/30)	0.9
MLV	PC/PS (70/30)	6.6

[a] Mice were injected intravenously with liposome preparations
containing radioiodinated phospholipids (2 μmol in 0.2 ml saline);
the radioactivity was monitored 4 hr later. From Fidler et al.,
1980; Schroit and Fidler, 1982.

In Situ Activation of Macrophages by Liposomes Containing
Immunomodulators.

 The intravenous injection of MLV composed of PC/PS (7:3) and
containing immunomodulators results in the in situ activation of
mouse alveolar macrophages (AM). AM obtained by lavage of lungs 24
hr after the administration of liposomes containing MDP are cytotoxic
against syngeneic melanoma cells (Table 2). MLV containing buffer

alone (control) in combination with unencapsulated MDP failed to
render AM tumoricidal (Fidler et al., 1981).

TABLE 2

Activation of mouse lung macrophages by intravenous administration
of liposomes containing MDP or MTP-PE[a]

Immunomodulator	% AM-mediated cytotoxicity	
(g/mouse)	MLV-MDP	MLV-MTP-PE
0.1	9.6 ± 2.3	15.4 ± 3.5
1.0	24.6 ± 3.8	31.5 ± 4.6
10.0	29.2 ± 3.8	44.6 ± 4.2
100.0	30.0 ± 4.2	49.2 ± 5.2

[a] C57BL/6 mice were given i.v. injections of 2.5)mol MLV
containing MDP or MTP-PE. AM were harvested 24 hr later, and
their tumoricidal activity was assayed against the syngeneic B16
melanoma. From Fidler et al., 1981.

To assess whether AM activated by liposomes containing MDP
maintain their tumoricidal properties over an extended period of
time, we harvested and assayed AM for their tumoricidal activity 1,
2, 3, or 4 days after the injection of a single dose of MLV
containing MDP, MTP-PE, or a combination of both agents. On day 1,
AM harvested from all the mice were cytotoxic against the syngeneic

tumor. By day 3, only AM harvested from mice that received MLV
containing MTP-PE exhibited significant levels of tumoricidal
activity (Fidler et al., 1982) (Table 3).

TABLE 3

Duration of tumoricidal properties in mouse lung macrophages after a
single intravenous injection of liposomes containing MDP and/or MTP-P

Treatment with	% AM-mediated cytotoxicity on:			
liposomes containing:	Day 1	Day 2	Day 3	Day 4
MDP (10 µg)	30.8	17.5	12.5	2.5
MDP (100 µg)	35.8	22.5	15.8	5.0
MTP-PE (10 µg)	52.5	40.0	30.8	10.8
MTP-PE (100 µg)	58.3	43.3	35.0	15.0
MDP (10 µg) plus MTP-PE (10 µg)	60.8	48.3	40.0	20.8

[a] Mice were injected i.v. with a single dose of MLV containing
various amounts of MDP or MTP-PE. AM were harvested at different
days thereafter, and tumoricidal activity was assayed in vitro.
From Sone and Fidler 1981; Fidler et al., 1982, 1983.

The in situ activation of macrophages by i.v. administration of liposomes containing macrophage activators resulted from the direct interactions of these liposomes with macrophages. It did not occur by an indirect action of the immunomodulator on T-cells leading to the release of lymphokines that activate macrophages (Fidler, 1981). This conclusion is based on data from experiments in which alveolar macrophages of mice with impaired T-cell function--mice exposed to UV radiation, thymectomized adult mice exposed to X-rays, and athymic nude mice--were rendered tumoricidal by the systemic administration of liposome-encapsulated MTP-PE but not by control liposome preparations (Fidler, 1981).

Stability of Liposomes

The stability of liposomes is crucial for the delivery of immunomodulators to macrophages. Two aspects of stability must be considered: (a) leakage problems associated with aqueous encapsulated compounds; and (b) the shelf life of liposomes containing immunomodulators.

Low-molecular-weight solutes such as MDP can rapidly leak from the liposome's aqueous compartment. This critical problem was obviated by the use of lipophilic MDP derivatives such as MTP-PE (Fidler et al., 1981), muramyldipeptide glyceryldipalmitate (MDP-GDP) (Phillips et al., 1985), and 6-0-stearoyl MDP (Lopez-Berestein et al., 1983). These compounds intercalate quantitatively into the MLV's phospholipid bilayer, resulting in greater than 90%

incorporation, as compared with 1%-3% incorporation obtained with
hydrophilic MDP (due to aqueous space limitations). In addition,
PC/PS/MTP-PE MLVs were stable for up to eight weeks when prepared
under sterile conditions and sealed in ampules under nitrogen (to
prevent lipid oxidation). In fact, eight-week-old or freshly made
MLV preparations produced similar levels of macrophage activation
whether they were assessed by in vitro or by in vivo techniques
(Schroit et al., 1983b). Similar results have been shown by Phillips
et al (1985) using lyophilized liposome preparations which, when
reconstituted immediately before use, yield reproducible vesicles
that efficiently activate macrophages in vitro and in vivo.

Therapy of Spontaneous Metastases by Repeated Intravenous Administrations of Immunomodulators Encapsulated in Liposomes

Numerous studies have shown that activation of macrophages can
be achieved by the intravenous administration of liposomes containing
various immunomodulators. Moreover, the efficiency of liposomes for
in situ activation was very similar to that observed under in vitro
conditions (Fidler, 1980, 1985, 1986). We therefore examined the
possibility that systemic administration of immunomodulators
encapsulated in liposomes could enhance host destruction of
metastases. To test this possibility, we injected liposomes
containing various immunomodulators into mice with spontaneous
metastases in their lungs and lymph nodes. The B16-BL6 melanoma cell
line, which is syngeneic to C57BL/6 mice, was used as the model to

determine the effectiveness of liposome-encapsulated materials in the
treatment of metastases. After implantation in the foot pad, this
tumor metastasizes to lymph nodes and the lungs in more than 90% of
the mice. Mice were given an intrafoot pad injection of B16 melanoma
cells, and four to five weeks later, when the tumors had reached a
size of 10 to 12 mm, the leg with the tumor was amputated at the mid
femur to include the popliteal lymph node. Three days after
resection of the primary tumor, liposomes containing immunomodulators
or placebo preparations were injected intravenously into the mice.
Both test and control groups were treated twice weekly for four weeks
(8 i.v. injections). We used MLVs consisting of PC/PS since they
have been shown to be nontoxic at the dose used and because they are
efficiently arrested in the lungs as well as in organs of the RES
following intravenous injection (Fidler et al., 1980).

Spontaneous metastases in the lungs and lymph nodes were well
established at the time liposome treatment began. Many individual
metastases could be seen macroscopically. The mice treated
intravenously with saline, with free MAF, with free MDP, or with
liposomes containing saline were dead by day 90 of the experiment.
In marked contrast, 66% and 60% of mice injected intravenously with
liposome-encapsulated MAF and with liposome-encapsulated MDP,
respectively, were alive when the experiments were terminated at 200
days. In this tumor system, we estimate that the metastases
contained 10^7 cells at the beginning of the liposome treatment.
Since the median survival time of mice injected with as few as 10

viable B16 cells (admixed with 10^6 dead cells) is 40 to 50 days (Fidler, 1980), we speculate that the residual tumor burden of the surviving mice must have been reduced to less than 10 viable cells (Figure 3).

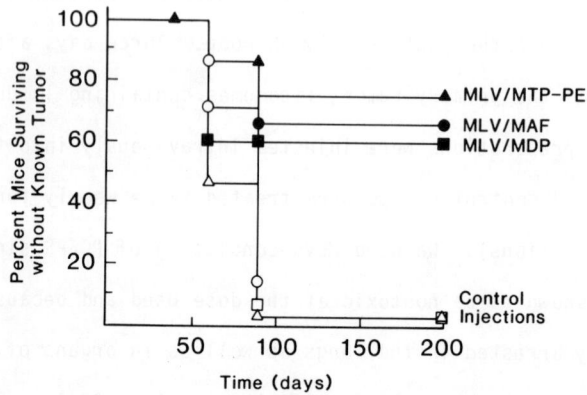

Figure 3. Treatment of spontaneous melanoma metastases by the systemic administration of lipsomes containing MAF, MDP or MTP-PE. Mice (20-30)/group) were injected with lipsomes 2 times per week for 4 weeks (from Fidler 1985, 1986).

Studies on mice treated with liposome-encapsulated MDP that had residual metastatic disease (albeit reduced compared with untreated control mice) have revealed that the tumor cells present in the lesions of those treatment "failures" were still fully susceptible to destruction by activated macrophages (Fogler and Fidler, 1985).

Similar data on the successful treatment of metastases by the

intravenous injections of liposomes containing immunomodulators have

been reported for several mouse fibrosarcomas (Deodhar et al.,

1982a,b; Lopez-Berestein et al., 1984) and melanomas (Phillips,

1985), colon carcinomas (Thombre and Deodhar, 1984), and spontaneous

mouse tumors induced by chronic UV irradiation (Talmadge et al.,

1986).

Synergistic Activation of Murine Macrophages by Combined
Immunomodulators Encapsulated within the Same Liposomes

Liposomes can be used to deliver more than a single agent to

macrophages. Recently, we have shown that the combination of MAF and

MDP entrapped in MLV at subthreshold doses leads to significant

synergistic activation of macrophages _in vitro_ (Sone and Fidler,

1980) and _in vivo_ (Fidler and Schroit, 1984). Moreover, combining

suboptimal concentrations of a lymphokine and MDP in the same

liposome preparations enhances the efficacy of macrophage-mediated

destruction of metastasis (Fidler and Schroit, 1984). For the _in_

vitro experiments, liposomes containing various dilutions of MAF,

various dilutions of MDP, combinations of MAF and MDP, or control

liposome preparations were injected intravenously into mice. AM were

harvested 24 hours later, and their antitumor activity was monitored

in vitro. Liposomes containing combinations of MAF and MDP were most

effective in generating cytotoxic properties in lung macrophages

(Sone and Fidler, 1980). Moreover, this therapeutic regimen was ver efficacious in the treatment of relatively large metastatic burdens (Fidler and Schroit, 1984). These therapy experiments began 7 days after the surgical removal of the subcutaneous B16 melanoma, when luny metastases were large, some even 1 mm in diameter. Liposomes were injected twice weekly for 4 weeks (total of 8 i.v. injections). The data demonstrated that the systemic administration of MLV containing subthreshold amounts of both MAF and MDP was associated with a significant increase in the long-term survival rat of mice and that the encapsulation of two distinct immunomodulators in the same liposome could lead to marked improvement in treatment o spontaneous metastases (Figure 4).

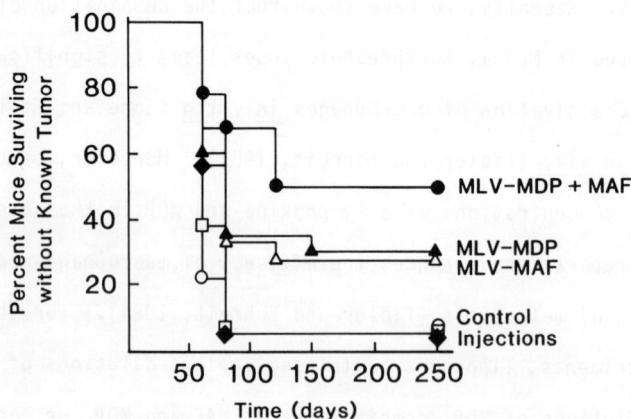

Figure 4. Treatment of large spontaneous melanoma metastases by the systemic administration of liposomes containing combinations of immunomodulators. Mice (18 to 20 mice per/group) received biweekly injections of 4 weeks of MLV containing 1:20 dilution MAF; MLV containing 0.3 µg MDP; MLV containing 1:2 dilution MAF; MLV containing 0.3 µg MDP and 1:20 MAF. (from Fidler and Schroit, 1984)

Synergistic activation of macrophages in vivo would be most advantageous if it could be accomplished with well-defined synthetic agents or highly purified lymphokines, such as IFN-γ. For this reason, we examined whether encapsulation of recombinant IFN-γ and MDP within the same MLV preparations generated synergistic activation of mouse macrophages (Saiki and Fidler, 1985) and human monocytes (Saiki et al., 1985). The delivery of recombinant IFN-γ and MDP to macrophages by liposomes bypassed the necessity of binding unencapsulated lymphokine to the cell surface and was associated with the abrogation of recombinant IFN-γ species specificity, i.e., once encapsulated, mouse IFN-γ or human IFN-γ activated both mouse and human macrophages (Fidler et al., 1985; Sone et al., 1986).

Treatment of Large Lung Metastases by Local Thoracic X-Irradiation Followed by Systemic Activation of Macrophages

There are at least two limiting factors in the treatment of metastases by activated macrophages. First, activated macrophages (blood monocytes) may not readily home to metastases and, second, the tumor burden in metastases may be too large for the available activated macrophages. For these reasons, we designed a study to determine whether the combination of local thoracic x-irradiation (LTI) with activation of blood monocytes would be beneficial for the treatment of large lung tumor colonies (Saiki et al. 1986). Ionizing radiation has been shown to produce a logarithmic decrease in tumor cell survival, and thus a reduction of metastatic burden. Radiation

also induces blood vessel damage and inflammation in irradiated
tissues, and this might promote the arrest and influx of macrophages
into the inflamed, damaged tissue. We combined LTI and systemic
activation of macrophages by i.v. administration of liposomes
containing MTP-PE to treat experimental lung metastases of a murine
fibrosarcoma.

In the first set of experiments, we determined the dose of LTI
that would not interfere with systemic activation of macrophages by
MLV-MTP-PE. Both 5- and 8-Gy LTI produced a marked decrease in the
number of recoverable lung macrophages. A 60% decrease in the number
of recoverable lung macrophages was found at 3 days after
irradiation. Thereafter, the number increased gradually but remained
below the control value even at 12 days after irradiation. These
data suggested that liposomes containing immunomodulators should be
administered at least 5 days after LTI, when recovery in the number
of harvested lung macrophages occurs.

In the next set of experiments, mice were given i.v. injections
of fibrosarcoma cells and 5 days later they were exposed to 8-Gy
LTI. After 5 additional days, the mice were given i.v. injections of
MLV containing saline or MTP-PE. The i.v. administration of only
MLV-MTP-PE activated tumoricidal properties in alveolar macrophages
of all groups of mice (normal, untreated, or irradiated mice with
lung tumor colonies). Thus, the presence of micrometastases in the
lung did not affect the magnitude of macrophage activation in normal
or irradiated mice.

In the final experiments, mice were given i.v. injections of fibrosarcoma cells. Five days later, treatment with LTI alone, liposomes alone or combinations of the treatments began. The mice were observed daily for up to 120 days. Moribund or dead mice were autopsied. The experiment was carried out twice with very similar results. Practically all the mice (24 of 25) with lung tumor colonies that received no treatment died. The median number of lung tumor colonies was 50 (range, 20-101) (Figure 5). Treatments with liposomes containing saline or MTP-PE did not decrease the median number of lung tumor colonies or significantly increase the overall survival of the mice. Treatment with irradiation alone followed by MLV containing saline prolonged the median survival and led to long-term survival of 8 of 25 mice. The most remarkable therapeutic results were achieved in mice given LTI and twice weekly (4 weeks) i.v. injections of liposomes containing MTP-PE. On day 120 of the experiment in this treatment group, 15 of 25 mice were alive. Thus, the combined therapeutic modalities of LTI with liposomes containing MTP-PE produced a highly significant survival rate of 60%. Mice surviving on day 140 of the study were killed and necropsied. No cancer was found in these mice (Saiki et al. 1986).

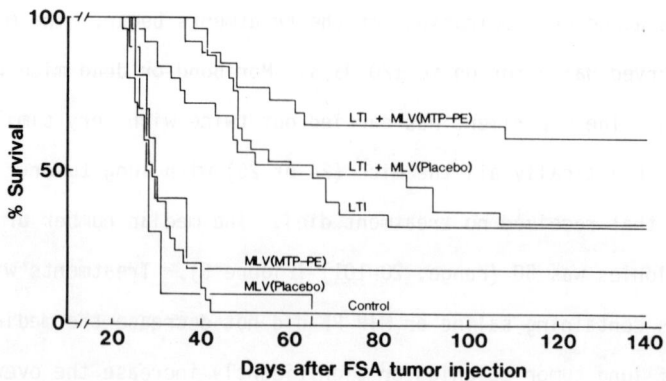

Figure 5. Therapy of large fibrosarcoma lung metastases by local thoracic irradiation and systemic activation of macrophages produced by liposomes containing MTP-PE (from Saiki et al., 1986.)

CONCLUSIONS

Macrophages can be activated to become tumoricidal by interaction with phospholipid vesicles (liposomes) containing various immunomodulators, such as lymphokines, bacterial products, and synthetic molecules. Tumoricidal macrophages can recognize and destroy neoplastic cells in vitro and in vivo, while leaving nonneoplastic cells unharmed. Intravenously administered liposomes are cleared from the circulation by phagocytic cells. This natural phenomenon allows delivery of various molecules to macrophages. Such an approach has the expected benefit of achieving high concentrations of drug at the appropriate site. Moreover, the endocytosis of lipsomes containing immunomodulators results in generating cytotoxic

properties in macrophages in situ. This process is independent of
the thymus and can be achieved in mice without functional T-cells
such as athymic nude mice or mice immunosuppressed by cyclosporin-
A. The multiple administration of liposomes containing
immunomodulators produces eradication of cancer metastases in several
rodent-tumor systems. The successful therapy of disseminated
metastases must circumvent the problem of neoplastic biological
heterogeneity and the development of resistance to therapy by tumor
cells. Since liposomes can deliver more than one agent to the
macrophages, synergistic activation of these cells can be
accomplished in situ. Macrophage destruction of cancer metastases is
limited by the ratio of effector to target cells. Thus, destruction
of small metastases is effective, but once metastases exceed a
certain number of cells, therapeutic efficacy is diminished.

REFERENCES

Allison, A.C. 1979. 'Mode of action of immunological adjuvants'.
J. Reticuloendothel. Soc. 26, 619-630.

Alving, C.R. 1983. 'Delivery of liposome-encapsulated drugs to
macrophages' Pharmacol. Thers. 22, 407-424.

Bucana, C., Hoyer, L.L., Hobbs, B., Breesman, S., McDaniel, M. and
Hanna, M.G. Jr. 1976. 'Morphological evidence for the translocation
of lysosomal organelles from cytotoxic macrophages into the cytoplasm
of tumor target cells'. Cancer Res. 36, 4444-4458.

Bucana, C.D., Hoyer, L.C., Schroit, A.J., Kleinerman, E. and Fidler,
I.J. 1983. 'Ultrastructural studies of the interaction between
liposome-activated human blood monocytes and allogeneic tumor cells
in vitro'. Am. J. Pathol. 112 101-111.

Chedid, L., Carelli, L., and Audibert, F. 1979. 'Recent
developments concerning muramyl dipeptide, a synthetic
immunoregulating molecule'. J. Reticuloendothel. Soc. 26, 631-641.

Cifone, M.A. and Fidler, I.J. 1981. Increasing metastatic potential is associated with increasing genetic instability of clones isolated from murine neoplasms. Proc. Natl. Acad. Sci. U.S.A. 78, 6949-6952.

Deodhar, S.D., Barna, B.P., Edinger, M. and Chiang, T. 1982a. 'Inhibition of lung metastases by liposomal immunotherapy in a murine fibrosarcoma model', J. Biol. Response Modifiers 1, 27-34.

Deodhar, S.D., James, K., Chiang, T., Edinger, M. and Barna, B. 1982b 'Inhibition of lung metastases in mice bearing a malignant fibrosarcoma by treatment with lipsomes containing human c-reactive protein. Cancer Res. 42, 5084.

Eppstein, D.A., Van Der Pas, M.A., Fraser-Smith, E.B., Kurahara, C.G., Felgner, P.L., Mathews, T.R., Waters, R.V., Venuti, M.C., Jones, G.H., Metha, R., Lopez-Berestein, G. 1986. 'Liposome-encapsulated muramyl dipeptide analogue enhances non-specific host immunity', Int. J. Immunotherapy 2, 115-126.

Fidler, I.J. 1978. 'Recognition and destruction of target cells by tumoricidal macrophages'. Isr. J. Med. Sci. 14, 177-191.

Fidler, I.J. 1980. 'Therapy of spontaneous metastases by intravenous injection of liposomes containing lymphokines'. Science 208, 1469-1471.

Fidler, I.J. 1981. The in situ induction of tumoricidal activity in alveolar macrophages by lipsomes containing muramyl dipeptide is a thymus-independent process. J. Immunol. 127, 1719-1720.

Fidler, I.J. 1984. 'The MAF dilemma'. Lymphokine Res. 3, 51-54.

Fidler, I.J. 1985. 'Macrophages and metastasis: A biological approach to cancer therapy'. Cancer Res. 45, 4714-4726.

Fidler, I.J. 1986a. 'Immunomodulation of macrophages for cancer and antiviral therapy. Site-Specific Drug Delivery, (Eds. Tomlinson, E. and Davis, S.S.), pp. 111-134 John Wiley and Sons, New York.

Fidler, I.J. 1986b. 'Optimization and limitations of systemic treatment of murine melanoma metastases with liposomes containing muramyl tripeptide phosphatidylethanolamine. Cancer Immunol. Immunother., 21, 169-173.

Fidler, I.J. and Balch, C.M. 1987. 'The biology of cancer metastasis and implications for therapy. 'The biology of cancer metastasis and implications for therapy. Current Problems in Surgery, (Eds Ravitch, M.M.), pp 137-209 Year Book Medical Publishers, Chicago.

Fidler, I.J., Barnes, Z., Fogler, W.E., Kirsh, R., Bugelski, P. and Poste, G. 1982. 'Involvement of macrophages in the eradication of established metastases following intravenous injection of liposomes containing macrophage activators'. Cancer Res. 42, 496-501.

Fidler, I.J., Fogler, W.E., Kleinerman, E.S., and Saiki, I. 1985. 'Abrogation of species specificity for activation of tumoricidal properties in macrophages by recombinant mouse or human gamma interferon encapsulated in liposomes'. J. Immunol. 135, 4289-4294.

Fidler, I.J., Fogler, W.E., Tarcsay, L., Schumann, G., Braun, D.G. and Schroit, A.J. 1983. 'Systemic activation of macrophages and treatment of cancer metastases by liposomes containing hydrophilic or lipophilic muramyl dipeptide'. Advances in Immunopharmacology, pp 235-253 Pergamon Press, Oxford.

Fidler, I.J. and Hart, I.R. 1982. 'Biological diversity in metastatic neoplasms: Origins and implications'. Science 217, 998-1003.

Fidler, I.J , Jessup, J.M., Fogler, W.E., Staerkel, R., Mazumder. 1986. Activation of tumoricidal properties in peripheral blood monocytes of patients with colorectal carcinoma. Cancer Res. 44, 994-998.

Fidler, I.J., Kleinerman, E.S. 1984. 'Lymphokine-activated human blood monocytes destroy tumor cells but no normal cells under cocultivation conditions'. J. Clin. Oncol. 2, 937-943.

Fidler, I.J. and Kripke, M.L. 1977. 'Metastasis results from preexisting variant cells within a malignant tumor'. Science 197, 893-895.

Fidler, I.J., Poste, G. 1982. 'Macrophage-mediated destruction of malignant tumor cells and new strategies for the therapy of metastatic disease', Springer Semin. Immunopathol. 5, 161-174.

Fidler, I.J. and Poste, G. 1985. 'The cellular heterogeneity of malignant neoplasms: Implications for adjuvant chemotherapy'. Semin. in Oncol. 12, 207-222.

Fidler, I.J. and Schroit, A.J. 1984. 'Synergism between lymphokines and muramyl dipeptide encapsulated in liposomes: In situ activation of macrophages and therapy of spontaneous cancer metastasis'. J. Immunol. 133, 515-518.

Fidler, I.J., Raz, A., Fogler, W.E., Kirsh, R., Bugelski, P. and Poste, G. 1980. 'The design of liposomes to improve delivery of macrophage-augmenting agents to alveolar macrophages'. Cancer Res. 40, 4460-4466.

Fidler, I.J., Sone, S., Fogler, W.E. and Barnes, Z.L. 1981. 'Eradication of spontaneous metastases and activation of alveolar macrophages by intravenous injection of liposomes containing muramyl dipeptide'. Proc. Natl. Acad. Sci. U.S.A. 78, 1680-1684.

Fidler, I.J., Sone S., Fogler, W.E., Smith, D., Braun, D.G., Tarcsay, L., Gisler, R.J. and Schroit, A.J. 1982. 'Efficacy of liposomes containing a lipophilic muramyl dipeptide derivative for activating the tumoricidal properties of alveolar macrophages in vivo'. J. Biol. Response Mod. 1, 43-55.

Fogler, W.E. and Fidler, I.J. 1984. "Modulation of the immune response by muramyl dipeptide'. Immune Modulation Agents and Their Mechanisms. (Eds. Chirigos, M.A. and Fenichel, R.L.), pp. 499-512 Marcel Dekker, Inc., New York.

Fogler, W.E. and Fidler, I.J. 1985. 'Nonselective destruction of murine neoplastic cells by syngeneic tumoricidal macrophages'. Cancer Res. 45, 14-18.

Fogler, W.E. and Fidler, I.J. 1986. The activation of tumoricidal properties in human blood monocytes by muramyl dipeptide requires specific intracellular interaction. J. Immun. 136, 2311-2317.

Fogler, W.E., Wade, R., Brundish, D.E., and Fidler, I.J. 1985. 'Distribution and fate of free and liposome-encapsulated [^{3}H]nor-muramyl dipeptide and [^{3}H]muramyl tripeptide phosphatidylethanolamine in mice'. J. Immunol. 135, 1372-1377.

Hart, I.R., Fogler, W.E., Poste, G., and Fidler, I.J. 1981. 'Toxicity studies of liposome-encapsulated immunomodulators administered intravenously into dogs and mice'. Cancer Immunol. Immunother. 10, 157-166.

Hibbs, J.B., Jr. 1974a. 'Discrimination between neoplastic and non-neoplastic cells in vitro by activated macrophages'. JNCI 53, 1487-1492.

Hibbs, J.B., Jr. 1974b. 'Heterocytolysis by macrophages activated by bacillus Calmette-Guerin: Lysosome exocytosis into tumor cells'. Science 184 468-471.

Key, M.E., Talmadge, J.E., Fogler, W.E., Bucana, C., and Fidler, I.J. 1982. 'Isolation of tumoricidal macrophages from lung melanoma metastases of mice treated systemically with liposomes containing a lipophilic derivative of muramyl dipeptide'. JNCI 69, 1189-1198.

Kleinerman, E.S., Schroit, A.J., Fogler, W.E. and Fidler, I.J. 1983a. 'Tumoricidal activity of human monocytes activated in vitro by free and liposome-encapsulated human lymphokines'. J. Clin. Invest. 72, 1-12.

Kleinerman, E.S., Erickson, K.L., Schroit, A.J., Fogler, W.E. and
Fidler, I.J. 1983b. 'Activation of tumoricidal properties in human
blood monocytes by liposomes containing lipophilic muramyl
tripeptide'. Cancer Res. 43, 2010-2014.

Kleinerman, E.S., Fogler, W.E. and Fidler, I.J. 1985. Intracellular
activation of human and rodent macrophages by human lymphokines
encapsulated in lipsomes. J. Leukocyte Biology 37, 571,-584.

Knight, C.G. (Ed.) 1981. Liposomes from Physical Structure to
Therapeutic Applications, Elsevier, Amsterdam.

Koff, W.C. and Fidler, I.J. 1985. The potential use of lipsome-
mediated antiviral therapy. J. Antiviral Res., 228, 495-497.

Kripke, M.L., Budmen, M.B. and Fidler, I.J. 1977. Production of
specific macrophage activating factor by lymphocytes from tumor-
bearing mice. Cell Immunol. 30, 341-352.

Lederer, E. 1980. 'Synthetic immunostimulants derived from the
bacterial cell wall'. J. Med. Chem. 23, 819-825.

Lopez-Berestein, G., Mehta, K., Mehta, R., Juliano, R.L. and Hersh,
E. M. 1983. 'The activation of human monocytes by liposome-
encapsulated muramyl dipeptide analogues'. J. Immunol. 130, 1500-
1504.

Lopez-Berestein, G., Milas, L., Hunter, N., Mehta, K., Eppstein, D.,
VanderPas, M.A., Mathews, T.R. and Hersh, E.M. 1984. 'Prophylaxis
and treatment of experimental lung metastases in mice after treatment
with liposome-encapsulated 6-0-steroyl-N-acetyl muramyl-L-
aminobutyryl-D-isoglutamine'. Clin. Exp. Metastasis 2, 366-367.

Mehta, K., Lopez-Berestein, G., Hersh, E.M., and Juliano, R.L.
1982. 'Uptake of liposomes and liposome-encapsulated muramyl
dipeptide by human peripheral blood monocytes'. J. Reticuloendothel.
Soc. 32, 155-164.

Nayar, R. and Schroit, A.J. 1985. 'Generation of pH-sensitive
liposomes: Use of large unilamellar vesicles containing
N-succinylphosphatidylethanolamine'. Biochemistry 24, 5967-5971.

Nicolau, C. and Paraf, A. (Eds.) 1981. Liposomes, Drugs and
Immunocompetent Cell Functions Academic Press, London.

Nicolson, G.L. 1987. 'Tumor cell instability, diversification and
progression to the metastatic phenotype: From oncogene to oncofetal
expression. Cancer Res. 47, 1473-1488, 1987.

Parant, M., Parant, F., Chedid, L., Yapo, A., Petit, J.F. and
Lederer, E. 1979. 'Fate of the synthetic immunoadjuvant, muramyl

dipeptide (^{14}C-labelled) in the mouse'. Int. J. Immunopharmacol. 1, 35-41.

Phillips, N.C., Mora, M.L., Chedid, L., Lefrancier, P. and Bernard, J.M. 1985. 'Activation of tumoricidal activity and eradication of experimental metastases by freeze-dried liposomes containing a new lipophilic muramyl dipeptide derivative'. Cancer Res. 45, 128-134.

Poste, G. 1979. 'The tumoricidal properties of inflammatory tissue microphages and multinucleate giant cells'. Am. J. Pathology 96, 595-606.

Poste, G. 1983. 'Liposome targeting in vivo: Problems and opportunities. Biol. Cell 47, 19-39, 1983.

Poste, G., Kirsh, R. and Bugelski, P. 1984. 'Liposomes as a drug delivery system in cancer therapy'. Novel Approaches to Cancer Chemotherapy (Ed. Sunkara, P.S.), pp. 166-221 Academic Press, New York.

Poste, G., Kirsh, R., Fogler, W. and Fidler, I.J. 1979. 'Activation of tumoricidal properties in mouse macrophages by lymphokines encapsulated in liposomes'. Cancer Res. 39, 881-892.

Poste, G., Bucana, C., Raz, A., Bugelski, P., Kirsh, R. and Fidler, I.J. 1982. 'Analysis of the fate of systemically administered liposomes and implications for their use in drug delivery'. Cancer Res. 42, 1412-1422.

Raz, A., Bucana, C., Fogler, W.E., Poste, G. and Fidler, I.J. 1981. 'Biochemical, morphological and ultrastructural studies on the uptake of liposomes by murine macrophages', Cancer Res, 41, 487-494.

Saiki, I. and Fidler, I.J. 1985. 'Synergistic activation by recombinant mouse interferon-gamma and muramyl dipeptide of tumoricidal properties in mouse macrophages'. J. Immunol. 135, 684-688.

Saiki, I., Sone, S., Fogler, W.E., Kleinerman, E.S., Lopez-Berestein, G. and Fidler, I.J. 1985. 'Synergism between human recombinant gamma-interferon and muramyl dipeptide encapsulated in liposomes for activation of antitumor properties in human blood monocytes'. Cancer Res. 45, 6188-6193.

Saiki, I., Milas, I., Hunter, N. and Fidler, I.J. 1986. Treatment of experimental lung metastases with local thoracic irradiation followed by systemic macrophage activation with liposomes containing muramyl tripeptide. Cancer Res. 46, 4966-4970.

Schroit, A.J. and Fidler, I.J. 1982. 'Effects of liposome structure and lipid composition on the activation of the tumoricidal properties of macrophages by liposomes containing muramyl dipeptide. Cancer Res. 42, 161-167.

Schroit, A.J., Hart, I.R., Madsen, J. and Fidler, I.J. 1983a. 'Selective delivery of drugs encapsulated in liposomes: Natural targeting to macrophages involved in various disease states'. J. Biol. Response Mod. 2, 97-100.

Schroit, A.J., Galligioni, E. and Fidler, I.J. 1983b. 'Factors influencing the in situ activation of macrophages by liposomes containing muramyl dipeptide'. Biol. Cell 47, 87-94.

Sone, S. and Fidler, I.J. 1980. 'Synergistic activation by lymphokines and muramyl dipeptide of tumoricidal properties in rat alveolar macrophages'. J. Immunol. 125, 2454-2460.

Sone, S. and Fidler, I.J. 1981. 'In vitro activation of tumoricidal properties in rat alveolar macrophages by synthetic muramyl dipeptide encapsulated in liposomes'. Cell Immunol. 57, 42-50.

Sone, S., Lopez-Berestein, G. and Fidler, I.J. 1986. Potentiation of direct cytotoxicity nd production of tumor cytolytic factors in human blood monocytes by human recombinant interferon-gamma and muramyl dipeptide derivatives. Cancer Immunol. Immunother. 2, 93-99.

Sone, S. and Tsubura, E. 1982. 'Human alveolar macrophages: Potentiation of their tumoricidal activity by liposome-encapsulated muramyl dipeptide'. J. Immunol. 129, 1313-1317.

Straubinger, R.M., Duzgunes, N. and Papahadjopoulos, D. 1985. 'Ph-sensitive liposomes mediate cytoplasmic delivery of encapsulated macromolecules'. FEBS Lett. 179, 148-154.

Sullivan, S.M. and Huang, L. 1985. 'Preparation and characterization of heat-sensitive immunoliposomes'. Biochim. Biophys. Acta 812, 116-126.

Talmadge, J.E., Lenz, B.F., Klabansky, R., Simon, R., Riggs, C., Guo, S., Oldham, R.K. and Fidler, I.J. 1986. 'Therapy of autochthonous skin cancers in mice with intravenously injected liposomes containing muramyltripeptide', Cancer Res. 46, 1160-1163.

Talmadge, J.E., Wolman, S.R. and Fidler, I.J. 1982. 'Evidence for the clonal origin of spontaneous metastases', Science 217, 361-363.

Thombre, P. and Deodhar, S.D. 1984. Inhibition of liver metastases in murine colon adenocarcinoma by liposomes containing human c-reactive protein or crude lymphokine. Cancer Immunol. Immunother. 16, 1984.

Xu, Z.L. and Fidler, I.J. 1984. 'The in situ activation of cytotoxic properties in murine Kupffer cells by the systemic administration of whole Mycobacterium bovis organisms or muramyl

Drug Carrier Systems
Edited by F.H.D. Roerdink and A.M. Kroon
© 1989 John Wiley & Sons Ltd.

Liposomes as Drug Carriers

in the Therapy of Infectious Diseases

by

R. L. Juliano, Ph.D.
Department of Pharmacology
School of Medicine
The University of North Carolina at Chapel Hill
Chapel Hill, NC   27599-7365

OUTLINE

1)   Background

        During the last dozen years there has been
considerable interest in the possible uses of lipid
vesicles (liposomes) as carriers for drugs
(Gregoriadis, 1981; Poznansky and Juliano, 1984;
Ostro, 1987; Raymond, 1987). Lately, much of this
interest has concerned the use of liposomal carriers
for drugs used in the therapy of infectious diseases,
especially those caused by obligate or facultative
intracellular pathogens. This interest derives from a
growing realization that liposome technology provides
an innovative but rational approach to the therapy of
such diseases. In this chapter we will briefly
examine the technical background for liposomal drug
delivery systems and then go on to review a number of
therapeutic applications of that technology in the
infectious diseases area. Thus, we will briefly
consider basic lipid vesicle chemistry in relation to
the problem of drug incorporation into liposomes. We
will then go on to describe the behavior of liposomal
drugs in vivo. Finally, we will examine applications
to the therapy of bacterial, fungal, viral and
protozoan pathogens. The first part of the chapter,
that dealing with liposome technology, will draw
heavily on a number of excellent recent reviews of
this subject. The therapeutics section will deal
mainly with developments since about 1983; previous
reviews are cited to deal with the earlier literature.
While intending to provide a detailed and coherent
review of this area, the author makes no claims as to
completeness. Rather, he apologizes in advance for
any errors of omission by which important
contributions to this ever-growing field are
overlooked. The recent reviews by Emmen and Storm
(1987), Bakker-Woudenberg and Roerdink (1986), and
Popescu, et al. (1987) complement the present work.

2)   Liposome Chemistry and Formulation

        As described in more detail elsewhere in this
volume, liposomes are closed structures composed of
bimolecular lipid membranes surrounding closed water-
filled compartments. The membrane structure can
include a variety of substances including natural or
synthetic phospholipids, sterols, fatty acids,
glycolipids or even proteins (Szoka and

Papahadjopoulos, 1980; Cullis et al., 1987).
Liposomes can be made to assume a variety of physical
configurations including large multilamellar vesicles
(MLVs) (≈1 micron), small unilamellar vesicles (SUVs)
(≈300 Angstroms) or large unilamellar vesicles (LUVs).
Both the size of the liposomes as well as their
chemical composition and physical form will affect the
characteristics of a particular vesicle population as
a drug carrier. Basically, drugs can be incorporated
into liposomal carriers in two distinct ways. Thus
highly polar, water soluble drugs can be entrapped or
encapsulated in the internal aqueous compartment(s) of
the liposome while lipophilic drugs can intercalate
into and become part of the liposome membrane. Thus
for carriage of water soluble drugs in lipid vesicles,
it is important that the vesicles have a large
internal volume per unit weight of lipid so that as
much drug as possible can be encapsulated; generally
speaking, LUVs are most suited to this purpose. By
contrast, for lipophilic drugs the amount of drug
incorporated is simply proportional to the total mass
of lipid, thus simple MLVs can often be effectively
used as carriers for apolar or amphiphilic drugs.
    A different situation arises when the drug of
interest has an intermediate oil/water partition
coefficient. In this case the drug does not bind
particularly well to the membrane lipid, nor is it
highly soluble in water and easily retained in the
aqueous compartment by the membrane barrier; thus
special approaches to liposome formulation must be
taken. One of the most interesting recent
developments in this regard is the work of Cullis and
collaborators (Mayer et al., 1985; Cullis et al.,
1987) who have used pH or ion gradients sustained
across the liposome membrane as a way of concentrating
weakly lipophilic cationic drugs (e. g.,
anthracyclines) within a liposomal carrier. These
workers showed that the preparation of liposomes with
a strong transmembrane potential (K diffusion
potential) or pH gradient could cause the
intraliposomal accumulation of a variety of lipophilic
cationic drugs, with interior/exterior accumulation
ratios of up to 100x in some cases. Besides high
efficiency drug entrapment, this approach can also
provide the opportunity for "remote loading" that is,
separate formulation of drug and liposome with loading
(pH driven) of the drug into the liposome at the time
of use (Cullis et al., 1987). A variety of other
procedures have been developed recently to permit
efficient entrapment of high concentrations of drugs
in liposomes. Among the most notable examples are
freeze-thaw procedures (Kirby and Gregoriadis, 1984;

Mayer et al., 1985b) high pressure extrusion
procedures (Hope et al., 1985) and procedures
involving forming liposomes on support material (Payne
et al., 1986). All of these approaches afford
efficient drug entrapment even when concentrated
suspensions of drug and lipid are used; presumably
multiple ruptures and resealing of the liposomes
caused either by freezing and thawing or by hydraulic
shear allow a much more efficient solvent and solute
capture than does simple hydration of the lipid.
     In thinking about liposomal drug-carrier
complexes, one must realize that, especially in the
cases of lipophilic compounds, the incorporation of
the drug into the liposome will change the physical
characteristics of the liposome membrane. In other
words, liposome plus drug implies an interactive
situation. Thus a number of years ago our laboratory
showed that the thermotropic phase transition profiles
of liposomes composed of a pure phospholipid could be
markedly perturbed by lipophilic antitumor drugs
(Actinomycin D, Vinblastine), but not by highly polar
antitumor drugs (fluorodeoxyuridine) (Juliano and
Stamp, 1979). In terms of drugs used in therapy of
infectious disease, this aspect of liposome-drug
interactive behavior has not been extensively explored
with the notable exception of the case of amphotericin
B, the polyene antifungal agent. These studies will
be discussed in the section on antifungal therapy
below.

3). Liposome Behavior In Vivo

     This topic has been the subject of an extremely
large number of investigations and here we can only
touch upon some of the highlights as they relate to
infectious disease. More detailed recent reviews
include those of Hwang (1987), Poznansky and Juliano
(1984), and Juliano (1987).
     It is important to remember that there is no such
thing as a "naked" liposome in vivo. Thus, immediately
upon entering the body, liposomes will adsorb a
coating of protein, either from the serum, if the
vesicles are administered intravenously, or from
lymph/extracellular fluid, if administered by another
route. A wide variety of blood proteins are known to
bind to lipid vesicle surfaces; this would include
albumin, alpha and beta globulins, high and low
density lipoproteins, fibronectin, immunoglobulin G,

certain clotting factors, and possibly complement
components (reviewed by Bonte and Juliano, 1986).
While most types of liposomes studied clearly tend to
adsorb albumin and IgG, the relative contribution of
the other proteins will depend on the chemical nature
of the liposome surface. The blood protein-liposome
interaction can have a number of biological
consequences. For example, certain vesicle types
interact strongly with and are destabilized by high
density lipoprotein; such vesicles would tend to
release their contents of encapsulated drugs upon
encountering HDL in the blood. Detailed studies of
serum lipoprotein-liposome interaction have been
published by a number of authors (for review see
Scherphof et al., 1984).
        Perhaps the most significant liposome-protein
interactions, in the context of this review, are those
which result in the adsorption of proteins with
opsonic activity - that is, molecules such as IgG,
fibronectin or C-reactive protein, for which receptors
exist on the cell surface of the macrophage. The
binding of such opsonic proteins will potentiate the
interaction of liposomes with the phagocytic
macrophages of the reticuloendothelial system (RES)
and thus enhance the rate of liposome clearance from
the blood and accumulation in tissues. Thus the well
known fact that negatively charged liposomes are
cleared from the blood much more rapidly than neutral
ones of the same size may be due to the fact that
negatively charged vesicles adsorb much greater
quantities of IgG than neutral (or positive) vesicles
(Bonte et al., 1987). The clearance kinetics and
tissue distribution of intravenously administered
liposomes have been studied extensively. Very early
on it became clear that large liposomes were cleared
more rapidly than small ones and that negatively
charged liposomes were cleared more rapidly than
neutral or positive ones (Juliano and Stamp, 1975;
Abra and Hunt 1981, reviewed by Gregoriadis et al.,
1983). Differences in liposome clearance rates may
relate, in part, to differential binding of opsonic
protein factors (see above). However, another
important aspect clearly is the in vivo stability of
the vesicle preparation. Thus use of long, fully
saturated PCs, use of sphingomyelin and inclusion of
cholesterol all help to stabilize vesicles in vivo
(Senior and Gregoriadis, 1982). It should be noted
that vesicles administered into the peritoneum also
eventually enter the blood and are then cleared in a
manner similar to those vesicles originally injected
i.v. (Ellens, et al., 1981). The dose of liposomes
used is also important since saturation of the RES can

occur (Bosworth and Hunt, 1982; Kao and Juliano, 1981).

The endothelial lining of the vasculature is a major barrier to the tissue uptake and distribution of liposomes (Juliano, 1987). Thus lipid vesicles can only leave the circulation at sites where the endothelium is open or fenestrated; this would include, most importantly, the liver and spleen, and to a lesser extent bone marrow and lymphoid organs (Poznansky and Juliano, 1984; Hwang, 1987). In these sites, liposomes interact with and are taken up by the phagocytic cells of the reticuloendothelial system (reviewed by Juliano, 1987; Poznansky and Juliano, 1984), a process greatly aided by the coating of opsonic factors acquired by some liposome types (see above). Thus liposome distribution in tissues is dominated by accumulation in the liver and spleen; however, one should keep in mind that there is always a diffuse distribution of liposomes in other organs and in the muscle mass, presumably due to nonspecific trapping of liposomes in capillary beds (Abra and Hunt, 1981; Hwang, 1987).

The in vivo clearance rates of liposomes can be modulated by actions or factors which affect liposome interaction with reticuloendothelial cells. Thus liposomes themselves provide an excellent means for inducing a transient, reversible blockade of the reticuloendothelial system (Kao and Juliano, 1981; Ellens, et al., 1982; Merion, 1985). Increasing the administered dose of liposomes will decrease the clearance rate, since the RES uptake capacity becomes saturated (Hwang, 1987;, Abra and Hunt, 1981). These observations, of course, raise the concern that liposomes may be quite toxic to the reticuloendothelial system. Clearly investigators have reported such toxicities (Allen et al., 1984), although these are usually reversible (Ellens et al., 1982; Merion, 1985). An interesting report from the perspective of this chapter, however, is that of Gilbreath et al., (1985) who reported that liposomes could reduce the microbicidal activity of macrophages.

Another approach to modifying liposome behavior has been to employ chemical modification of the liposome surface with the intent of reducing interaction with reticuloendothelial cells. That this may be possible is supported by the work of Illum et al. (1986) who prolonged the circulation lifetime of microparticles by coating them with certain types of polymers. Several groups are now actively studying the possibility of coupling high densities of sugar groups or of certain polymers to the liposome surface so as to hinder RE cell interaction, but the success

of these approaches is yet to be clearly validated.
Figure 1 summarizes the behavior of liposomal drugs _in_
vivo.

Figure 1

Behavior of Liposomal Drugs In Vivo

Phase I- binding of opsonic and destabilizing blood proteins and release of drug	Phase II- uptake by reticuloendothelial cells & diffusional transfer of drug in the extracellular space	Phase III-release of drug within the cell

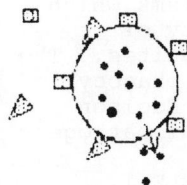

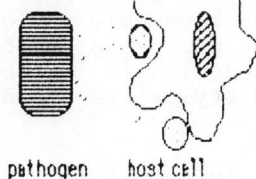

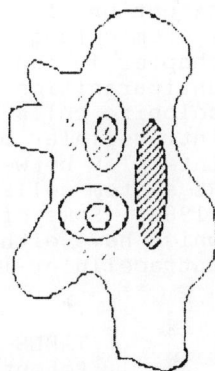

pathogen   host cell

See text for detailed explanations.

4)   Pathogenesis of Intracellular Infections;
     A Pharmacological Sanctuary.

     Microbes have adopted a variety of strategies to
colonize their hosts. It is beyond the scope of this
chapter to present a detailed consideration of these
strategies. Rather, it will suffice to make the
simple observation that liposomes, when used as drug
carriers in infectious diseases, have displayed the
most uniform successes against diseases caused by
facultative or obligate intracellular pathogens,
especially those which tend to colonize the cells of
the RE system. This is not hard to comprehend since
liposomes offer a very effective, highly targeted
means for concentrated delivery of drug to the
phagocytic cells of the RE system. With the possible
exception of use of liposomal polyenes for therapy of
systemic fungal disease (see below), the brightest
chapter in the history of liposomal antibiotics and
antiparasitics involves use against organisms which
colonize cells. There exist a wide variety of
intracellular bacterial and protozoan parasites. The
interplay between these organisms and the phagocytic
RE system cells has been ably reviewed by Edelson
(1982). Some of the protozoan and bacterial pathogens
which have either facultative or obligate
intracellular life styles are shown in TABLE 1.

TABLE 1.   Intracellular Infectious Disease
Potentially Treatable with Liposomal Drugs

Bacterial	Protozoan	Fungal
Tuberculosis	Leishmaniasis	Histoplasmosis
Leprosy	Toxoplasmosis	Cryptococcosis
Brucellosis	Malaria	
Salmonellosis		

Adapted from Fidler, 1987

     It has been known for some time that
intracellular pathogens occupy a pharmacological
sanctuary. Thus early on it was shown that penicillin
G and streptomycin are both ineffective in killing
intracellular staphylococci in polymorphonuclear
leukocytes (Holmes et al., 1966) as was methicillin;
however, the lipophilic antibiotic rifampin was more

effective in this regard (Mandell and Vest, 1972).
Subsequent studies have confirmed the concept that the
efficacy of certain antibiotics in treating pathogens
capable of intracellular colonization is related to
the ease of penetration of the antibiotic into the
host cell. Thus in alveolar macrophages (Johnson et
al., 1980) and in polymorphonuclear leukocytes
(Prokesch and Hand, 1982) a variety of beta lactam
antibiotics penetrated poorly, gentamicin, isoniazid
and tetracycline penetrated moderately well, while the
more lipophilic drugs rifampin, lincomycin and
chloramphenicol penetrated and were accumulated within
host cell; further, erythromycin and clindamycin were
markedly accumulated. Other studies have also
confirmed the relatively poor uptake of beta lactams
and greater uptake of chloramphenicol (Jacobs et al.,
1982) or erythromycin (Martin et al., 1985) by
mammalian cells. Tetracyclines have also been shown,
in model systems, to diffuse rather readily through
lipid bilayer membranes (Argast and Beck, 1984) thus
presumably accounting for their efficacy against
intracellular pathogens. By contrast, aminoglycoside
antibiotics, besides being quite polar, are known to
strongly bind to certain negatively charged lipid
constituents of cell membranes (polyphosphoinositols)
(Au et al., 1986; Brasseur et al., 1984) a situation
which might further reduce their efficacy in the
intracellular environment. Besides basic
considerations of membrane permeability, another
aspect which might be of importance in antibiotic
therapy of intracellular infections is the activity of
the antibiotic in the intracellular environment. For
example, there is some evidence that rifampin, while
active against intracellular pathogens, is less active
in the low pH environment of the endosome than it
would otherwise be (Lam and Mathison, 1983).

5)   Antibacterial Actions

     As discussed above, important classes of
antibiotics including the aminoglycosides and many
members of the beta lactam group penetrate poorly into
the intracellular environment. Thus there is a good
rationale in using liposomes to deliver these drugs to
tissues infected with intracellular pathogens. A
number of in vitro and in vivo studies indicate that
this approach can be at least modestly successful.
The use of liposomes in antibacterial chemotherapy has

been reviewed previously by Richardson (1983) and by
Bakker-Woudenberg and Roerdink (1986).

In a very early report, Bonventre and Gregoriadis
(1978) showed that dihydrostreptomycin entrapped in
PC/cholesterol/PA vesicles could promote the killing
of S. aureus within phagosomes of peritoneal
macrophages. Similar in vitro studies have also been
made by other groups. Thus Stevenson et al. (1983)
showed that the apparent activity of both streptomycin
and chloramphenicol for in vitro killing of E. coli
within phagosomes of the J774.2 mouse macrophage cell
line was increased approximately 10 fold by
incorporation of the drug in large unilamellar
liposomes. These investigators found a marked
dependence on liposome composition for this effect.
Interestingly, the liposomal forms of the drug were
inactive in simple tube dilution assays against E.
coli, indicating that the macrophages probably caused
release of the drug in the phagosome compartment.
Fountain et al. (1985) have also shown enhanced
effectiveness of liposomal streptomycin in killing
intracellular B. canis or B. abortus organisms in
mouse or guinea pig peritoneal macrophages, while Dees
et al. (1985) have made similar findings in bovine
macrophages with B. abortus. In these studies several
types of liposomes seemed effective, in contrast to
the report of Stevenson et al. (1983). Working with
Listeria infections of mouse peritoneal macrophages,
Bakker-Woudenberg et al. (1986) showed that a
liposomal formulation of ampicillin killed
intracellular organisms whereas free ampicillin with
or without co-administration of "empty" liposomes only
slowed bacterial growth.

        Thus a number of investigators have demonstrated
increased effectiveness of several different
antibiotics against intracellular pathogens during in
vitro experiments when the drugs where given in
liposomal form. Although most investigations in this
area have concentrated on the liposome mediated
delivery of drugs to intracellular sites, a few
studies have shown altered effectiveness of liposomal
antibiotics against bacteria in simple in vitro assays
not involving bacterial uptake by mammalian cells.
Thus Chowdury et al. (1981) showed increased
effectiveness of liposomal penicillin against several
penicillin resistant bacteria; the type of liposome
used here did not seem to be important. In a similar
vein, Nacucchio et al. (1985) showed that the
incorporation or bonding of piperacillin to liposomes
protected the drug against beta lactamases and thus
enhanced in vitro antistaphylococcal activity. The
mechanism by which the liposomal beta lactam drug

interacts with the bacterial cells is unclear from the
studies cited here.

A number of investigators have also reported on
the successful therapy of facultative or obligate
intracellular bacterial pathogens with liposomal
antibiotics. Thus Desiderio and Campbell (1983)
showed an improved action of the beta lactam drug
cephalothin against an experimental murine
salmonellosis when the drug was given in liposomal
form. Tadakuma et al. (1985) have also examined
effects of liposomal streptomycin in experimental
murine salmonella infections, finding that single dose
treatments with the liposomal drug prolonged survival
whereas similar treatment with free streptomycin did
not. Prolonged survival was attributed to suppression
of bacterial growth in liver and spleen which are, of
course, the major sites of liposome accumulation.
Fountain et al. (1985) examined effects of liposomal
aminoglycosides on Brucella infections in mice and
guinea pigs; they found a reduction in the number of
Brucella organisms in several organs (not only liver
and spleen) when the liposomal forms of the drug were
used but not when free drug was used.
Bakker-Woudenberg et al. (1985) studied effects of
free or liposomal ampicillin in therapy of Listeria
infections in normal or athymic (nude) mice. The
liposomal form of the drug displayed an 80 fold
increase in therapeutic potency as compared to free
drug in normal mice. Thus liposomal ampicillin
treatment produced a marked reduction in liver and
spleen viable bacteria in normal mice. However, in
nude mice the effect was much less dramatic,
indicating that the liposomal drug delivery strategy
does not obviate the need for a strong host response
to pathogen. Anderson and Kirby (1986) have studied
effects of cloxacillin liposomes in experimental
mastitis. Interestingly, thus far there is no
evidence of enhanced immune responses against
penicillins incorporated into liposomes, despite the
fact that liposomes usually have an adjuvant action
(de Haan et al., 1986).

A number of workers have begun to explore the use
of liposomal antibiotics in therapy of mycobacteria
infections including tuberculosis and leprosy. Thus
Wasserman et al. (1986) and Orozco et al. (1986) have
developed methods for loading liposomes with
rifampicin and isoniazid and have used these to treat
mice with severe tuberculosis. The results indicated
that the group of animals treated with the liposomal
drug showed the best survival as well as the least
lung inflammation. Similar results were reported by

Vladmirsky and Ladigina (1982) using streptomycin to
treat an animal model of tuberculosis.

6)   Antifungal Actions

        Systemic fungal infections are a severe and
increasingly prevalent clinical problem. These
infections are relatively rare in individuals with
normal immune function and tend to occur primarily in
patients who are immunocompromised in one way or
another (Bodey, 1986). The numbers of
immunocompromised individuals are growing for a
variety of reasons including the spread of AIDS,
aggressive use of immunosuppressive chemotherapy and
radiation therapy in cancer, intentional
immunosuppresion of organ transplant recipients, and
growth of the aged and debilitated population.
Although there has been a concerted effort to develop
new antifungal drugs in recent years, particularly
agents of the imidazole and triazole classes, the
mainstay of systemic antifungal therapy continues to
be amphotericin B, a polyene antibiotic which has been
available for about thirty years. Amphotericin B
(AMB) is also used as a second line drug in the
therapy of Leishmaniasis, a disease caused by a
protozoan parasite common in parts of Africa and the
Middle East. AMB is an extremely effective, broad
spectrum anti-fungal agent; however its use is fraught
with problems due to the extreme toxicity of the drug.
These toxicities include acute side effects such as
nausea, hypotension, chills and fever, whereas chronic
use of the drug (as is commonly required) almost
inevitably leads to substantial nephrotoxicity (Pratt
and Fekety, 1986).
        AMB is a well studied drug and there is a good
deal known about its mechanism of action
(Hamilton-Miller, 1973). AMB is a very lipophilic
molecule and readily binds to a variety of cellular
(or artificial) membranes. If the membranes contain a
sterol, the AMB can form a stoichiometric complex with
sterol molecules giving rise to a transmembrane
channel or pore; the characteristics of AMB-sterol
pores have been quite well defined. However, as
recently reviewed by Bolard (1986), AMB may have a
variety of other actions on cell membranes unrelated
to its pore forming ability, and some of these actions
may play a role in the therapeutic effectiveness of
the drug. Nonetheless, our ability to use AMB in

therapy of fungal infections probably rests primarily
on the fact that AMB complexes very effectively with
ergosterol, the primary sterol of fungal membranes,
but somewhat less well with cholesterol the major
sterol of mammalian cells (Pratt and Fekety, 1986).
Despite this degree of selectivity, many of the toxic
effects of AMB no doubt arise from its interactions
with cholesterol in critical host cell populations,
for example, in proximal tubule cells or in cardiac
cells.

The clinically utilized form of AMB [Fungizone
(R)] is a micelle of AMB complexed with deoxycholate
as a solubilizing agent. Some time ago, several
laboratories began to evaluate the question of whether
reformulation of AMB in liposomes could improve its
therapeutic ratio. The first observations were due to
R. New and his colleagues (1981) who showed that a
liposomal form of AMB could be used to treat an animal
model of Leishmaniasis and that the drug was less
toxic in liposomal form . A number of other
laboratories then investigated the use of various
formulations of liposomal AMB in animal models
(usually murine) of systemic fungal infections,
including histoplasmosis (Taylor et al., 1982),
cryptococcosis (Graybill et al., 1983), and
candidiasis (Tremblay et al., 1984; Lopez-Berestein et
al., 1983). The work of Lopez-Berestein et al. (1983)
seemed especially promising, since this group reported
a major (approx. 20 fold) reduction in the in vivo
toxicity of AMB with full retention of antifungal
potency in a model of candidiasis. The liposomes used
in this study (MLVs 7:3 DMPC/DMPG), as we will see
below, had a near optimal composition for protecting
the host against AMB toxicity. Interestingly,
liposomal AMB has also been considered for treating
fungal disease of the eye. Thus Barza et al.
(1985a,b) showed that liposomal AMB exhibits reduced
toxicity to monkeys or rabbits when administered by
intravitreal injection.

Work on the DMPC/DMPG formulation of AMB resulted
in a number of important observations, including: (a)
the liposomal formulation essentially retained the
broad spectrum antifungal characteristics of free AMB
(Hopfer et al., 1984); (b) the liposomal AMB
maintained its effectiveness in treatment of fungal
infections in neutropenic animals (Lopez-Berestein et
al., 1984); (c) liposomal AMB could work
synergistically with immunomodulators in prophylaxis
of candida infections (Mehta et al., 1985).
Encouraged by these results, Lopez-Berestein and
colleagues engaged in preliminary clinical trial of
liposomal AMB in a series of cancer patients with

systemic fungal disease who had failed all
conventional antifungal therapy (Lopez-Berestein et
al., 1985). The results were quite encouraging in
that toxicity of the drug was clearly reduced when it
was given in liposomal form; further, the drug seemed
to have some efficacy against disease which had not
responded to other treatment. More recently,
additional studies have been reported in patients with
hepatosplenic candidiasis (Lopez-Berestein, et al.,
1987; Shirkoda, et al., 1986) and the issues of
clinical treatment with liposomal AMB have been
discussed (Lopez-Berestein, 1986).

The observations discussed so far leave open the
question of the mechanism whereby liposomal
incorporation increases the therapeutic index of AMB.
The first insights came with a study which showed,
using erythrocytes as a "model" mammalian cell, that
whereas free AMB was toxic in vitro to both fungal
cells and mammalian cells, the liposomal form of the
drug was toxic only to fungal cells (Mehta, et al.,
1984). This sort of observation has also been
confirmed in studies of the action of liposomal AMB on
macrophages and lymphoid cells (Mehta et al., 1985),
that is, the liposomal drug was far less toxic than
free AMB. Recent studies (Juliano, et al., 1987) have
partially delineated the basis for the selective
toxicity of liposomal AMB. These studies indicate
that for liposomes of an appropriate composition, the
AMB can rapidly and effectively transfer from the
liposomal carrier to fungal cell membranes but not to
mammalian cell membranes. Alteration of the liposomal
lipid constituents, including modification of
acylchain unsaturation as well as modification of the
polar head group, can markedly affect the ability to
selectively transfer AMB to fungal cells Although
these observations are rather unorthodox, they are
consistent with a growing body of information
concerning the lipid phase partitioning behavior of
AMB. Thus Bolard (1986) as well as Clejan and Bittman
(1985) and Vertut-Croquin et al. (1984) point out that
binding of AMB to membranes depends not only on the
sterol content but on the overall physical
characteristics of the membrane. Rao et al. (1985)
have also pointed out that the phospholipid and
protein composition of the "target" cell membrane as
well as its sterol content can affect polyene action.
Further, it has been shown (Bolard et al., 1981;
Bolard, 1986) that the transfer rate of AMB between
vesicle populations will depend on the physical state
of both donor and recipient vesicles; for example, AMB
will transfer rapidly from fluid vesicles to other
fluid vesicles, but not from gel phase vesicles to

fluid ones. This suggests the theoretical basis for
the construction of vesicles which can transfer or
donate AMB to one type of cell membrane (fungal cells)
but not to another type of cell membrane (mammalian
cells) which has very different membrane physical
characteristics. A more detailed study of the
biophysics of AMB transfer processes between membranes
is needed, as well as studies to determine if other
types of membrane active drugs also display the
selective partitioning characteristics shown by the
polyene antibiotics.

7)   Antiparasitic Actions

      Some of the most successful early work on the use
of liposomes in the treatment of infectious disease
involved therapeutic approaches to Leishmaniasis, a
parasitic disease common in tropical and subtropical
regions. The disease is caused by a protozoan
organism which is transmitted by the bite of
sandflies. There are several forms of Leishmaniasis
including the visceral form (L. donovani), the
cutaneous form (L. tropica) and the mucocutaneous form
(L. braziliensis). These organisms share a very
unusual life style in that they colonize the
phagocytic vacuoles of macrophages and related cells;
thus the Leishmania parasites infect one of the body's
key host defense cells. The drugs used most widely
against Leishmaniasis have been pentavalent antimony
compounds such as sodium stibogluconate, while
aminoquinolines and amphotericin B have been used as
secondary drugs (Pratt and Fekety, 1986).
      In a dramatic early demonstration of the value of
liposomal carriers in treatment of parasitic disease,
Alving et al. (1978) showed the incorporation of
antimonial compounds into liposomes produced a
remarkable enhancement of their potency against
Leishmaniasis in an animal model. The toxicity of the
drug was also increased, but a substantial net
improvement in the therapeutic index was attained.
The key issue here is that the use of a liposomal
carrier produced accumulation of the drug at the
precise site where parasite colonization occurred,
namely the phagocytic vacuoles of macrophages. Early
studies on the use of liposomal antimony drugs in
therapy of protozoal disease have been thoroughly
reviewed by Alving (1983). Another interesting study
was the work of New et al. (1981) who used liposomal

amphotericin B to treat an animal model of
Leishmaniasis. As with the application of liposomal
amphotericin B to antifungal therapy discussed above,
this application gave rise to enhanced results due, at
least in part, to the reduced toxicity of the drug to
the host.

More recent work in the antiprotozoal area has
concentrated on understanding in more detail the
mechanism of action of the liposomal drug
formulations. Thus Weldon et al. (1983) and Heath et
al. (1984) have explored the intracellular
distribution of antileishmanial drugs after
administration in liposomes. A recent study has also
demonstrated that liposomal amphotericin B is
effective in treatment of visceral Leishmaniasis in
hamsters and in monkeys at doses which are known to be
virtually nontoxic in man. This looks very promising
as a useful and minimally toxic approach to treatment
of Leishmaniasis and may well set the stage for a
clinical trial (Berman et al., 1986).

Liposomes have also been used as carriers for
antimalarial drugs. Thus the toxicity of the
important antimalarial compound primaquine was reduced
about 4 fold when the drug was incorporated into lipid
vesicles; this occurred with no loss of therapeutic
efficacy (Pirson et al., 1980). This same group has
also studied the biodistribution of free and liposomal
antimalarial drugs (Pirson et al., 1983). Since
plasmodia, the malarial parasites, seem to colonize
both phagocytic liver sinusoidal cells and non-
phagocytic hepatocytes (Pratt and Fekety, 1986), the
rationale for liposomal drug delivery in malaria is
not as clear as for Leishmaniasis.

8)   Immunomodulation

Macrophages are key cells in the host defense
systems against both cancer and infectious disease.
It has been known for some time that various
immunomodulating substances including lymphokines and
bacterial cell wall products such as muramyl dipeptide
(MDP) and lipopolysaccharide can activate macrophages
thus enhancing their tumoricidal capabilities as well
as their ability to destroy pathogens (Poste and
Fidler, 1979). In general, in vivo activation of
macrophages through the administration of soluble
immunomodulators has met with limited success due to
the fact that these substances are rapidly degraded

and/or excreted. Some time ago Fidler and his
colleagues demonstrated that the macrophage activating
effects of lymphokines and of muramyl dipeptide could
be markedly enhanced via incorporation in liposomes
(Fidler, 1987). The lipid vesicles tended to promote
binding and uptake of concentrated "packets" of
immunomodulators in vitro, while in vivo they reduced
excretion and metabolism and promoted uptake of the
encapsulated immunomodulators by circulating monocytes
and by other cells of the reticuloendothelial system.
This approach was first applied to therapy of
metastatic cancer in animal models with very
impressive results (Fidler, et al., 1981, 1982). More
recently Fidler's group as well as a number of other
investigators have used liposomal immunomodulators in
the prophylaxis and treatment of infectious disease in
animals. Results on immunomodulators in antiviral
therapy are discussed immediately below; here we will
concentrate on studies of treatment of fungal and
parasitic diseases.
     In an early study, Osada et al. (1982) showed
that a stearic acid analog of muramyl dipeptide could
stimulate the resistance of immunocompromised mice to
challenge with E. coli or with Candida albicans.
Since this C18 derivative of MDP likely forms a
micelle, this study can be viewed as a forerunner of
studies with liposome encapsulated immunomodulators.
A water soluble analog of MDP was studied by
Fraser-Smith et al. (1983) administered in free form,
encapsulated within liposomes and co-mixed with
liposomes. The infectious challenge in this case was
an intravenous dose of Candida albicans.
Interestingly, the MDP derivative was far more
effective when either encapsulated within liposomes or
co-mixed with liposomes than when given alone. There
was an apparent synergistic stimulation of the
reticuloendothelial system by the liposomes and by the
MDP derivative; the liposomes used in this case
contained phosphatidyl serine and phosphatidyl
choline. In another study with liposomal MDP
derivatives, Lopez-Berestein et al. (1985) studied the
joint effects of liposomal amphotericin B and of a
stearic acid analog of MDP in liposomes on the
prophylaxis of Candida albicans infections in mice.
They found a substantial synergistic effect of this
combination of cytotoxic drug and immunomodulator over
the effect of either agent given alone.
     In addition to work on bacterial and fungal
infections, liposome encapsulated immunomodulators
have been examined in the context of protozoal
infections. Thus Gupta et al. (1986) observed that a
tuftsin derivative encapsulated in liposomes provided

good protection against <u>Plasmodium</u> <u>berghei</u> infections
in an animal model. Reed <u>et</u> <u>al.</u> (1984) have
encapsulated crude lymphokines in liposomes and
studied effects on visceral Leishmaniasis. In another
study of the effect of immunomodulators and
"conventional" drugs used together, Adinolfi <u>et</u> <u>al.</u>
(1985) used an antimony drug in combination with a
stearic acid analog of MDP in liposomes to treat
visceral Leishmaniasis in mice and hamsters.
Immunostimulation, prophylactically or
therapeutically, or both, enhanced the effectiveness
of the antimonial drug.
     Thus in a variety of infectious disease states,
the use of immunomodulators incorporated into
liposomes has provided an effective means of
immunostimulation and has resulted in enhanced
prophylaxis or therapy.

9)   Antiviral Actions

     There has been but a limited amount of work on
the use of liposomes in antiviral therapy (Streissle,
<u>et</u> <u>al.</u>, 1985). Basically, two experimental approaches
have been followed. First, some investigators have
utilized liposomes in conjunction with drugs which
directly inhibit steps in the virus replicative cycle.
Thus Snolin <u>et</u> <u>al.</u> (1985) have explored the use of
IUdR in therapy of ocular Herpes infections. The drug
ribavirin has also been used in liposomal form in
therapy of Rift Valley Fever Virus infection in mice
with some success (Kende <u>et</u> <u>al.</u>, 1985). A more
complete account of experiences with direct antiviral
therapy is given in Popesco, <u>et</u> <u>al.</u> (1987).
     Second, there has been a greater amount of work
on the use of liposomal immunomodulators to enhance
the host response to viral infection, primarily
through macrophage activation. Since activated
macrophages have the ability to recognize and destroy
virus infected cells (Mogenson, 1985), the use of
liposomes to target immunomodulating substances
directly to macrophages seems a very reasonable
approach. A comprehensive discussion of this approach
and of the underlying pathophysiology is given by Koff
and Fidler (1985). Perhaps the most striking example
of the use of immunomodulators in liposomes to attain
prophylaxis against viral challenge is the work of
Koff <u>et</u> <u>al.</u> (1985). These investigators use MTP-PE, a
lipid analog of muramyl dipeptide, incorporated into

liposomes to protect mice against a potentially lethal
challenge with Herpes simplex type 2 virus. Another
group has recently studied the joint use of liposomal
ribavirin and liposomal MTP-PE in protecting mice
against pneumonitis caused by influenza virus or
Herpes simplex type 1 (Gangemi et al., 1987). The two
drugs, used in conjunction, were more effective than
single agent therapy with either drug. In another
recent report Knudsen et al. (1986) have used liposome
incorporated Avridine, an immunostimulating lipophilic
amine, to protect guinea pigs against foot and mouth
disease virus; this therapy seemed to block spread of
the virus but to permit the establishment of titres of
antiviral antibodies sufficient to immunize against
reinfection.

Although there has been relatively little work
done on liposome incorporated antiviral drugs to date,
the realization that macrophages may be a major
reservoir of HIV (Streicher and Joynt, 1986) and thus
that they may contribute significantly to the
pathogenesis of AIDS, suggests that we can anticipate
numerous attempts to use liposomes to target anti-HIV
drugs to macrophages. This will no doubt add a very
significant chapter to the development of drug carrier
systems for therapy of viral disease.

10)   Conclusions

Lipid vesicle carrier systems are beginning to
make an important contribution to experimental therapy
of infectious disease. The most promising application
seems to be in connection with obligate and
facultative intracellular pathogens, especially those
afflicting the RE system, where liposomes can
effectively deliver therapeutic agents directly to the
site of pathogenic infection. Another promising,
albeit specialized, application is in the therapy of
systemic fungal infections using polyene antibiotics.
Here, use of liposomes seems to prevent host toxicity
by permitting selective transfer of the drug to the
fungal pathogen but not to the host cells.
Presumably, other interesting applications will emerge
as the technology for liposomal drug delivery matures
further.

11) <u>References</u>

Abra, R.M. and Hunt, C.A. (1981). `Liposome
disposition <u>in</u> <u>vivo</u> III. dose and vesicle-size
effects', Biochim. Biophys. Acta, 666, 493-503.

Adinolfi, L.E., Bonventre, P.F., Vander Pas, M., and
Eppstein, D.A. (1985). `Synergistic effect of
glucantime and a liposome-encapsulated muramyl
dipeptide analog in therapy of experimental visceral
leishmaniasis', Infect. Immun., 48, 409-416.

Allen, T.N., Murray, L., MacKeigan, S., and Shah, M.
(1984). `Chronic liposome administration in mice:
effects on reticuloendothelial function and tissue
distribution', J. Pharmacol. Exp. Ther. 229, 267-
275.

Alving, C.R., (1983). `Delivery of liposome-
encapsulated drugs to macrophages', Pharmacol.
Ther., 22, 407-424.

Alving, C.R., Steck, E.A., Chapman, W.L., Waits, V.B.,
Hendricks, L.D., Swartz, G.M., and Hanson, W.L.
(1978). `Therapy of leishmaniasis: superior
efficacies of liposome encapsulated drugs', Proc.
Natl. Acad. Sci., USA, 75, 2959-2963.

Anderson, J.C. and Kirby, C.J. (1986). `The effect
of incorporation of cloxacillin in liposomes on
treatment of experimental staphylococcal mastitis in
mice', J. Vet. Pharmacol. Ther., 9, 303-309.

Argast, M. and Beck, C.F. (1984). `Tetracycline
diffusion through phospholipid bilayers and binding
to phospholipids', Antimicrob. Agents Chemother.,
26, 263-265.

Au, S., Weiner, N., and Schacht, J. (1986).
`Membrane effects of aminoglycoside antibiotics
measured in liposomes containing the fluorescent
probe, 1-anilino-8-naphthalene sulfonate', Biochim.
Biophys. Acta, 6, 205-210.

Bakker-Woudenberg, I.A. and Roerdink, F.H. (1986).
`Antimicrobial chemotherapy directed by liposomes',
J. Antimicrob. Chemother., 17, 547-549.

Bakker-Woudenberg, I.A., Lokerse, A.F., Roerdink, F.H., Regts, D., and Michel, M.F. (1985). `Free versus liposome-entrapped ampicillin in treatment of infection due to Listeria monocytogenes in normal and athymic (nude) mice', J. Infect. Dis., 151, 917-924.

Bakker-Woudenberg, I.A., Lokerse, A.F., Vink-van den Berg, J.C., Roerdink, F.H., and Michel, M.F. (1986). `Effect of liposome-entrapped ampicillin on survival of Listeria monocytogenes in murine peritoneal macrophages', Antimicrob. Agents Chemother., 30, 295-300.

Barza, M., Baum, J., and Szoka, Jr., F. (1984). `Pharmacokinetics of subjunctive liposome-encapsulated gentamicin in normal rabbit eyes', Invest. Ophthalmol. Vis. Sci., 25, 486-490.

Barza, M., Baum, J., Tremblay, C., Szoka, F., and D'Amico, D.J. (1985b). `Ocular toxicity of intravitreally injected liposomal amphotericin B in rhesus monkeys', Am. J. Ophthalmol., 15, 259-263.

Berman, J.D., Hanson, W.L., Chapman, W.L., Alving, C.R., and Lopez-Berestein, G. (1986). `Antileishmanial activity of liposome-encapsulated amphotericin B in hamsters and monkeys', Antimicrob. Agents Chemother., 30, 847-851.

Bodey, M.D. (1986). `Fungal infection and fever of unknown origin in neutropenic patients', Am. J. Med, 80, 112-119.

Bolard, J. (1986). `How do the polyene macrolide antibiotics affect the cellular membrane properties?', Biochim. Biophys. Acta, 864, 257-304.

Bolard, J., Vertut-Croquin, A., Cybulska, D.E., and Gary-Bobo, C.M., (1981). `Transfer of the polyene antibiotic amphotericin B between single-walled vesicles of dipalmitoylphosphatidycholine and egg-yolk phosphatidylcholine', Biochim. Biophys. Acta, 647, 241-248.

Bonte, F. and Juliano, R.L. (1986). `Interactions of liposomes with serum proteins', Chem. Phys. Lipids, 40, 359-372.

Bonte, F., Hsu, M.J., Papp, A., Wu, K., Regen, S.L., and Juliano, R.L. (1987). `Interactions of polymerizable phosphatidylcholine vesicles with blood components: relevance to biocompatibility', Biochim. Biophys. Acta, 900, 1-9.

Bonventre, P.F. and Gregoriadis, G. (1978). `Killing of intraphagocytic Staphylococcus aureus by dihydrostreptomycin entrapped within liposomes', Antimicrob. Agents Chemother., 13, 1049-1051.

Bosworth, M.E., Hunt, C. (1982). `Liposome disposition in vivo. II. Dose dependency', J. Pharm. Sci., 71, 100-104.

Brasseur, R., Laurent, C., Ruysschaert, J.M., and Tulkens, P. (1984). `Interactions of aminoglycoside antibiotics with negatively charged lipid layers. Biochemical and conformational studies', Biochem. Pharmacol., 33, 629-637.

Chowdhury, M.K.R., Goswami, R., and Chakrabarti, P. (1981). `Liposome-trapped penicillins in growth inhibition of some penicillin-resistant bacteria', J. Applied Bacteriology, 51, 223-227.

Clejan, S. and Bittman, R. (1985). `Rates of amphotericin B and filipin association with sterols. A study of changes in sterol structure and phospholipid composition of vesicles', J. Biol. Chem., 260, 2884-2889.

Cullis, P., et al. (1987). `Liposomes as Pharmaceuticals' in Liposomes: From Biophysics to Therapeutics, (M. Ostro, ed.). Marcel Dekker, NY, 39-72.

Dees, C., Fountain, M.W., Taylor, J.R., and Schultz, R.D. (1985). `Enhanced intraphagocytic killing of Brucella abortus in bovine mononuclear cells by liposome-containing gentamicin', Vet. Immunol. Immunopathol., 8, 171-182.

deHaan, P., Claasen, E., and vanRooijen, N. (1986). `Liposomes as carrier for antibiotics: a comparative study on the immune response against liposome-encapsulated penicillin and other penicillin preparations', Int. Archs, Allergy appl. Immun., 81, 186-188.

Desiderio, J.V. and Campbell, S.G. (1983).
`Intraphagocytic killing of Salmonella typhimurium
by liposome-encapsulated cephalothin', J. Infect.
Dis., 148, 563-570.

Desiderio, J.V. and Campbell, S.G. (1983).
`Liposome-encapsulated cephalothin in the treatment
of experimental murine salmonellosis', J.
Reticuloendothel. Soc., 34, 279-287.

Edelson, P.J. (1982). `Intracellular parasites and
phagocytic cells: cell biology and
pathophysiology', Rev. Infect. Diseases, 4, 124-153.

Ellens, H., Mayhew, E, and Rustum, Y.M. (1982).
`Reversible depression of the reticuloendothelial
systems by liposomes', Biochim. Biophys. Acta, 714,
479-485.

Ellens, H., Morselt, H., and Scherphof, G. (1981).
`In vivo fate of large unilamellar sphingomyelin-
cholesterol liposomes after intraperitoneal and
intravenous injection into rats', Biochim. Biophys.
Acta, 674, 10-18.

Emmen, F. and Storm, G. (1987). `Liposomes in
treatment of infectious diseases', Pharmaceutisch
Weekblad Scientific Edition, 9, 162-171.

Fidler, I.J. (1987). `Immunomodulation of
macrophages for cancer and antiviral therapy', in
Site Specific Drug Delivery (E. Tomlinson and S.
Davis, editors), John Wiley, Chichester, 111-134.

Fidler, I.J., Barnes, Z., Fogler, W.E., Kirsch, R.,
Bugeski, P., and Poste, G. (1982). `Involvement of
macrophages in the eradication of established
metastasis folling intravenous injection of
liposomes containing macrophage activators', Cancer
Res., 42, 496-501.

Fidler, I.J., Sone, S., Fogler, W.E., Barnes, Z.L.
(1981). `Eradication of spontaneous metastases and
activation of alveolar macrophages by intravenous
injection of liposomes containing muramyl
dipeptide', Proc. Natl. Acad. Sci., USA, 78, 1680-
1684.

Fountain, M.W., Weiss, S.J., Fountain, A.G., Shen, A., and Lenk, R.P. (1985). `Treatment of Brucella canis and Brucella abortus in vitro and in vivo by stable plurilamellar vesicle-encapsulated aminoglycosides', J. Infect. Dis., 152, 529-535.

Fraser-Smith, E.B., Eppstein, D.A., Larsen, M.A., and Matthews, T.R. (1983). `Protective effect of a muramyl dipeptide analog encapsulated in or mixed with liposomes against Candida albicans infection', Infect. Immun., 39, 172-178.

Gangemi, J.D., Nachtigal, M., Barnhart, D., Krech, L., and Jani, P. (1987). `Therapeutic efficacy of liposome-encapsulated ribavirin and muramyl tripeptide in experimental infection with influenza or Herpes simplex virus', 155, 510-517.

Gilbreath, M.J., Swartz, G.M., Alving, C.R., Nacy, C.A., Hoover, D.L., and Meltzer, M.S. (1985). `Differential inhibition of macrophage microbial activity by liposomes', Infect. Immun., 47, 567-569.

Graybill, J.R., Craven, P.C., Taylor, R.L., Williams, D.M., and Magee, W.E. (1982). `Treatment of murine cryptococcosis with liposome-associated amphotericin B', J. Infect. Dis., 145, 748-752.

Gregoriadis, G. (1981). `Targeting of drugs: implications in medicine', Lancet, 2, 241-246.

Gregoriadis, G., Kirby, C., and Senior, J. (1983). `Optimization of liposome behavior in vivo', Biol. Cell, 47, 11-18.

Gupta, C.M., Puri, A., Jain, R.K., Bali, Al, and Anand, N. (1986). `Protection of mice against Plasmodium berghei infection by a tuftsin derivative', FEBS Lett. 205, 351-354.

Hamilton-Miller, J.M.T. (1973). `Chemistry and biology of the polyene macrolide antibiotics', Bact. Rev. 37, 166-196.

Heath, S., Chance, M.L., and New, R.R.C., (1984). `Quantitative and ultrastructural studies on the uptake of drug loaded liposomes by mononuclear phagocytes infected with Leishmania donovani', Molecular and Biochemical Parasitology, 12, 49-60.

Holmes, B., Quie, P.G., Windhorst, D.B., Pollara, B.,
and Good, R.A. (1966). 'Protection of
phagocytized bacterial from the killing action of
antibiotics', Nature, 210, 1131-1132.

Hope, M.J., Bally, M.B., Webb, G., and Cullis, P.R.
(1985). 'Production of large unilamellar vesicles
by a rapid extrusion procedure. Characterization of
size distribution, trapped volume and ability to
maintain a membrane potential', Biochim. Biophys.
Acta, 812, 55-65.

Hopfer, R.L., Mills, K., Mehta, R., Lopez-Berestein,
G., Fainstein, V., and Juliano, R.L. (1984). 'In
vitro antifungal activities of amphotericin B and
liposome-encapsulated amphotericin B', Antimicrob.
Agents Chemother., 25, 387-389.

Hwang, C. (1987) 'Liposome Pharmacokinetics' in
Liposomes: From Biophysics to Therapeutics, (M.
Ostro, ed.). Marcel Dekker, NY, 109-156.

Illum, L., Hunneyball, I.M., and Davis, S.S. (1986).
'The effect of hydrophilic coatings on the uptake of
colloidal particles by the liver and by peritoneal
macrophages', Int. J. Pharmaceutics, 29, 53-66.

Jacobs, R.F., Wilson, C.B., Laxton, J.G., Haas, J.E.,
and Smith, A.L. (1982). 'Cellular uptake and
intracellular activity of antibiotics against
Haemophilus influenzae type b', J. Infect. Dis.,
145, 152-159.

Johnson, J.D., Hand, W. L, Francis, J.B.,
King-Thompson, N., and Corwin, R.W. (1980).
'Antibiotic uptake by alveolar macrophages', J. Lab.
Clin. Med., 95, 429-439.

Juliano, R.L. (1987). 'Microparticulate Drug
Carriers: Liposomes Microspheres and Cells. In:
Sustained and Controlled Release Drug Delivery
Systems', (J. Robinson and V. Lee, eds.) Marcel
Dekker, NY.

Juliano, R.L. and Stamp, D. (1975). 'Effect of
particle size and charge on the clearance rates of
liposomes and liposome encapsulated drugs', Biochim.
Biophys. Res. Commun., 63, 651-658.

Juliano, R.L. and Stamp, D. (1979). `Interaction of drugs with lipid membranes: characteristics of liposomes containing polar or non-polar antitumor drugs', Biochim. Biophys. Acta, 586, 137-145.

Juliano, R.L., Grant, C.W., Barber, K.R., and Kalp, M.A. (1987). `Mechanism of the selective toxicity of amphotericin B incorporated into liposomes', Mol. Pharm., 31, 1-11.

Kao, Y.J. and Juliano, R.L. (1981). `Interactions of liposomes with the reticuloendothelial system: effects of blockade on the clearance of large unilamellar vesicles', Biochim. Biophys. Acta, 677, 453-461.

Kende, M., Alving, C.R., Rill, W.L., Swartz, Jr., G.M., and Canonico, P.G. (1985). `Enhanced efficacy of liposome-encapsulated ribavirin against rift valley fever virus infection in mice', Antimicrob. Agents Chemother., 27, 903-907.

Kirby, C. and Gregoriadis, G. (1984). Liposome Technology, Volume 1, (G. Gregoriadis, ed.), CRC Press, Boca Raton, FL, 79-107.

Knudsen, R.C., Card, D.M., and Hoffman, W.W. (1986). `Protection of guinea pigs against local and systemic foot-and-mouth disease after administration of synthetic lipid amine (Avridine) liposomes', Antiviral Res., 6, 123-133.

Koff, W.C. and Fidler, I.J. (1985). `The potential use of liposome-mediated antiviral therapy', Antiviral Res., 5, 179-190.

Koff, W.C., Showalter, S.D., Hampar, B., and Fidler, I.J. (1985). `Protection of mice against fatal Herpes simplex type 2 infection by liposomes containing muramyl tripeptide', Science, 228, 495-497.

Lam, C. and Mathison, G.E. (1983). `Effect of low intraphagolysosomal pH on antimicrobial activity of antibiotics against ingested staphylococci', J. Med. Microbiol, 16, 309-316.

Lopez-Berestein, G. (1986). `Liposomal amphotericin b in the treatment of fungal infections', Annals of Internal Medicine, 105, 130-131.

Lopez-Berestein, G., Bodey, G.P., Frankel, L.S., and Mehta, K. (1987). `Treatment of hepatosplenic candidiasis with liposomal amphotericin B', J. Clin. Oncol., 5, 310-317.

Lopez-Berestein, G., Fainstein, V., Hopfer, R., Mehta, K., Sullivan, M.P., Keating, M., Rosenblum, M.G., Mehta, R., Luna, M., and Hersh, E.M. (1985). `Liposomal amphotericin B for the treatment of systemic fungal infections in patients with cancer: a preliminary study', J. Infect. Dis., 151, 704-710.

Lopez-Berestein, G., Hopfer, R.L., Mehta, R., Mehta, K, Hersh, E.M., and Juliano, R.L. (1984). `Liposome-encapsulated amphotericin B for treatment of disseminated candidiasis in neutropenic mice', J. Infect. Dis., 150, 278-284.

Lopez-Berestein, G., Hopfer, R.L., Mehta, R., Mehta, K., Hersh, E.M., and Juliano, R.L. (1984). `Prophylaxis of Candida albicans infection in neutropenic mice with liposome-encapsulated amphotericin B', Antimicrob. Agents Chemother., 25, 366-367.

Lopez-Berestein, G., Mehta, R., Hopfer, R.L., Mills, K., Kasi, L., Mehta, K., Fainstein, V., Luna, M., Hersh, E.M., and Juliano, R.L. (1983). `Treatment and prophylaxis of disseminated infection due to Candida albicans in mice with liposome-encapsulated amphotericin B', J. Infect. Dis.,147, 939-945.

Mandell, G.L. and Vest, T.K. (1972). `Killing of intraleukocytic Staphylococcus aureus by Rifampin: in vitro and in vivo studies', J. Infect. Dis., 125, 486-490.

Martin, J.R., Johnson, P., and Miller, M.F. (1985). `Uptake, accumulation, and egress of erythromycin by tissue culture cells of human origin', Antimicrob. Agents Chemother., 27, 314-319.

Mayer, L.D., Bally, M.B., Hope, M.J., and Cullis, P.R. (1985a). `Uptake of antineoplastic agents into large unilamellar vesicles in response to a membrane potential', Biochim. Biophys. Acta, 816, 294-302.

Mayer, L.D., Hope, P.R., Cullis, P.R., and Janoff, A.S., (1985b). `Solute distributions and trapping efficiencies observed in freeze-thawed multilamellar vesicles', Biochim. Biophys. Acta, 817, 193-196.

Mehta, R., Lopez-Berestein, G., Hopfer, R., Mills, K., and Juliano, R.L. (1984). `Liposomal amphotericin B is toxic to fungal cells but not to mammalian cells', Biochim. Biophys. Acta, 14, 230-234.

Mehta, R.T., Lopez-Berestein, G., Hopfer, R.L., Mehta, K., White, R.A., and Juliano, R.L. (1985). `Prophylaxis of murine candidiasis via application of liposome-encapsulated amphotericin B and a muramyl dipeptide analog, alone and in combination', Antimicrob. Agents Chemother., 28, 511-513.

Merion, R.M. (1985). `Measurements of reticuloendothelial system phagocytic activity in the rat after treatment with silica, liposomes, and cyclosporine', Transplantation, 40, 86-90.

Mogensen, S.C. (1985). `Genetic aspects of macrophage involvement in natural resistance to virus infections', Immunol. Lett., 11, 219-224.

Nacucchio, M.C., Bellora, M.J.G., Sordelli, D.O., and D'Aquino, M. (1985). `Enhanced liposome-mediated activity of piperacillin against staphylococci', Antimicrob. Agents Chemother., 27, 137-139.

New, R.R.C., Chance, M.L., and Heath, S. (1981). `Antileishmanial activity of amphotericin and other antifungal agents entrapped in liposomes', J. Antimicrob. Chemother., 8, 371-381.

Orozco, L.C., Quintana, F.O., Beltran, R.M., deMoreno, I., Wasserman, M., and Rodriguez, G. (1986). `The use of rifampicin and isoniazid entrapped in liposomes for the treatment of Murine tuberculosis', Tuvercle, 67, 91-97.

Osada, Y., Mitsuyama, M., Matsumoto, K., Une, T., Otani, T., Ogawa, H., and Nomoto, K. (1982). `Stimulation of resistance of immunocompromised mice by a muramyl dipeptide analog', Infect. Immun., 37, 1285-1288.

Ostro, M. (ed.). (1987). Liposomes: From Biophysics to Therapeutics, Marcel Dekker, NY.

Payne, N.I., Timmins, P., Ambrose, C.V., Ward, M.D., and Ridgway, F. (1986). `Proliposomes: a novel solution to an old problem', J. Pharm. Sci., 75, 325-329.

Pirson, P., Steiger, R.F., and Trouet, A. (1980). `Primaquine liposomes in the chemotherapy of experimental murine malaria', Annals of Tropical Medicine and Parasitology, 74, 383-391.

Pirson, P., Steiger, R.F., Trouet, A., Gillet, J., and Herman, F., (1980). `Primaquine liposomes in the chemotherapy of experimental murine malaria', Annals of Tropical Medicine and Parasitology, 74, 384-391.

Popesco, M., Swenson, C.E., and Ginsberg, R.S. (1987). `Liposome mediated treatment of viral, bacterial and protozoal infection', in Liposomes: From Biophysics to Therapeutics (M. Ostro, ed.), Marcel Dekker, NY, 219-252.

Poste, G and Fidler, I.J. (1979). `The pathogenesis of cancer metastasis', Nature, 283, 139-146.

Poznansky, M.J. and Juliano, R.L. (1984). `Biological approaches to the controlled delivery of drugs: a critical review', Pharmacol. Reviews, 36, 278-325.

Pratt, W.B. and Fekety, R. (1986). The Antimicrobial Drugs, Oxford University Press, New York.

Prokesch, R.C. and Hand, W.L. (1982). `Antibiotic entry into human polymorphonuclear leukocytes', Antimicrob. Agents Chemother., 21, 373-380.

Rao, T.V., Das, F., and Prasad, R. (1985). `Effect of phospholipid enrichment on nystatin action: differences in antibiotic sensitivity between in vivo and in vitro conditions', Microbios, 42, 145-53.

Raymond, C.A. (1987). `Liposomes embark on rescue mission to make highly toxic drugs more useful', JAMA, 257, 1143-1144.

Reed, S.G., Barral-Netto, M., and Inverso, J.A. (1984). `Treatment of experimental visceral leishmaniasis with lymphokine encapsulated in liposomes', J. Immunol., 132, 3116-3119.

Richardson, V.J. (1983). `Liposomes in antimicrobial chemotherapy', J. Antimicrob. Chemother., 12, 532-534.

Scherphof, G., Damer, J., and Huekstra, D. (1981).
Liposomes from Physical Structure to Therapeutic
Application, (G. Knight, ed.), Elsevier, Amsterdam,
299-322.

Senior, J. and Gregoriadis, G. (1982). `Stability of
small unilamellar liposomes in serum and clearance
from the circulation: the effect of the
phospholipid and cholesterol components', Life
Sciences, 30, 2123-2136.

Shirkhoda, A., Lopez-Berestein, G., Holbert, J.M., and
Luna, M.A. (1986). `Hepatosplenic fungal
infection: CT and pathologic evaluation after
treatment with liposomal amphotericin B', Radiology,
159, 349-353.

Smolin, G. Okumoto, M., Feiler, S., nand Condon, D.
(1981). `Idoxuridine liposome therapy for Herpes
simplex keratitis', Am. J. Opthamol., 9, 220-225.

Stevenson, M., Baillie, A.J., and Richards, R.M.E.
(1983). `Enhanced activity of streptomycin and
chloramphenicol against intracellular Escherichia
coli in the J774 macrophage cell line mediated by
liposome delivery', Antimicrob. Agents Chemother.,
24, 742-749.

Streicher, H.Z. and Joynt, R.J. (1986). `HTLV-
III/LAV and the monocyte/macrophage', JAMA,
256,2390-2391.

Streissle, G., Paessens, A., and Oediger, H. (1985).
`New antiviral compounds', Adv. Virus Res., 30, 83-
138.

Szoka, F., Jr. and Papahadjopoulos, D. (1980).
`Comparative properties and methods in preparation
of lipid vesicles (liposomes)', Ann. Rev. Biophys.
Bioeng., 9, 467-508.

Tadakuma, T., Ikewaki, N., Yasuda, T., Tsutsumi, M.,
Saito, S., and Saito, K. (1985). `Treatment of
experimental salmonellosis in mice with streptomycin
entrapped in liposomes', Antimicrob. Agents
Chemother., 28, 28-32.

Taylor, R.L. (1982). `Amphotericin B in liposomes:
a novel therapy for histoplasmosis', Am. Rev.
Respir. Dis., 125, 610-611.

Tremblay, C., Barza, M., Fiore, C., and Szoka, F.
(1984). `Efficacy of liposome-intercalated
amphotericin B in the treatment of systemic
candidiasis in mice', Antimicrob. Agents Chemother.,
26, 170-173.

Vertut-Croquin, A., Bolard, J., and Gary-Bobo, C.M.
(1984). `Enhancement of amphotericin B selectivity
by antibiotic incorporation into gel state vesicles.
A circular dichroism and permeability study',
Biochim. Biophys. Acta, 30, 360-366.

Vladimirsky, M.A. and Ladigina, G.A. (1982).
`Antibacterial activity of liposome-entrapped
streptomycin in mice infected with mycobacterium
tuberculosis', Biomedicine, 36, 375-377.

Wasserman, M., Beltran, R.M., Quintana, F.O., Mendoza,
P.M., Orozco, L.C., and Rodriguez, G. (1986). `A
simple technique for entrapping rifampicin and
isoniazid into liposomes', Tubercle, 67, 83-90.

Weldon, J.S., Munnell, J.F., Hanson, W.L., and Alving,
C.R. (1983). `Liposomal chemotherapy in visceral
leishmaniasis: an ultrastructural study of an
intracellular pathway', 69, 415-424.

Drug Carrier Systems
Edited by F.H.D. Roerdink and A.M. Kroon
© 1989 John Wiley & Sons Ltd.

TARGETING OF LIPOSOMES TO LIVER CELLS

Gerrit L. Scherphof, Halbe H. Spanjer, Johannes
T.P. Derksen, George Lázár and Frits H. Roerdink
Laboratory of Physiological Chemistry,
University of Groningen,
Bloemsingel 10, 9712 KZ Groningen, The Netherlands

One of the several drug carrier systems which have been intensively studied in the past decades are the liposomes. Liposomes are particles formed from amphiphilic lipids such as phosphoglycerides and cholesterol, consisting of arrays of bilayered lipid membranes encapsulating as many aqueous sheels as there are bilayers. Drugs can either be entrapped as solutes in these aqueous spaces or they can, if their lipophilic character allows that, be accommodated in the hydrophobic moiety of the bilayer. Advantages of liposomes as a drug carrier system include low or no toxic effects, biodegradability, low or not immuno- genicity, a high degree of versatility allowing easy manipulation of size and surface characteristics, including covalent coupling of ligands for specific cell-surface receptor recognition. Every investigator studying the in vivo fate of liposomes will experience that it is virtually impossible to avoid uptake of liposomes by liver cells. Targeting of liposomes to liver cells is therefore no big deal. The question of interest then is: which cell type(s) in the liver is/are responsible for this uptake and can we perhaps manipulate the relative contribution of each cell type to total hepatic uptake? Our group has furnished a number of contributions with respect to these questions over the past few years and our major observations will be summarized in this chapter.

The mammalian liver consists mainly of the parenchymal cells or hepatocytes which are reponsible for typical hepatic functions such as bile formation, lipid and lipoprotein synthesis, regulation of blood glucose levels, detoxification, and synthesis of numerous plasma proteins. They are arranged in plates, two cells thick, separated by the smallest branches of the hepatic microcirculatory system, the so-called sinusoids, which carry the blood from arterial and portal origin to the terminal branches of the hepatic venous system. The sinusoidal walls are formed by flattened endothelial cells which are characterized by the presence of numerous pores or fenestrations of about 100 nm in diameter (Wisse, 1970). Within the sinusoids, in direct contact with whole blood, we find the liver macrophages or Kupffer cells. Hepatocytes and endothelial and Kupffer cells are the three major hepatic cell types with a relative abundancy of approximately 60, 30 and 10%, respectively. In terms of mass the hepatocytes constitute more than 80% of the total liver mass due to their relatively large size. In addition to endothelial and Kupffer cells two other non-parenchymal or sinusoidal cell types have been described, the pit-cell and the fat-storing cell (Wisse, 1977). The former is probably identical to the large granular lymphocyte and involved in immune function (Bouwens et al., 1987) and the latter plays a role in fat-soluble vitamin metabolism and collagen synthesis. Methods have been described to isolate purified liver cell populations. Hepatocytes can be isolated in high yield and purity following perfusion of the liver with a collagenase solution. Non-parenchymal (i.e. sinusoidal) cells are usually isolated after pronase treatment of the liver which destroys the hepatocytes. Further fractionation of the non-parenchymal cells can be achieved by means of elutriation centrifugation (Knook and Sleyster, 1976). In the work described below non-parenchymal liver cells were fractionated in endothelial and Kupffer cell fractions without taking into account the presence of pit cells or fat-storing cells. Occasionally, the Kupffer cell fraction was subfractionated according to size, also by elutriation centrifugation.

The first evidence we obtained on intrahepatic localization of intravenously administered liposomes was in a collaborative study with Dr. Eddie Wisse, at that time still in Leiden. By encapsulating horse radish peroxidase in so-called multilamellar liposomes, prepared by brief sonication of a dispersion of phosphatidylcholine, cholesterol and dicetylphosphate, we could visualize the liposomes, or rather their contents, by a histochemical staining procedure on electronmicrographs of thin sections of (rat) liver after intravenous injection (Roerdink et al., 1977). The stain was mostly concentrated in vacuoles of the Kupffer cells, but significant staining was also observed in endothelial cells. There was no detectable staining of hepatocytes. Some years later, when we had learned to isolate the various cell populations from the liver we repeated these experiments with a few modifications. First, we incorporated a larger proportion of cholesterol into the liposomal bilayers in order to reduce the chance of release of encapsulated solute from the liposomes during their presence in the blood compartment. In our earlier experiments we could not exclude the possibility that peroxidase, leaked out of the liposomes, had been taken up by the endothelial cells by means of (adsorptive) pinocytosis. Second, we used a different liposome marker, ^{125}I-labeled polyvinylpyrrolidone instead of horse radish peroxidase, which would allow us to determine uptake in isolated cell fractions by radioactivity measurements. The results confirmed the Kupffer cells as the main sites of liposome uptake (Roerdink et al., 1981) but two discrepancies with the morphological observations became apparent: we found virtually no radioactivity associated with the endothelial cell fraction and small but significant amounts of radioactivity were recovered in the parenchymal cell fraction. The lack of participation of the endothelial cells was confirmed in a number of later studies in which we used different liposome types (e.g. small unilamellar vesicles, SUV) and compositions (e.g. SUV, containing the glycosphingolipid lactosylceramide) as well as different liposome labels (e.g. ^{3}H-inulin). These observations strengthened our conviction that the earlier observed localization of liposomal peroxidase in endothelial cells had to be attributed to uptake of the peroxidase by these cells, after release from the liposomes during their stay in the bloodstream.

The presence of minor amounts of ^{125}I-PVP in isolated hepatocytes is considered to represent reality. It is likely that in MLV preparations a fraction of the liposomes is small enough to have access to the hepatocytes via the 100-um fenestrations (see also below). The reason why this did not show up in the electron-micrographs is probably a matter of "dilution". The 5-fold excess of hepatocytes over Kupffer cells and the approximately 10-fold larger volume of the former would conceivably cause a small fraction of the total liposome dose to be "diluted" in such a mass of parenchymal cells that it would remain unnoticed in the electronmicrographs.

As already alluded to above, the limited participation of hepato-cytes in liposome uptake has to be attributed to the presence of the endothelial lining of the sinusoids, which separates the hepatocytes from the sinusoidal blood. The fenestrations in the endothelial cells are of insufficient dimensions to allow passage to the relatively large liposomes. That the hepatocytes do possess the capacity to take up liposomes in contrast to the endothelial cells which fail to take up liposomal label despite direct contact with the blood compartment, can be demonstrated by the use of SUV instead of MLV. Of these smaller liposomes a much larger proportion was found to become associated with the hepatocyte fraction (Roerdink et al., 1984). Endothelial cells, however, were still not participating in liposome uptake to any detectable extent. Attempts to further increase uptake of SUV by the hepatocytes by means of incorporation into the liposomes of the galactose-exposing lipid lactosylceramide were only partly successful or even not successful at all, depending on the bulk lipid composition of the liposomes. When the synthetic lipid dimyristoylphosphatidylcholine (DMPC) was the main phospholipid constituent of the vesicles, incorporation of a small proportion of lactosylceramide led to a substantial increase in the amount of liposomes taken up by the hepatocytes, presumably by means of interaction with the galactose receptor on the surface of these cells (Spanjer and Scherphof, 1983). This increase in hepatocyte uptake was not effected at the expense of Kupffer cells uptake, however, since we found a small increase in uptake by this cell fraction as well. A more pronounced increase in Kupffer cell uptake was found, however, when the lactosylceramide was incorporated in SUV with sphingomyelin as a major phospholipid constituent (Spanjer et al, 1984). Similar results were obtained in a collaborative study with Drs. Herman-Jan Kempen and Theo van Berkel at Leiden and Rotterdam, respectively, using a tris-galactosyl derivative of cholesterol (TGC) as a liposomal ligand for the galactose receptor (Spanjer et al., 1985). In fact, the results of these experiments were even more prominent in that many-fold higher rates of blood elimination and liver uptake were observed in response to the TGC incorporation and also an almost 10-fold increase in Kupffer cell uptake as compared to only a 1.2-fold increase in hepatocyte uptake. The reported galactose receptor on Kupffer cells (Kolb-Bachofen et al., 1982) was believed to be involved in this galactose-induced increase in Kupffer cell uptake. Questions have been raised, however, concerning the nature of this receptor and it has been suggested that the structure recognizing the galactose moiety is, in fact, a galactose-binding plasma protein (Kolb-Bachofen, personal communication), which would serve to opsonize particles such as liposomes for Kupffer cell uptake. If this were true, the difference observed by us between DMPC and sphingomyelin liposomes could be explained by assuming that differences in fluidity determines the extent of opsonization and thereby the rate of uptake of the liposomes by Kupffer cells.

The mere finding that a liposomal label is associated with the hepatocyte fraction does not necessarily prove that these cells are able to internalize the liposomes. These lines of evidence indicate, however, that internalization by hepatocytes takes place. First, we observed extensive conversion of liposomal ^{14}C-choline-labeled sphingomyelin into phosphatidylcholine (Roerdink et al., 1984), a process which was inhibited by prior injection of chloroquin. This indicates that the liposomes must have been internalized and degraded in the lysosomal system upon which radiolabeled products such as choline became available for phosphatidylcholine by synthesis. Second, we showed that inulin label encapsulated in intravenously injected SUV followed the distribution pattern of the lysosomal enzyme acid phosphatase upon subcellular fractionation of the hepatocytes (Spanjer, 1985). Finally, with cholesterol or cholesterol oleate as liposomal markers we found rapid biliary excretion of the label as bile salts (Kuipers et al., 1986; Scherphof et al., 1987; Kuipers et al., 1987) again indicating that the SUV not only arrive at and bind to the hepatocytes, but are internalized and processed as well.

Intracellular stability of liposomes, following their endocytic uptake by cells is an important aspect of their potential as a drug carrier system. The rate at which an encapsulated biologically active agent is released into the cell may determine to a large extent the effectiveness of the substance and it is therefore of relevance to gain insight into the rates at which liposomes, once taken up by endocytosis, are degraded intracellularly. Obviously, the pertinence of this concept only applies to agents that are resistant to lysosomal hydrolases or whose biological activity is at least not influenced by such enzymes; in addition, the active agent should have a sufficiently low molecular weight to allow it to travel from the lysosomal compartment to its intracellular site of action.

In early experiments we observed already indications of rapid intracellular degradation of liposomes. With encapsulated ^{125}I-labeled albumin release of acid-soluble radioactivity into the medium of cultured hepatocytes or into the blood following intravenous injection (Roerdink et al., 1981), was observed within 10-15 min after the start of the incubation or the intravenous injection, respectively. In later studies by Jan Dijkstra in our laboratory detailed information was obtained on intracellular liposome degradation by liver macrophages in vitro (Dijkstra et al., 1984a, 1984b). Again, encapsulated radiolabeled albumin was rapidly released from the cells in an acid-soluble form. Within 2 h the cells degraded and released over 90% of the total amount of albumin taken up in liposome-encapsulated form during a 1-h preincubation in presence of ammoniumchloride. When the ammonium-chloride was left in the medium after the 1-h preincubation period, virtually no release of label was observed, while nearly all cell-associated radioactivity was acid-insoluble. The ammonium-chloride was apparently acting as a lysosomotropic agent, dissipating the pH-gradient over the lysosomal membrane and thus inhibiting the acid hydrolases. Remarkably, liposomal phospho-lipid, labeled in the choline moiety of egg-phosphatidylcholine, seemed to be degraded to much lower extents than the encapsulated albumin, as judged by the limited release of water-soluble radio-activity from the cells. In a later study (Dijkstra et al., 1985), however, it was shown that such degradation products are efficiently reutilized by the cells to synthesize their own phosphatidylcholines. By using ^{14}C-choline-labeled sphingomyelin or dimyristoylphosphatidylcholine, both of which can be separated by simple thin-layer chromatography from cellular and egg-phosphati-dylcholine, it was possible to show an extensive conversion of the choline label from the liposomal sphingomyelin or dimyristoyl-phosphatidylcholine to the cellular phosphatidylcholine. Also this conversion was efficiently inhibited by lysosomotropic agents, confirming the lysosomal localization of at least part of the converting process, presumably the degradation step, mediated by sphingomyelinase and phospholipase A activities, respectively.

Recently, we undertook a combined in vitro and in vivo study on degradability of liposomes of variable lipid composition (Roerdink et al., 1986). Liposomes with high cholesterol content were found to be much more resistant to degradation by macrophages of liver and spleen than low-cholesterol vesicles. This was confirmed in an in vitro system with isolated Kupffer cells in maintenance culture and in a cell-free system with a lysosomal fraction from rat liver at pH 5.5. In a collaborative study with Drs. Daan Crommelin and Gert Storm in Utrecht a good correlation was found between the susceptibility of liposomes towards intracellular phospholipases and the rate of release of liposome-encapsulated adriamycin from the cells (Storm et al., 1987, 1988). In these studies we used another parameter to assess the extent of liposome degradation; the liposomes were doubly labeled, in their aqueous space with the metabolically inert label ^{3}H-inulin and in their membranes with a trace of cholesterol-^{14}C -oleate. The latter is susceptible to hydrolytic cleavage by cholesterolesterase, leading to the liberation of the ^{14}C -oleate moiety. In vivo the labeled oleate is readily released from the cells, only a minor fraction becoming incorporated into cellular lipids such as triglycerides. In vitro, efficient release of oleate occurs only if an adequate acceptor such as albumin is present in the medium; in vitro incubations were done, therefore, in serum-containing media. The encapsulated inulin, as a metabolically inert susbstance, remains to a large extent cell-associated once taken up in the form of liposomes. Thus, liposome degradation will result in a gradual increase in the ^{3}H/^{14}C ratio. As for the liver, one has to take into consideration, however, that oleate released from the Kupffer cells is efficiently taken up by the hepatocytes so that the measurement of the isotopic ratio on whole liver tissue will imply a gross underestimation of the extent of liposome degradation. For more accurate estimations it will be necessary to either isolate the macrophage fraction or extract the tissue sample and analyze the extract by means of thin-layer chromatography for the amount of undegraded cholesteryl-oleate.

A recent study in our laboratory on the comparison of a number of liposomal labels, with respect to their usefulness as markers for cell uptake and degradation, revealed the superiority of yet another label, cholesterol hexadecylether. This compound shares the metabolic inertia with inulin but it was shown to be retained by liver macrophages much more efficiently than inulin, which, presumably by way of exocytosis, is released with an initial rate of approximately 5-10% per hour when taken up in liposome-encapsulated form. In addition, the cholesterol ether label is not subject to cell-surface-induced release from the liposomes prior to uptake.

Several years ago we attempted to shift the intrahepatic distribution of liposomes from the Kupffer cells towards the hepatocytes by treating the rats, prior to injection with liposomes, with GdCl$_3$ or La(NO$_3$). These rare earth metals had been shown to block preferentially the phagocytic capacity of liver macrophages (Husztik et al., 1980). Although administration of the lanthanides caused a retardation in the rate of elimination of liposomes from the blood and also gave rise to a shift from liver uptake to splenic uptake, we failed to observe an increase in the amount of liposome uptake by the hepatocytes. In retrospect this is, obviously, readily explained by the presence of the 100-nm fenestrations in the hepatic endothelium which will prevent access to the hepatocytes of particles larger than 100 nm, which the liposomes we used in fact were. More recently we repeated these experiments with small unilamellar liposomes (SUV) of approximately 30 nm in diameter. By choosing a lipid composition which we previously found to direct the SUV to about equal extents towards Kupffer cells and hepatocytes (Spanjer et al., 1986), i.e. sphingomyelin/cholesterol/phosphatidylserine in a 4:5:1 molar ratio, we were able to observe a considerable shift in liposome uptake from the Kupffer cells to the hepatocytes. Interestingly, with large multilamellar liposomes, we noticed also a Gd-induced shift in the distribution of liposome label within the Kupffer cell fraction. The lanthanide caused a very substantial inhibition in the uptake of liposomes by the large Kupffer cells, which was in part compensated for by an increased uptake by the much more abundant small cells, which, normally, participate only to a minor extent in liposome uptake (Daemen et al., 1988).

Applications

Both hepatocytes and Kupffer cells have been shown to be potential targets for systemically administered liposomes. For hepatocytes we have to deal with the restriction of limited liposome dimensions in view of the diameter of the sinusoidal endothelium. This implies that only limited use can be made of water-soluble agents encapsulated in the aqueous liposomal space since the encapsulated volume of such SUV is extremely small. SUV obtained by ultrasonication with a typical diameter of approximately 30 nm have an internal aqueous volume in the order of 0.3 l/mol of lipid, i.e. an order of magnitude lower than that of a typical MLV preparation. This means that only small volumes of

lipid-encapsulated solutions can be applied; e.g. with a lipid dose
as high as 100 umol per kg, an injected volume of only as little as
30 ul is available. If a 100-mM drug solution is used to
encapsulate, the amount of drug that can be administered will be
limited to about 3 umol/kg. It will therefore require drugs which
display activity at low concentrations to apply such liposomal
carriers successfully. Certain antiviral drugs may be considered
for this purpose. Perspectives for the use of SUV are much better
when lipophilic agents are considered which can be incorporated in
the liposomal bilayer. Either the drug itself or lipophilic
derivatives can be used for that objective. In recent years a
number of such lipophilic drug derivatives have been described.
Incorporation of 10-30 mol % of such derivatives in the liposomal
membrane should be feasible, allowing an injected dose of 10-30
umol/kg. In addition, the use of membrane-associated drug
derivatives brings along the advantage of much more efficient
incorporation. The theoretical value of 100% should be closely
approached, whereas encapsulation efficiencies of less than 1% are
obtained during ultrasonic formation of SUV. Obviously, the design
of liphophilic membrane-anchored drug derivatives will require the
chemical link between drug and lipophilic anchor to be susceptible
to intracellular cleavage, e.g. by lysosomal hydrolytic enzymes, in
order for the drug to become available in an active form. The
recent development of techniques to prepare large or small unilamellar vesicles in very high concentrations, by freeze-thawing
and high-pressure extrusion, may, however, provide a useful
alternative in this respect.

The massive uptake of most types of liposomes by the macrophages
in general and by liver macrophages in particular can be exploited
in several ways. The first ideas in this connection were directed
towards the combat of intracellular parasites such as the
Leishmaniasis parasite. The well-known studies of Alving and
coworkers (Alving et al., 1978), Black et al. (1977) and New et al.
(1978) demonstrated the tremendous potentiating effect of liposome
encapsulation of known anti-parasitic drugs as a result of which
they could be concentrated at the very site of infection, i.e. the
lysosomal system of the liver macrophages. More recently also the
encapsulation within liposomes of a number of antibiotics resulted
in a strongly enhanced therapeutic efficacy towards intracellular
bacterial infections (Bakker-Woudenberg et al., 1985; Bakker-
Woudenberg and Roerdink, 1986).

The work of Fidler and his associates on the use of liposomes to
potentiate biological response modifiers such as muramyl dipeptide
(MDP) and lymphokines was also based on the notion that liposomes
tend to be taken up by cells of the mononuclear phagocyte system.
Thus, impressive results were obtained in animal models, showing
the enormous efficacy with which macrophages, activated by
liposome-encapsulated immunomodulators, were able to eradicate
metastatic growth in, particularly, the lung (Fidler et al., 1982).
Following a similar approach we showed that also liver macrophages
can be activated to tumorcytotoxicity with liposomal MDP and that
this leads to a very substantial reduction in metastatic growth
from intrasplenically administered colon adenocarcinoma cells in
the liver of mice and to significant increase in survival time of

the animals (Daemen et al., 1986).

Acknowledgements
   The authors gratefully acknowledge the expert help of Bert
Dontje, Jan Wijbenga, Mieke van Galen, Joke Regts and Henriëtte
Morselt and the secretarial help of Rinske Kuperus.
   The stay of G.L. in the laboratory of Physiological Chemistry was
made possible through a foreign visitor's grant from the
Netherlands Organization for Pure Scientific Research (NWO).

References

Alving, C.R., Steck, E.A., Chapman Jr., W.L., Waits, V.B.,
   Hendricks, L.D., Swartz Jr., G.M. and Hanson, W.L. (1978).
   'Therapy of Leishmaniasis: superior efficacies of
   liposome-encapsulated drugs', Proc. Natl. Acad. Sci. USA, 75,
   2959-2963.

Bakker-Woudenberg, I.A.J.M., Lokerse, A.F., Roerdink, F.H., Regts,
   D. and Michel, M.F. (1985). 'Free versus liposome-entrapped
   ampicillin in the treatment of infection due to Listeria
   monocytogenes in normal and athymic (nude) mice', J. Infect.
   Dis., 151, 917-924.

Bakker-Woudenberg, I.A.J.M. and Roerdink, F.H. (1986). 'Leading
   article, Antimicrobial chemotherapy directed by liposomes', J.
   Antimicrob. Chemother., 17, 547-549.

Black, C.D.V., Watson, G.J. and Ward, R.J. (1977). 'The use of
   pentostam liposomes in the chemotherapy of experimental
   Leishmaniasis', Trans. Roy. Soc. Trop. Med. Hyg., 71, 550-552.

Bouwens, L., Remels, L., Backeland, M., Van Bossuyt, H. and Wisse,
   E. (1987). 'Large granular lymphocytes or "Pit cells" from rat
   liver: isolation, ultrastructural characterization and natural
   killer activity', Eur. J. Immunol., 17, 37-42.

Daemen, T., Veninga, A., Roerdink, F.H. and Scherphof, G.L. (1986).
   'In vitro activation of rat liver macrophages to tumoricidal
   activity by free or liposome-encapsulated muramyl dipeptide',
   Cancer Res., 46, 4330-4335.

Daemen, T., Veninga, A., Roerdink, F.H. and Scherphof, G.L. (1988).
   'Endocytic and tumoricidal heterogeneity of rat liver macrophage
   populations, implications for drug targeting', Cancer Drug
   Delivery, in press.

Dijkstra, J., Van Galen, W.J.M., Hulstaert, C.E., Kalicharan, D.,
   Roerdink, F.H. and Scherphof, G.L. (1984a). 'Interaction of
   liposomes with Kupffer cells in vitro', Exp. Cell Res., 150,
   161-176.

Dijkstra, J., Van Galen, M. and Scherphof, G. (1984b). 'Effects of
   ammonium chloride and chloroquine on endocytic uptake of

liposomes by Kupffer cells in vitro', Biochim. Biophys. Acta, 804, 58-67.

Dijkstra, J., Van Galen, M., Regts, J. and Scherphof, G. (1985). 'Uptake and processing of liposomal phospholipids by Kupffer cells in vitro', Eur. J. Biochem., 148, 391-397.

Fidler, I.J., Barnes, Z., Fogler, W.E., Kirsh, R., Bugelski, P. and Poste, G. (1982). 'Involvement of macrophages in the eradication of established metastases following intravenous injection of liposomes containing macrophage activators', Cancer Res., 42, 496-501.

Husztik, E., Lázár, G. and Párducz, A. (1980). 'Electron microscopic study of Kupffer cell phagocytosis blockade induced by gadolinium chloride', Br. J. Exp. Pathol., 61, 624-630.

Knook, D.L. and Sleyster, E.C. (1976). 'Separation of Kupffer and endothelial cells of the rat liver by centrifugal elutriation', Exp. Cell Res., 99, 444-449.

Kolb-Bachofen, V., Schlepper-Schäfer, J., Vogell, W. and Kolb, H. (1982). 'Electron microscopic evidence for an asialoglycoprotein receptor on Kupffer cells: localization of lectin-mediated endocytosis', Cell, 29, 859-866.

Kuipers, F., Spanjer, H.H., Havinga, R., Scherphof, G.L. and Vonk, R.J. (1986). 'Lipoproteins and liposomes as in vivo cholesterol vehicles in the rat: Preferential use of cholesterol carried by small unilamellar liposomes for the formation of muricholic acids', Biochim. Biophys. Acta, 876, 559-566.

Kuipers, F., Derksen, J.T.P., Gerding, A., Scherphof, G.L. and Vonk, R.J. (1987) 'Biliary lipid secretion in the rat. The uncoupling of biliary cholesterol and phospholipid secretion from bile acid secretion by sulfated glycolithocholic acid', Biochim. Biophys. Acta, 922, 136-144.

New, R.R.C., Chance, M.L., Thomas, S.C. and Peters, W. (1978). 'Antileishmanial activity of antimonials entrapped in liposomes', Nature 272, 55-56.

Roerdink, F.H., Wisse, E., Morselt, H.W.M., Van der Meulen, J. and Scherphof, G.L. (1977). 'Cellular distribution of intravenously injected protein-containing liposomes in the rat liver', in Kupffer cells and other liver sinusoidal cells (Eds. E. Wisse and D.L. Knook), pp. 263-272, Elsevier/North-Holland Biomedical Press.

Roerdink, F., Dijkstra, J., Hartman, G., Bolscher, B. and Scherphof, G. (1981), 'The involvement of parenchymal, Kupffer and endothelial cells in hepatic uptake of intravenously injected liposomes. Effects of lanthanium and gadolinium salts', Biochim. Biophys. Acta, 677, 79-89.

Roerdink, F.H., Regts, J., Van Leeuwen, B. and Scherphof, G.L. (1984). 'Intrahepatic uptake and processing of intravenously injected small unilamellar vesicles in rats'. Biochim. Biophys. Acta, 770, 195-202.

Roerdink, F., Regts, D., Daemen, T., Bakker-Woudenberg, I. and Scherphof, G.L. (1986). 'Liposomes as drug carriers to liver macrophages: fundamental and therapeutic aspects', in Targeting of drugs with synthetic systems (Eds. G. Gregoriadis, G. Poste, J. Senior and A. Trouet), pp. 193-206, Plenum Press.

Scherphof, G.L., Derksen, J.T.P., Kuipers, F. and Vonk, R.J. (1987). 'Intrahepatic processing of liposomes; conversion of liposomal cholesterol to bile salts', Biochem. Soc. Trans., 15, 62S-68S.

Spanjer, H. and Scherphof, G. (1983). 'Targeting of lactosyl-ceramide containing liposomes to hepatocytes in vivo', Biochim. Biophys. Acta, 734, 40-47.

Spanjer, H., Morselt, H. and Scherphof, G.L. (1984). 'Lactosyl-ceramide-induced stimulation of liposome uptake by Kupffer cells in vivo', Biochim. Biophys. Acta, 774, 49-55.

Spanjer, H.H., Scherphof, G.L., Van Berkel, T.J.C. and Kempen, H.J.M. (1985). 'The effect of a water-soluble tris-galactoside terminated cholesterol derivative on the in vivo fate of small unilamellar vesicles in rats', Biochim. Biophys. Acta, 816, 396-402.

Spanjer, H.H. (1985). 'Targeting of liposomes to liver cells in vivo', Ph.D. Thesis, State University Groningen.

Spanjer, H.H., Van Galen, M., Roerdink, F.H., Regts, J. and Scherphof, G.L. (1986). 'Intrahepatic distribution of small unilamellar liposomes as a function of liposomal lipid composition', Biochim. Biophys. Acta, 863, 224-230.

Storm, G., Roerdink, F.H., Steerenberg, P.A., De Jong, W.H. and Crommelin, D.J.A. (1987). 'Influence of lipid composition on the antitumor activity exerted by doxorubicin-containing liposomes in a rat solid tumor model', Cancer Res., 47, 3366-3372.

Storm, G., Nässander, U.K., Roerdink, F.H., Steerenberg, P.A., De Jong, W.H. and Crommelin, D.J.A. (1988). 'Studies on the mode of action of doxorubicin-liposomes', in Liposomes in the therapy of infectious diseases and cancer. UCLA Symposia on Molecular and Cellular Biology, Vol. 89 (Eds. G. Lopez-Berestein and I. Fidler), Alan R. Liss, Inc. New York, in press.

Wisse, E. (1970). 'An electron microscopic study of the fenestrated endothelial lining of rat liver sinusoids', J. Ultrastruct. Res., 31, 125-150.

Wisse, E. (1977). 'Ultrastructure and function of Kupffer cells and
other sinusoidal cells in the liver', in Kupffer cells and other
liver sinusoidal cells (Eds. E. Wisse and D.L. Knook), pp. 33-60,
Elsevier/North-Holland Biomedical Press.

Drug Carrier Systems
Edited by F.H.D. Roerdink and A.M. Kroon
© 1989 John Wiley & Sons Ltd.

IMPLANTABLE INFUSION PUMPS FOR DRUG DELIVERY IN MAN:
THEORETICAL AND PRACTICAL CONSIDERATIONS

Perry J. Blackshear, M.D., D. Phil. and
    Thomas D. Rohde*, M.S., C.A.S.

Duke University Medical Centre,
P.O.Box 3297,
Durham, NC 27514, USA

A. Introduction.
In 1969, my colleagues and I at the University of Minnesota
designed and began animal experimentation with a totally
implantable drug delivery pump. This device was first implanted
in man in 1975 and since that time has been implanted in more than
20,000 human patients worldwide for a variety of drug delivery
applications. It is now known as the Infusaid Implantable
Infusion Pump, manufactured at the present time by the
Shiley-Infusaid Corporation of Norwood, MA. The purpose of this
chapter is to discuss some aspects of implantable pumps of this
type as they relate to innovative drug delivery systems in
general, citing as examples specific experiences which we have had
over the past 17 years with this device. We do not intend to
discuss other types of implantable drug delivery systems such as
polymeric systems or liposomes or other drug carrier systems, since
these have been amply discussed in previous chapters.
B. Rationale for and advantages of totally implantable drug
infusion systems.
Anyone who has received intravenous drugs in hospital by means
of external or extracorporeal infusion systems can appreciate the
potential advantages of totally implantable infusion systems,
especially for the long-term delivery of parenteral drugs. The
major advantages of such totally implantable systems for drug
delivery can be summarized in approximately six major categories.
The first compares these types of delivery systems to more
conventional types of drug therapy, that is, oral or other
conventional dosage forms. For most implantable infusion systems,
although not necessarily in all cases, the drug to be delivered is
most optimally a drug which can only be given by the parenteral
route, that is, by injection or infusion. This is true of drugs
such as many protein hormones, of which insulin is a good example,
which are either inactivated by proteases in the gastrointestinal
tract, or are not absorbed from the gastrointestinal tract into the
circulation, or both. In addition, some drugs, even when taken
orally, have such a very short half-life in the circulation as to
make their repeated dosing impractical. This is especially true in
the case of drugs which must be given parenterally but which have
very short serum half-lives because of rapid clearance. Again,
insulin is an excellent example of such a drug which can only be

given parenterally and when injected intravenously has a half-life of approximately four minutes in the circulation. Obviously, for insulin to be maintained at a constant concentration in the plasma by means of repeated intravenous injections, the injections would have to be made at very frequent intervals which would be highly impractical. The same goes for other types of drugs such as protein hormones, anticoagulants such as heparin and others which when injected as an intravenous bolus, cause peak concentrations which may be dangerous, whereas during the nadir of their serum concentration following the bolus, inadequate therapeutic effect may be maintained. The avoidance of this peak and valley effect of rapidly metabolized parenteral drugs is one of the main advantages of totally implantable drug infusion systems.

A second major advantage of totally implantable infusion systems over conventional drug therapy is that rate control of drug delivery can be improved. Simple single rate infusion systems such as the Infusaid pump are capable of mimicking the type of continuous single rate infusion readily provided by the familiar bedside intravenous apparatus. However, modifications to this pump, and more elaborate pumps developed by other groups now under initial trials, can provide very complex rate control with almost unlimited varieties of drug delivery profiles. In general, once the systems have been completely implanted, it is impractical to change the pump itself, which obviously requires a surgical procedure; in addition, it is also inconvenient to change the concentration of drug in the infusate more frequently than every two to three weeks during pump refills. Therefore, changes in drug delivery rate can readily be accomplished in sophisticated devices by changing the flow rate from the pump. In addition, as exemplified in several recent clinical applications of the Infusaid pump, drugs can be delivered by more than one catheter to more than one site, so that combinations of infusion, for example, into the arterial and venous circulation at the same time are now possible.

One of the greatest advantages of totally implantable infusion systems for long-term drug delivery lies in in the fact that the entire device is located under intact skin, a location which provides much greater protection against infection than can be achieved in practice with any kind of percutaneous foreign material. Numerous examples of foreign bodies located in the percutaneous position for long periods of time are available in the literature, and all of them are subject to problems with bacteria and other organisms getting from the external environment into the body around the non-physiological material of the apparatus. As one can imagine, an infection around a totally implanted device is often a disastrous occurrence, usually necessitating removal of the device before the infection can be completely controlled. This has been amply demonstrated with implanted cardiac pacemakers, prosthetic hips and other devices, in which the presence of the foreign body makes infection eradication much more difficult. In addition to troublesome local infections around the device, any such bacterial invasion can result in infection of the blood stream and septicemia, especially if a cannula is permanently installed in the venous or arterial system; this is obviously a serious situation, one which can have fatal consequences. With totally implantable infusion systems, these problems of infection are largely obviated by the intact skin covering the entire device; if suitable skin cleansing precautions are taken during the refill procedure, there is an extremely low rate of pump pocket infections in patients in whom Infusaids have been implanted.

Another major advantage of this type of drug delivery system is that the presence of one or more delivery cannulas, small tubes through which the drug flows immediately before reaching the body site, allows the skilled surgeon to insert the cannulas directly into a wide variety of body sites or body cavities leading to a site-specific drug delivery. In combination with rate control, as noted above, this capability of site-specific drug delivery over long periods of time offers immense advantages over conventional systemic drug administration. Two examples from the Infusaid experience will suffice to make this point. One is in the treatment of refractory osteomyelitis with aminoglycoside antibiotics. Osteomyelitis is a bacterial infection of bone which, when allowed to go untreated for long periods of time, can become extremely refractory to treatment with conventional antibiotics. Many such infections are treated most appropriately with aminoglycoside antibiotics, which have severe toxicity if the serum concentrations are allowed to remain too high for too long: these toxicities include deafness as well as renal damage sometimes resulting in kidney failure. This is a good example of the so-called therapeutic window, in which drug concentration in the plasma needs to be high enough to maintain the therapeutic effect but not so high as to cause toxicity. In the case of refractory osteomyelitis treated with aminoglycoside antibiotics, this therapeutic window can be non-existent, since systemic serum concentrations high enough to eradicate the infection would cause severe and irreversible toxicity. This can be readily overcome by the use of site-specific antibiotic delivery, that is, one or more delivery cannulas placed at or near the site of infection so that extremely high local concentrations of the drug can be achieved while minimizing the amount of drug that gets into the systemic circulation, thus minimizing systemic toxicity.

Another example of the advantages of site-specific drug delivery is the use of intrathecal or subdural infusions of preservative-free morphine in the treatment of refractory pain of malignancy. In this case, the analgesic narcotic is infused directly into the spaces surrounding the spinal cord, providing analgesia to the areas of the body supplied by nerves emanating from that region of the cord. This provides marked pain relief to patients suffering from cancer pain in those areas of the body, while at the same time minimizing the effects of high concentrations of narcotics in the systemic circulation, which can involve respiratory depression to the point of respiratory arrest, and central nervous system depression leading to confusion, sleepiness and even coma in high doses.

These are just two examples of the advantages of site-specific drug delivery in preventing systemic toxicity. Interestingly, this advantage can be combined with some of the other advantages of

implantable infusion systems to provide truly unique forms of drug
delivery. For example, in some diseases of the brain and meninges,
infusion of drug directly into the cerebral ventricles can be
beneficial. However, in most cases surgeons have been loathe to
place delivery cannulas directly into the cerebral ventricles as
long as there is a percutaneous access site, since infection of the
cerebrospinal fluid in the presence of a foreign body is usually a
serious complication. However, using a totally implantable
infusion device, with its inherent low risk of infection, catheters
can and have been placed in this site and connected to a completely
implanted pump, so that infusions can be delivered into this
specific body site, at the same time providing high concentrations
of drug in the local area, minimizing concentrations of drug in the
systemic circulation, and virtually preventing the infection which
might develop from use of a percutaneous device.

A fifth major advantage of totally implantable infusion pumps
as opposed to many of the other innovative drug delivery systems
reviewed in this volume is that these represent simple infusion
devices of relatively large useable infusate volume. This means
that, in many cases, commercially available, FDA-improved drugs can
be used directly in the pump, although obviously regulatory
clearance must be obtained for every new drug-pump combination.
This advantage is particularly important in diseases for which the
numbers of patients involved might not be sufficient to justify the
expense to a drug company of starting a whole new regulatory
program involving approval of a drug-pump combination. The
Infusaid pump which we will describe in some detail below has a
useable internal volume of about 50 ml, allowing many commercially
available approved drugs to be delivered at flow rates of one-half
to two milliliters per day. This does not mean, as we will
indicate below, that any commercial drug can be used in this way;
several instances of pump-drug incompatibility and other problems
will be discussed in section D.

Finally, a major advantage to patients who need to undergo
drug infusions for long periods of time is in the fact that the
device is completely implanted under the skin, and is not generally
detectable in a fully clothed patient. This is in contrast to
almost all types of external infusion systems, including some of
the very small pumps which have been developed in recent years for
insulin delivery; by virtue of the holster or harness used for
holding the external pumps and maintaining them close to the site
of drug infusion, these devices can be detected readily in most
patients. Thus, the ability of the devices to be totally implanted
is advantageous in several ways: it allows patients to go about
daily activities such as athletics, showering or bathing or
swimming, without risk of infection through the percutanous access
site; it allows them a major cosmetic advantage, in that the drug
delivery system is usually not detected; and together these two
advantages give many patients an enormous psychological boost when
compared to the use of external devices. Many have volunteered
that the pumps are completely forgettable under most situations,
and allow patients to avoid what they have termed the

stigmatization of being seen to require continuous out-patient drug delivery.

Many other advantages of totally implantable drug delivery systems could be envisioned; however, as indicated below, these types of systems also have their drawbacks. An obvious drawback is that once the pump is implanted it is not a trivial matter to remove it for servicing, battery replacement if batteries are used in the device, major changes in flow characteristics, etc. However, a review of the clinical applications of such a device indicates that this potential obstacle to use of these devices has actually not been a serious problem.

C. Design aspects of the Infusaid pump: central concepts.

The overall design and operational principles of the Infusaid pump have been described in detail elsewhere (Blackshear et al., 1972, Blackshear, 1979, Blackshear et al., 1979). Briefly, the pump is a simple device composed of a disk-shaped titanium cannister separated internally into two chambers by collapsible welded titanium bellows (Fig. 1). The inner chamber contains the liquid to be delivered or the infusate; the outer chamber contains a vapor-liquid mixture of a chemical power source. At the virtually constant temperatures which prevail under the skin of a human or animal body, the vapor in equilibrium with its liquid phase exerts a constant vapor pressure on the bellows, regardless of the enclosing chamber volume. This vapor pressure extrudes the infusate through a series of filters and flow regulating resistance elements into a delivery cannula and then into a body cavity, artery or vein.

One of the central concepts which made this device possible was that a vapor in equilibrium with its liquid phase exerts a constant vapor pressure at a given temperature, regardless of the enclosing volume. This means that as long as some of propellant liquid is still available at any part of the bellows stroke cycle, it could still evaporate to provide the constant vapor pressure necessary for a constant flow rate. It was absolutely critical for us in the early days to find a suitable vapor-liquid combination, which would provide a fairly high vapor pressure of approximately one atmosphere over atmospheric pressure within the pump, but which would be non-toxic, non-inflammable, non-explosive, and in other ways not cause a major disaster if the external pump cannister were to be ruptured during some accident. In the early days of the pump design, we experimented with a variety of compounds which seemed to fit the bill, including diethyl ether and a rather toxic compound known as tetramethylsilane, both of which appeared to provide suitable vapor pressures. However, with the development of various fluorocarbons used as propellants in other commercial products, we determined that one of these would be the most appropriate propellant for this type of device. The final compound used in these studies is a fluorocarbon propellant which exerts a vapor pressure of approximately 440 mm of mercury at 37°C, the average body temperature. It has the additional advantages of being colorless, odorless, non-corrosive to pump internal components, and non-toxic if there should be accidental leakage under the skin. This chemical power source is really the central concept behind the

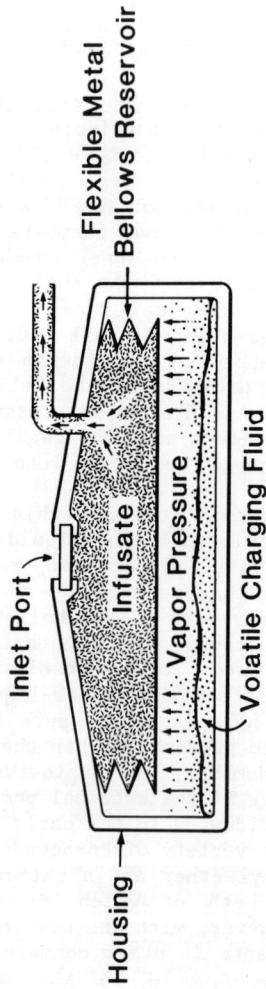

Fig. 1. Diagrammatic cross-section of implantable infusion pump.

pump design; not only does it provide for continuous, constant pressure during all phases of the pump cyle, but the same physical principle allows for recycling of the pump. In other words, by increasing pressure in the pump's inner chamber, by pressure on the hand-held or mechanically assisted syringe filling the inner chamber, the pressure exerted on the driving vapor in the outer chamber can be increased, causing it to condense and to recycle to the liquid phase. An alternative possibility is the use of cold refilling solutions which could equally well condense the driving vapor; however, our early experience with this approach led to actual negative pressures being applied to the delivery cannula, with the disastrous result of blood being aspirated into the delivery cannula, clotting, and complete cannula occlusion. Therefore, refills now rely solely on the use of external pressure to cause condensation and recycling of the driving vapor.

The inner chamber and outer chamber of the pump needed to be separated by a collapsible or displaceable membrane which could be deformed in such a way as to not change the effective pressure exerted on the inner chamber. Our initial idea was to have some sort of thin rubber or plastic membrane separating the two chambers which could be readily collapsible during the flow cycle. However, it was clear that most thin membranous barriers would eventually be permeable to the vapor enclosed within the pump's outer chamber. This would not only cause leaching of the possibly harmful and non-therapeutic driving vapor into the infusate, but it would also result in gradual loss of molecules from the external driving vapor chamber, with gradual decreases in flow rate and eventually complete cessation of flow. Therefore, it was apparent to us from the earliest days of pump development that a completely impermeable barrier was needed between the vapor-liquid-containing outer chamber and the liquid-containing inner chamber. A central concept which allowed the final device to be possible was that of the collapsible welded metal bellows, which at that time was manufactured by the Metal Bellows Corporation of Sharon, MA. Relatively few types of barriers met our specifications, but this was the one that seemed most appropriate. By making the bellows of stainless steel in our initial trials, or later of titanium, we could be assured of complete impermeability to liquids and gases, and the use of noncorrosive materials like titanium assured us of reasonably good compatibility with most driving vapors and infusates. However, most commercially available bellows at that time were cast rather than welded which meant that during the stroke cycle of the bellows there was a gradual but quite marked increase in the pressure needed to collapse the bellows. This problem was largely obviated by the development and use of welded bellows by the Metal Bellows Company, in which each flange of the bellows was welded to its partner in such a manner as to make the spring rate of the bellows very low and certainly much lower than in the case of the cast devices. We therefore initially used welded stainless steel bellows in our early models of the pump, but rapidly switched to titanium when it became clear that this metal was somewhat lighter and somewhat more tissue and bio-compatible than stainless steel. The titanium bellows now used in this device do indeed provide a completely impermeable barrier, so that the theoretical life span of the pump after implantation is indefinite; studies in the pre-market approval phase for several drug applications have suggested that many thousands of pump cycles are possible in such a device before fatigue at the welded bellows joints begins to set in. However, even the welded bellows have a finite spring rate which will, indeed, alter drug delivery rate to some extent. In the welded titanium bellows currently in use in the Infusaid pump, we have estimated that during the working stroke volume of the device, the spring rate changes the drug delivery rate by approximately 5 or 6% over the complete flow cycle. In other words, as the flow cycle progresses, there is a gradual decrease in nominal flow rate of approximately 6% at the end of the cycle when compared to the beginning of the cycle. This could possibly be decreased by further refinements in bellows fabrication technology, but we suspect that it is impossible to completely eliminate the bellows spring rate as a cause of changes in flow rate.

The next central concept which made this device possible was the use of a long, narrow-bore stainless steel and later titanium capillary tube as a flow-regulating resistance element. Flow through this type of tube is governed by the Poisseuille equation which is

$$Q = \frac{\pi D^4 \Delta P}{128 \, \mu L}$$

in which ΔP equals the difference in pressure between the pump's drug chamber and the drug delivery site, μ is the viscosity of the solution, D is the internal diameter of the capillary tube, Q is the flow rate and L is the length of the capillary tube. It can be seen that the flow through such a capillary tube is inversely proportional to the first power of the length of the tube and the viscosity of the solution, the latter of which will be relevant to our discussion of individual drugs. However, it is also apparent that fluid flow rate is directly proportional to the fourth power of the internal diameter of the tube. It is this very powerful fact, in combination with the length and viscosity factors, which makes very dramatic changes in flow rate possible in a given device. We were initially attracted to the use of a capillary tube as a flow regulator because its behavior was governed by this well-defined mathematical equation, in which all of the variables could be controlled quite specifically. However, at the relatively high pressure heads that we were using, and the very low flow rates which we wanted to achieve, on the order of one ml per day, it became necessary to use very long lengths of very narrow bore stainless steel and later titanium capillary tubing to achieve these ends. One can see that with the substantial pressure head generated by the fluorocarbon vapor pressure power source, to achieve a flow rate of one ml per day using a capillary tube of tiny internal diameter, it was necessary to use a coil of a long length of this narrow bore tubing. Initially, we had problems in obtaining capillary tubing of reproducible internal diameter, especially when we changed over to titanium. It was difficult in the early days to weld the opening of such a fine bore capillary tube to the cannister of the pump, and special processes needed to be evolved to make this connection possible. However, although we included bubble trap filters in our earliest versions of the device, and still include them in most useable clinical models, extensive de-gassing of solutions for use in the device has not been necessary, since we have had few problems with flow rate stoppage by surface tension artifacts or minute bubbles which could have evolved from the liquid during the course of the pump cycle. Although somewhat unwieldy in the manufacturing process, the capillary tubing flow regulators have not been supplanted by any other type of flow regulating resistance element because of some of the advantages enumerated above. At the present time, using commercially available titanium or stainless steel tubing, virtually any flow rate that is necessary for a drug or liquid of given viscosity can be achieved, down to about 0.1 ml per day. At some point, when the flow rate becomes vanishingly low, problems with clotting in the venous system begin to occur and it is probably necessary to maintain the fluid flow rate above a certain point with most drugs to prevent this from happening.

Another crucial aspect of the design of the device was the refill septum by which means the pump's inner chamber or drug chamber could be refilled by percutaneous injection. This septum needed to be relatively easy to penetrate with a needle from outside the body, and therefore needed to be of a size sufficiently large so it could be located under the skin of even obese patients. However, even with this large area through which a needle might be inserted, the septum had to completely reseal itself to prevent drug leakage into the subcutaneous tissues, be usable for many hundreds of needle sticks over the lifetime of the septum, be composed of non-toxic materials so that it would not induce an inflammatory reaction under the skin of the patients, and be made of materials such that cores or other pieces of the septum would not be introduced by the needles into the pump's inner chamber. After a long period of trial and error we eventually used a combination silicone rubber and titanium septum which seemed to fit most of the design requirements. This septum does not allow drug leakage when refills are done in an appropriate manner using a special Huber point needle with the orifice on the side of the shaft of the needle rather than on the needle tip, allowing the needle to be introduced through the skin and through the septum without a core of either substance being introduced into the pump's inner chamber, possibly with disastrous results of infection or device occlusion. This combination septum has also proved strong enough to withstand many, many refills at fairly high vapor pressure, and prevent leakage of the pump's fluid into the subcutaneous tissues. Obviously, with some potent drugs such as heparin or insulin, a significant leak in this refill septum could cause a major disaster, so that it was extremely important that this septum be relatively failure proof when used with the appropriate type of needles. Throughout the rather long and broad clinical experience with this device, several problems have arisen in which inexperienced personnel have introduced ordinary large bore needles through the septum, resulting in leakage of drug into the subcutaneous tissues.

Finally, the entire outer surface of the pump needed to be compatible with body tissues so as not to provoke local inflammatory reaction or cause sloughing of the skin overlying the pump. For this reason, the device needed to be relatively small in size so as not to cause undue pressure on the skin over the top of the device. We relied on data from early pacemaker implants to construct a device that was large enough to contain a useful volume of infusate, on the order of 50 ml for most models, and a size that would not be cosmetically offensive or cause skin sloughing. The eventual design was a disk shaped device as pictured in Fig. 2 and is now composed almost entirely on its external surfaces of biocompatible titanium. Other components in contact with subcutaneous tissues or blood include teflon suture loops for fixing the device to the underlying fascia, a silicone rubber band around the exterior of the device to keep the flow regulating capillary tube in place, the silicone rubber and teflon refill septum alluded to above, and the radio-opaque silicone rubber cannula for eventual placement in a vein, artery, or body cavity. This last substance is probably the least thrombogenic material which can be incorporated into a vein or artery, and its properties were particularly useful in this application because it is relatively flexible, does not injure blood vessels because of its rigidity, while at the same time being difficult to kink if the wall thickness is appropriately large. Although there is often a gradual development of a thin fibrous capsule around the pump after many years of implantation, this is not inflammatory in nature and does not lead to any harmful effects. Indeed, at times when pump replacement was necessary, new pumps could be inserted into the same subcutaneous capsule.

D. Problems with the implantable pump: theoretical and practical.

One theoretical problem which is apparent from a consideration of the design of the implantable pump is that increases in ambient temperature could cause increases in the vapor pressure of the pump's outer compartment, resulting in a proportional increase in flow rate (Blackshear et al., 1979). For the fluorocarbon vapor liquid chemical power source currently in use in the pump, it is estimated that the greatest commonly encountered increase in body temperature of about $4^{\circ}C$ could result in an increase of approximately 40% in drug flow rate. With some drugs with a very narrow therapeutic window, this might be expected to cause problems, although this has not been a problem in practice to date. One possible reason for this is that most common fevers can be readily controlled with anti-pyretic agents such as aspirin or acetaminophen, even when the diseases causing the fever are serious bacterial infections. Secondly, fevers of that magnitude are very rare in a adult population without concurrent chronic diseases leading to decreased functioning of the immune system. Third, in many cases it appears that the increased body metabolism which occurs in the setting of most febrile illnesses may increase the turnover of the delivered drugs, so that major net increases in

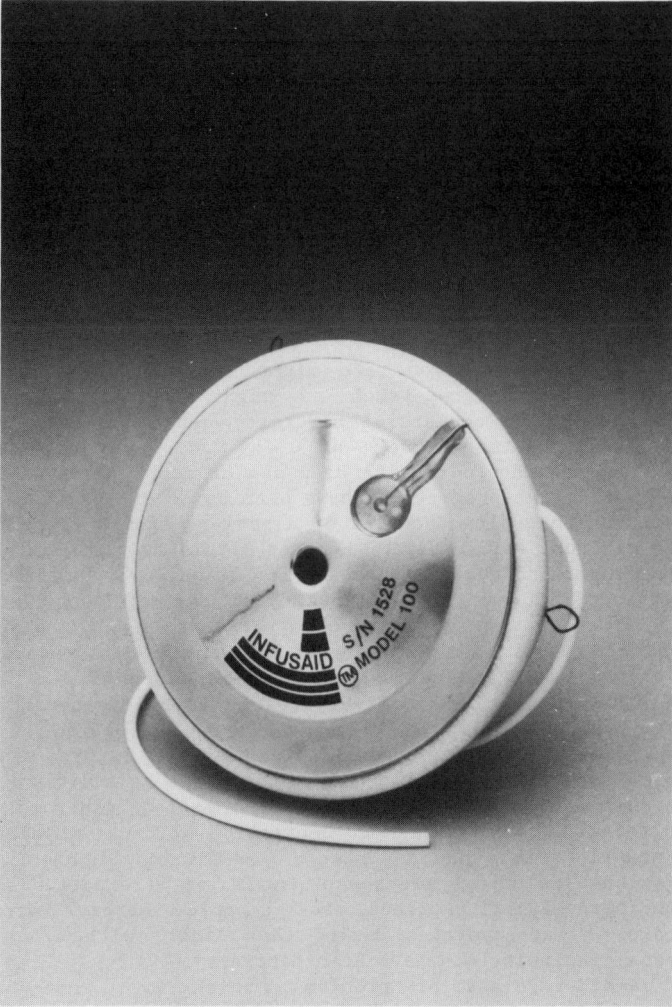

Fig. 2. External appearance of implantable infusion pump.

plasma concentration do not occur; this is still speculative, but anecdotal reports suggest that it might be the case. Finally, as we determined many years ago in dog studies at the University of Minnesota, routine changes in ambient air temperature do not affect the subcutaneous temperature significantly. However, prolonged immersion in a very hot tub or ice bath, for example, or application of very hot or cold packs to the area immediately above the pump, can and has resulted in minor changes in flow rate.

Another environmental variable which might be predicted to lead to changes in fluid flow rate is the atmospheric pressure surrounding the body of the pump's recipient (Blackshear et al., 1979). This is a possible problem because of the elastic nature of the human body which transmits freely changes in air and water pressure which occur, for example, during ascension to high altitudes or deep sea diving, where extremes in ambient pressure can be noted. Since the pump is completely self contained within a rigid cannister, these changes in ambient environmental pressure affect the second part of the Δ P in the Poisseuille equation so that increases in flow rate might be expected to occur at low atmospheric pressures such as might be encountered at high altitudes or in commercial aircraft, and decreases in flow rate would be expected in very high environmental pressures such as might be encountered during deep sea or scuba diving. Once again, the theoretical predictions have been borne out by observed changes in flow rate during transfer of a patient from a low altitude to a high altitude environment. However, it is fairly simple to predict the changes and quantitate them; for example, it has been estimated that every increase in altitude of approximately 6,000 feet leads to an increase in flow rate of approximately 30%. When a patient wants to move from sea level to Denver, for example, it is relatively simple to adjust the concentration of drug in the pump so that constant drug delivery rates are maintained. Similarly, if a patient needs to fly for an extended period of time in a commercial air craft, where cabin pressure is the equivalent of approximately 6,000 feet of altitude, it is relatively easy to fill the pump with a lower amount of drug immediately before the flight, and refill it with a high concentration of drug immediately after. In practice, this has not been a problem in flights of reasonable duration including transcontinental or transatlantic flights. Insulin-receiving patients, for example, merely increase their intake of carbohydrates during the flight; with other drugs, no special manipulations have been necessary.

Another class of problems, which might not be entirely predictable on theoretical grounds, has to do with problems encountered with specific interactions of individual drugs with the infusion pumps (Blackshear, 1984). Some of these could be predicted from the known properties of the drugs: for example, some antibiotics such as penicillin are known to be relatively unstable at 37°C, as are other compounds of potential therapeutic interest such as dopamine, which might be useful in the treatment of Parkinson's disease. Problems with many drugs were encountered by individual trial and error. For example, heparin is an extremely valuable anticoagulant which can only be given parenterally, and which was used in early human studies in the treatment of refractory venous thromboembolic disease or refractory vertebrobasilar transient ischemic attacks. However, as the concentration of heparin increases in solution, the viscosity of that solution also increases, although not in a simple proportional way. Therefore, when changes in heparin dose were required, a complicated calculation involving the viscosity changes as well as the drug concentration changes was necessary. From other points of view, however, heparin was a very desirable drug to deliver by this means, since it kept the tips of the delivery cannulas patent, its very high concentration in commercially available form allowed for very low infusion rates of less than one ml/day, and the drug is entirely stable at 37°C.

Other problems were encountered in the commercially available concentrations of drugs (Blackshear, 1984). For example, the drug 5-fluorouracil, which has been given by intravenous injection and infusion for many years in the treatment of bowel carcinoma metastatic to the liver, was too dilute in commercially available form to be of practical use in the pump. In other words, use of this drug at maximally available concentrations would still require very frequent refill intervals which would be largely impractical for outpatients. Luckily, an analog of this drug, 5-fluorodeoxyuridine, is available in high concentrations which can be used for intra-arterial infusion chemotherapy for the same disease.

Many other problems with other drugs have been encountered. For example, deferoximine, an iron chelating drug which might useful in the treatment of iron overload states such as thalassemia, when delivered by continuous infusion appeared to precipitate and plug the pump flow passages when incubated at 37°C at the required concentrations. The cancer chemotherapeutic agent, adriamycin, which would be desirable to deliver intravenously for a number of malignancies, appeared to cause local thromboembolic phenomena in the venous system. The anti-arrhythmic agent lidocaine, which might be useful to infuse chronically for the prevention of ventricular arrhythmia, appeared to set up a galvanic current in the pump, and actually dissolved some of the welds between the pump cannister and the capillary tube. And finally, the hormone insulin, which is a desirable drug to deliver in the treatment of various types of diabetes, has a number of problems which have made its use in this device particularly difficult (Blackshear and Rohde, 1983). For example, conditions known to cause precipitation of insulin in solution include acidification of the insulin solutions to the isoelectric point of the protein, about 5.5 to 5.8; introduction of divalent cations into the solution; chelation of the available zinc, which is used to maintain the stability of the zinc insulin crystals; and a process known as heat fibrillation in which insoluble fibrils of protein form in response to heating, shaking, shear, or a variety of other denaturing stresses. Finally, even in the absence of precipitation or fibril formation, under certain conditions insulin is prone to form soluble, but inactive, oligomeric species representing several combined oligomers of the insulin dimer (Blackshear et al., 1983).

All of these problems have prevented the widespread clinical use of this and similar devices for the delivery of insulin in diabetic patients. Clinical studies have been made possible by the use of high concentrations of glycerol, which has been known to protein chemists for many years as a means of protecting proteins from denaturation from a variety of stresses. The introduction of insulin mixed with 80% glycerol prevented the various types of insulin precipitation which plagued our early studies with the pump, especially the formation of insoluble insulin fibrils through the process of heat fibrillation. However, glycerol in our early studies appeared to promote the formation of soluble insulin oligomers or polymers, and decrease the amount of bioavailable insulin in the pump at the end of each delivery cycle (Blackshear et al., 1983, Blackshear et al., 1985). In addition to this practical problem of decreased insulin potency, there was a theoretical problem of increased antigenicity of the insulin oligomers, which have been demonstrated by direct measurement to exist in the circulation of patients receiving insulin by this means (Blackshear et al., 1985). However, in recent years we have determined that further purification of the glycerol through a series of ion exchange columns can decrease the formation of these insulin oligomers to a very great extent, and may make glycerol insulin solutions of more practical use in future patient studies (Blackshear et al., 1987, Rohde et al., 1987b). It is still an unwieldy preparation to use because of its extreme viscosity, and ongoing development of insulin protected from denaturation by a variety of surfactant compounds is eagerly anticipated.

It can be seen from the above that each new drug tested requires individual validation before it can be used in clinical studies in the pump. Therefore, the simple conception that this is merely an implantable version of the familiar bedside intravenous apparatus is clearly an oversimplification, since drug-pump compatibility must be studied and tested in each case before any individual application. Obviously, this potential interaction, when combined with site-specific aspects of drug delivery into veins, arteries, or other body cavities, means that drug delivery from devices of this type is clearly more complex than a simple peripheral intravenous infusion.

E. Current clinical uses of the Infusaid Pump.

We have recently described the current clinical uses of the pump, and some practical aspects of clinical use in a number of recent reviews (Blackshear, 1979; Blackshear, 1985; Blackshear et al., 1985; Blackshear and Rohde, 1983; Rohde et al., 1987a). Figure 3 shows the approximate number of patients who have been implanted with this device for various indications, compiled in 1984. Current estimates are that over 20,000 devices have been shipped for clinical use, of which an estimated 16,000 were for cancer chemotherapy for tumors involving the liver, 2,500 were for intraspinal analgesia, 300 for heparin infusion and 200 for insulin infusion.

It can be seen that a wide spectrum of diseases is represented, and the use of the device for site specific delivery in a number of applications can be readily appreciated.

PUMP IMPLANTATIONS: 1984

Disease	Drug	Site	Number of Patients
1. FDA Approved Uses			
LIVER METASTASES	FUDR	1A	7200
	FUDR	1A + IV	20
	5-FU	1A or IV	15
THROMBEMBOLIC DISORDERS	HEPARIN	IV	80
PAIN OF MALIGNANCY	MORPHINE	SUBDURAL OR INTRATHECAL	450
CA HEAD & NECK	FUDR	1A	50
CA CERVIX	METHOTREXATE	1A	3
CA MENINGES	METHOTREXATE	INTRATHECAL	2
2. Experimental Uses			
HEPATOMA	ADRIAMYCIN	1A	9
CA BRAIN	BLEOMYCIN	INTRAVENTRICULAR	4
PAIN OF MALIGNANCY	CLONIDINE	INTRATHECAL	5
	D-ALANINE-D-LEUCINE (DADL)	INTRATHECAL	2
DIABETES	INSULIN	IV or IP	115
ALZEIMER'S DISEASE	UROCHOLINE	INTRAVENTRICULAR	4
CHF	DOBUTAMINE	IV	3
OSTEOMYELITIS	AMINOGLYCOSIDES	LOCAL INFUSION (DUAL CATHETER)	15
		TOTAL PATIENTS	7977
		TOTAL CENTERS	1440

Fig. 3. Clinical uses of Infusaid implantable pump. Abbreviations used: FDA, U.S. Food & Drug Administration; CA, cancer; CHF, congestive heart failure; FUDR, 5-fluorodeoxyuridine; 5-Fu, 5-fluorouracil; IA, intra-arterial; IV, intravenous; IP, intraperitoneal.

F. Recent developments and prospects for the future.

This device was the first of its type to be designed and used clinically, and certainly has the longest track record and broadest number of applications of any implantable infusion pump in the world. However, it is a relatively simple device, as can be appreciated from the above description of its design, and should probably be viewed as a first generation or prototype device for other more complicated types of devices which have been developed since that time. We have reviewed the several types of devices developed since our initial invention in some detail in several recent reviews (Blackshear and Rohde, 1983; Rohde et al., 1987a). These encompass devices with a variety of pumping mechanisms, with design features making them suitable for use in certain specific applications more than others. For example, the programmable implantable medication system (PIMS) developed at the Applied Physics Lab at Johns Hopkins University has been developed largely for use as an insulin delivery vehicle, and current ongoing clinical trials of this device are testing it in this application. We anticipate that simple devices such as the Infusaid pump which have demonstrated themselves to be rather foolproof and safe for long-term use will continue to have clinical applications for conditions requiring simple basal rate infusions, with drug concentration alterations at refill periods. However, for more complex rate control, the simple Infusaid pump will need to be modified, or one of the more complicated devices will be required. It remains to be seen whether the introduction of increased complexity in these newer devices makes them more prone to failure, since, as discussed above, the fact that the device is completely implanted makes it much more difficult to replace or alter if any sort of device malfunction occurs. We feel that devices of this type have introduced a new type of drug delivery into the physicians' and pharmacologists' therapeutic armamentarium, and are of such practical advantage in certain clinical circumstances that they will remain with us for the foreseeable future.

Acknowledgments
         We thank our many colleagues who contributed to the work
cited   in   this   chapter,   and   Lessie   Detwiler   for   typing   the
manuscript.   We are grateful to Mr.   Vin Bucci of Shiley-Infusaid
Corporation for the data contained in Fig. 3.   Perry J. Blackshear
is an Investigator of the Howard Hughes Medical Institute.

References
Blackshear, P. J. (1979).'Implantable drug delivery systems',
     Scientific Amer., 241, 66-73.
Blackshear, P. J.   (1984).'Long-term parenteral drug delivery with
     an   implantable infusion pump: Progress and problems', in:
     Proceedings Second World Conference on Clinical Pharmacology
     and Therapeutics (Eds. M. M. Reidenberg, L. Lemberger) pp.
     459-468, ASPET, Bethesda.
Blackshear, P. J. (1985). 'Implantable infusion pumps: clinical
     applications', in Methods in Enzymology, Drug and Enzyme
     Targeting (Eds. S. P. Colowick and N. O. Kaplan) pp. 520-530,
     Vol. 112, Academic Press, New York.
Blackshear, P.J., Dorman, F. D., Blackshear P. L., Jr., Varco, R.
     L. and Buchwald, H. (1972). 'The design and initial testing
     of an implantable infusion pump', Surg. Gynecol. Obstet.,
     134, 51-57.
Blackshear, P. J., Robbins, D. C., Rohde, T. D., Langer, R. S.,
     Moses, A.C. and Massey, E. H. (1987). 'Insulin Replacement:
     Current Concepts', ASAIO Vol. 9, 646-655.
Blackshear, P. J. and Rohde, T. D. (1983). 'Artificial devices for
     insulin infusion in the treatment of patients with diabetes
     mellitus', in Controlled Drug Delivery, (Ed. S. D. Bruck)
     pp.111-147, Vol II, CRC Press, Inc., Boca Raton.
Blackshear, P. J., Rohde, T.D., Palmer, J. L., Wigness B. D., Rupp
     W. M. and Buchwald H. (1983) 'Glycerol prevents insulin
     precipitation and interruption of flow in an implantable
     insulin infusion pump', Diabetes Care, 6: 387-392.
Blackshear, P. J., Rohde, T. D., Prosl, F. and Buchwald, H. (1979).
     'The implantable infusion pump: A new concept in drug
     delivery', Med. Prog. Technol., 6, 149-161.
Blackshear, P. J., Shulman, G. I., Roussell, A. M., Nathan, D. M.,
     Minaker, K. L., Rowe, J. W. Robbins, D. C. and Cohen, A. M.
     (1985). 'Metabolic response to three years of continuous
     basal rate intravenous insulin infusion in Type II diabetic
     patients', J. Clin. Endocrinol. & Metab., 61, 753-760.
Rohde, T. D., Buchwald, H. and Blackshear, P. J. (1987a).
     'Implantable infusion pumps', in Drug Delivery by Devices:
     Pharmaceutical and Biological Applications (Ed. P. Tyle)
     Marcel Dekker, Inc., in press, New York.
Rohde, T. D., Kemp, M. S. and Kernstine, K. H., et al. (1987b).
     'Improved glycerol/insulin formulation for use in implantable
     pumps'. ASAIO in press.

Index

lysosomes, 3.12,13,60,63-80.89,91,140,156-
158,162,284,287
lysosomotropic agents, 285
lysozyme, 90,161

M-cells, 139
M5076, 79,83
macromolecular carriers, 20
macromolecular conjugates, 137
macromolecularization, 34,51
macrophage activation, 9,216-219,228,231,235-240,266
macrophage activation factors ,see MAF.
macrophages, 1,4,9,12,14,16,17,61,96,138,140,143,158,
162-164,170,189,196,214-241,253-258,262-266,281,284-287
MAF, 217,223,233-236
magnetic particles, 145
magnetic resonance imaging(MRI), 19
malaria, 10,256
mammary carcinoma, 76,84,113,121-126,186
mannose, 11,12,61
mannose receptor, 12,61
mannosylated albumin, 17
mastitis, 259
matrix degradation, 140
MC540, 162
melanoma, 86-88,225,228,232,233-236
melittin, 171
melphalan, 86
membrane asymmetry, 162
membrane fusion, 159-161
membrane integrity, 139
membrane modification, 170,171,174
membrane permeability, 159,257
membrane receptors, 69
membrane-mediated toxicity, 76
metabolic conversion, 59
metabolic pathways, 197
metal storage disease, 12,14
metastases, 8,9,93,95,141,186,190,195,200,213,214,219,
226,232-241,265,287,305,307
metastatic potential, 224
methicillin, 256
methotrexate, 63,79-82,90,91,96,164,189-191,200,307
methotrexate gamma-aspartate, 192,200
micelles, 97
microbial infections, 10
microbicidal, 12,254
microcapsules, 61,131,132,139
microenvironment, 3,13
microfluidisation, 142
micrometastases, 141,238
microparticles, 113
microporosity, see porosity.